ABOUT THE AUTHOR

James H. Johnson received his Ph.D. from Northern Illinois University in 1976. He has previously served on the faculty of the University of Texas Medical Center at Galveston (Division of Child Psychiatry) and as Assistant Professor of Psychology at the University of Washington, Seattle. At present he is an Associate Professor of Clinical Psychology at the University of Florida, Gainesville, where he is Director of Training in the clinical-child area. He is currently Associate Editor of the *Journal of Clinical Child Psychology* and is president of the American Psychological Association's Section on Clinical Child Psychology. He has published numerous articles dealing with life stress in children, adolescents, and adults, and has been instrumental in developing measures for assessing both adult and child life changes. He is coauthor (with S. Schwartz) of *Psychopathology of Childhood: A Clinical-Experimental Approach* (2nd ed., Pergamon, 1985), and is senior author (with W. Rasbury and L. Siegel) of forthcoming *Approaches to Child Treatment: Introduction to Theory, Research, and Practice* (Pergamon, 1986).

Toxic Injustice

The publisher gratefully acknowledges the generous support of the Humanities Endowment Fund of the University of California Press Foundation.

Toxic Injustice

A TRANSNATIONAL HISTORY OF
EXPOSURE AND STRUGGLE

Susanna Rankin Bohme

UNIVERSITY OF CALIFORNIA PRESS

University of California Press, one of the most distinguished university presses in the United States, enriches lives around the world by advancing scholarship in the humanities, social sciences, and natural sciences. Its activities are supported by the UC Press Foundation and by philanthropic contributions from individuals and institutions. For more information, visit www.ucpress.edu.

University of California Press
Oakland, California

Library of Congress Cataloging-in-Publication Data

Bohme, Susanna Rankin, 1973– author.
    Toxic injustice : a transnational history of exposure and struggle / Susanna Rankin Bohme.
        p.   cm.
    Includes bibliographical references and index.
    ISBN 978-0-520-27898-1 (cloth : alk. paper)
    ISBN 978-0-520-27899-8 (pbk. : alk. paper)
    ISBN 978-0-520-95981-1 (ebook)
    1. Dibromochloropropane—Toxicology.   2. Dibromochloropropane—Health aspects—Law and legislation.   3. Fruit trade—Health aspects—Law and legislation.   4. Agricultural laborers—Health and hygiene.
5. Environmental justice.   I. Title.
    RA1270.P4B586   2015
    363.738′4—dc23   2014016968

Manufactured in the United States of America

24   23   22   21   20   19   18   17   16   15
10   9   8   7   6   5   4   3   2   1

In keeping with a commitment to support environmentally responsible and sustainable printing practices, UC Press has printed this book on Natures Natural, a fiber that contains 30% post-consumer waste and meets the minimum requirements of ANSI/NISO Z39.48–1992 (R 1997) (*Permanence of Paper*).

CONTENTS

ILLUSTRATIONS

ACKNOWLEDGMENTS

While the story I tell here is by turns outrageous, enraging, and hopeful, I am lucky to be able to say that my experience in writing it has been often challenging, sometimes frustrating, but usually deeply gratifying. All that has been positive in researching, writing, and revising over the years, I attribute to the many people—from strangers to loved ones—who have helped and supported me along the way. *Toxic Injustice* began as my dissertation in the Department of American Studies at Brown University, and I was fortunate to have a committee outstanding in both intellect and generosity. Karl Jacoby, Josie Saldaña, Mari Jo Buhle, and David Egilman each provided incisive feedback and a model of rigorous scholarship. I owe David special thanks, as he has provided inspiration and support, as well as a political and practical education in the contested territory that is occupational health.

I am particularly indebted to those who provided the many documents I relied on to write this history that were not available in libraries or traditional archives. Litigation files—including the documents produced as evidence as well as the motions, filings, transcripts (and so on) that constitute lawsuits' working parts—are not easy to come by without the assistance of those involved in the process. I am grateful to a number of individuals who shared these files with me, including attorneys and others who had, in turn, received documentation from other lawyers. These include Vicent Boix, Christian Hartley, Scott Hendler, Jacinto Obregón, Carolina Quintero, Mark Sparks, and Lori Ann Thrupp. Writing about activist organizations and governmental efforts also at times posed challenges in locating primary sources. I am also grateful to Vicent Boix, Victorino Espinales, Jason Glaser, Giorgio Trucchi, and Ineke Wesseling for providing me with key documents in this vein. Without their willingness to share this documentation, much of the DBCP story would remain hidden. Thank you, Manuel Ángel Esquivel and Giorgio Trucchi, for agreeing to share your moving photographs. Finally, I thank the many people in the United States, Costa Rica, and Nicaragua who spent time speaking with me about DBCP history, litigation, and pesticide policy. Although not all these interviews or more-informal conversations have been cited in the book, each has informed my understanding of this history.

This book has benefitted enormously from the careful attention paid it by reviewers and colleagues. John Soluri and an anonymous reviewer for the University of California Press each provided astute and detailed feedback that guided me in transforming this from a dissertation to a book and expanded and refined my thinking on banana production and transnational politics. I am grateful to Chris Sellers and Jo Melling for providing their example, feedback, and venues for scholarship on "Dangerous Trade," and to participants in the conference of that name for the useful exchanges we had there and since. I am also grateful to Douglas Barraza, Douglas Murray, Susan Craddock, and Sarah Wald for smart feedback on ideas, conference papers, and chapters. Vicent Boix deserves another mention here, as he has been exceedingly generous with me, and our conversations and work together—as well as his own accomplished body of journalism on the DBCP issue—have been key in shaping my understanding of this history. Thanks also to my fellow members of the Trade and Health Forum of the American Public Health Association, who throughout the years I worked with them always helped me understand the broader issues at stake.

As this project took me far from home, I am especially grateful to those men and women in Costa Rica and Nicaragua who were generous with their time, homes, memories, and knowledge. I cannot name them all here, but do want to thank a few people who went well out of their way to help me with my research, including Ineke Wesseling and many others at the Instituto Regional de Estudios en Sustancias Tóxicas at the Universidad Nacional, in Heredia, Costa Rica; Isabel MacDonald, Francisco Cordero, and others at the Centro de Amigos para la Paz, in San José, Costa Rica; Belinda Forbes and Gerardo Gutiérrez in Managua, Nicaragua; and the members of the executive committee of PROSSTRAB (Promoción de la Salud de los Trabajadores), who have provided an inspiring example of a union–academic collaboration to improve occupational health, and also assisted me in making intellectual and personal connections that I might have missed otherwise. In Costa Rica, Carlos Arguedas Mora provided me with a place to stay in

Siquirres, introduced me to other unionists and DBCP-affected workers, and spent many hours talking with me about DBCP and about the current system of banana production and its ramifications on the local, national, and international scales. His death in 2010 was a sad loss for many.

Finally, thanks to Jessica Adams and Mary Sokolowski for incisive editing; Juan Quintana for careful copyediting; Judith Gallegos for transcribing a number of interviews; Maria Palmucci for help with citation management; and to Cari Lora, who started as a tutor and became a friend.

While I am indebted to all these people (and many others) for their help over the years, I of course take full responsibility for any errors or shortcomings. I anticipate that several of the people who have helped me will find points of analysis (perhaps several) that they disagree with (perhaps strongly): I would like to thank them most particularly and invite their input on issues we disagree on.

To my dearest friends and family who have sustained me over the course of this process, including those of you who have read and commented on chapters or on the entire manuscript, I am grateful for your intellectual engagement and loving support. Thank you from the bottom of my heart, Dick Adams, Charlotte Biltekoff, Winhard Bohme, Mimi Budnick, Liz Collins, Dayton Deighan, Dwyer Deighan, Hudson Deighan, Jonna Eagle, Dora Fisher, Lexi Matza, Margot Meitner, Mo Moulton, Ann Marie Nafziger, Wendy Rankin, Scout, Aliza Shapiro, Sarah Sharp, Wendy Sternberg, and Pamela Whitefield. Finally, love and thanks to Anne Griepenburg, whose insights on life and writing have immeasurably improved mine.

Introduction

IN A LOS ANGELES COURTROOM over the summer of 2007, twelve Nicaraguans brought suit against Dow Chemical and Dole Food Companies, alleging that the use of Dow's pesticide dibromochloropropane (DBCP) on Dole's banana plantations had rendered them sterile.[1] DBCP, a nematicide meant to prevent damage from the tiny, wormlike soil-dwelling creatures called nematodes, had been used on Nicaraguan and other Central American banana plantations in the 1970s and 1980s, even after it had been decisively linked to sterility in U.S. production workers. By 2007, tens of thousands of former Central American banana workers had reported health problems linked to their exposure—in Central America they were known as *los afectados,* the affected. But the 12 Nicaraguans—known as the *Téllez* group, after lead plaintiff José Adolfo Téllez—were the first to successfully bring the fruit and chemical corporations to trial.[2]

As the courtroom confrontation stretched out over more than four months, defendants' and plaintiffs' lawyers argued about the legacy of the chemical. Some of the central questions at issue were scientific and medical: What harm, exactly, did DBCP cause? Could it be blamed for each of these workers' sterility? Another set of questions considered by the jury went beyond the scientific, dealing instead with corporate decision making and labor practices: Had DBCP manufactured in the United States been sent to and used on Nicaraguan farms? Did the fruit and chemical companies know its dangers? Did workers? Did Dole hide information on those dangers from the workers who came into contact with the chemical? And beyond these questions was yet another level of conflict between the workers and corporations: What was the best forum for deciding the scientific and social questions about the effects of DBCP? More pointedly, should and could Central

Americans be allowed to bring these lawsuits in the United States in the first place? This 2007 trial was significant not only because of the possibility of monetary awards, but also because it was the first time—after 25 years of trying—that such a case had made its way into a U.S. courtroom. Did the victory promise that other similarly affected Nicaraguans might finally have a chance of receiving compensation from the corporations they held responsible for their illnesses?

The urgency of these questions is evidenced by the now widespread recognition of dangers faced by workers and others in the developing world. Some of the most recent (but hardly novel) examples include the high suicide rates among Foxconn workers producing electronics for Apple, environmental disease among indigenous populations in the Ecuadorean Amazon, and the death of more than 1,000 Bangladeshi garment workers at the collapsed Rana Plaza building.[3] For many workers like these, the lived experience of globalization includes poverty, ill health, violently limited life choices, and premature death. Understanding—and more so, changing—this brutal reality is daunting. Recent popular and scholarly efforts to make sense of globalization and blunt or rectify its injustices have often sought to trace the transnational movement of commodities, making visible long chains of production, distribution, use, and waste disposal and showing the social relations and material inequalities that emerge at and between various links in the chain.[4] Activist efforts by people in the developed and developing world, sometimes working together in solidarity, have included unionization, boycotts, factory inspections, the establishment of "fair trade" enterprises, and transnational litigation, famously including suits against Dow for the deadly 1984 gas plant explosion in Bhopal, and against ChevronTexaco for decades of dangerous practices in Ecuador.[5] In a political and scholarly sense, these efforts have focused on the development of and challenges faced by transnational social and labor movements, which often seek to link consumers and workers in various parts of the world in an effort to reverse abuses or demand accountability for past offenses.

Toxic Injustice tells a story of lived globalization through the history of one pesticide—DBCP—tracing not just its creation, circulation and use in the 1950s–1980s, but also the efforts since the 1980s by workers and others to demand justice for the harms associated with it. In the history of DBCP, corporate and state power in the fields of science, regulation, and law combined to create uneven geographies of exposure, resulting in disproportionate risk for both Central American agricultural workers and their (often immigrant) counterparts in the United States. Affected workers and their allies in Costa Rica and Nicaragua organized and engaged with both science and law to demand some modicum of justice, both at the national and transnational levels.

Looking at DBCP use and accountability together does more than expand the chronological frame of scholarly work on DBCP—which overwhelmingly focuses primarily on either use or accountability but rarely brings these two "moments" into dialog in a sustained way.[6] Considering the full sweep of DBCP's history, my inception-to-accountability approach helps us understand globalization better by looking at how the conditions of production, distribution, and use shape the terrain for the development of movements and strategies; and how resistance movements may challenge those conditions with varying degrees of success. David Harvey has argued that the current moment of globalization provides an opening to critically question the spatial organization of capitalism, an "opportunity to seize the nettle of capitalism's geography, to see the production of space as a constitutive moment within ... the dynamics of capital accumulation and class struggle."[7] For Harvey, understanding the terrain of capitalism is essential to "better speculate on how to exploit the weakest link and so explode the worst horrors of capitalism's penchant for violent though 'creative' destruction."[8] My aim here is both to show what the "nettle of [DBCP's] geography" looked like and to show how workers, lawyers and others tried to seize it, with varying degrees of success in obtaining both monetary compensation and a more abstract justice through legal recognition of their suffering.

The DBCP story has usually been told as—and indeed it is—a story of corporate malfeasance. Pesticide producers—including Dow, Shell, Occidental, Amvac, and others—sold the chemical even though they knew it was toxic to experimental animals, and they failed to provide a realistic warning to any user. Agriculturists, including corporate banana growers United Fruit and Standard Fruit, exposed workers to the chemical. In Central America, virtually no protections were provided to laborers. After the chemical was clearly shown to produce sterility in exposed industrial workers, Standard Fruit continued to use it on their plantations, and some chemical companies continued to produce and sell it outside the United States. And finally, chemical and fruit companies have strenuously fought banana workers' efforts to win compensation for the harms they have suffered.

The misdeeds of transnational fruit and chemical corporations are central to the most cited versions of the DBCP story. David Weir and Mark

Schapiro's influential investigative journalism in 1979–1981 focused on this nematicide (among a few other chemicals) as an example of what they dubbed the "Circle of Poison." While the journalists emphasized shared interests between developing-world workers applying pesticides and U.S. residents eating residue-laced produce, historian Angus Wright and others have since complicated such a neatly geometrical formulation of the transnational pesticide relation.[9] But Weir and Schapiro's work aptly framed the pesticide problem as one of corporate accountability; their "list of companies selling hazardous pesticides to the third world" included Dow, Shell, Chevron, Bayer, Monsanto, and many others.[10] DBCP remains a classic example of the phenomenon that has come to be known as toxic trade, and has been widely written about. Probably the most influential academic account is Lori Ann Thrupp's excellent 1991 article focusing blame primarily on "the dominance of short-term profit motives, and the control over information and technology [by] the manufacturers . . . [and] the banana producer companies."[11]

But the uneven landscape of DBCP damage was not shaped by corporations alone—national states also played key roles in shaping processes of both exposure and accountability. Most accounts of DBCP, including Thrupp's and Weir and Schapiro's, acknowledge the role of regulatory "loopholes" and failed enforcement in creating inequalities in DBCP exposure transnationally. But at the same time they do not account for the full complexity of the roles of state actors in DBCP's history. Throughout the chemical's history, state actors and institutions in both the U.S. and Central American banana-growing nations served as important sites of influence, debate, and contestation, profoundly shaping DBCP exposures and modes of justice-seeking for DBCP harms. Accordingly, I also focus on the state, looking at how corporations and workers engaged state actors in struggle over DBCP use and accountability, and how these struggles were shaped by national borders and laws, domestic democratic traditions, interstate power dynamics, and the changing role of the nation state in the architecture of neoliberal economy. The state maintains the social and economic order necessary for the functioning of even the "free"-est markets, and state actions have served to blunt the health effects of industrial capitalism by instituting (albeit incremental and contested) health and safety protections for workers and consumers since the late nineteenth century, and later for the environment.[12] Health and safety regulations have constituted an arena of struggle, as business interests—including corporations and trade associations—have managed to exercise a large degree of influence over regulations, whereas other actors—

including workers, public health advocates, and environmentalists—have sought to increase protections from a growing list of industrial dangers. Over the last three decades of the twentieth century and into the twenty-first, challenges to state regulatory capacity have mounted with the ascendance of a neoliberalism in which social policies are increasingly subordinated to the interests of capital, sacrificing welfare and labor protections, rolling back social and environmental regulations, and privatizing public assets. This has led many observers on both the left and the right to claim a diminished and diminishing role for the state.

While the ideology of neoliberalism holds that states are irrelevant or antagonistic to human well-being, people concerned with righting the health and environmental abuses of unfettered global capitalism have also increasingly turned to non-state targets and spaces for change and protest, including market mechanisms such as fair trade, boycotts, and certification campaigns, often coordinated by nongovernmental organizations. Such efforts are often based on the idea that a transnational network of local groups of "workers, citizens, church and other concerned groups all around the world" can improve corporate behavior through "constant monitoring and public exposure of wrongdoing,"[13] as has been the aim, for example, of campaigns against Dow, Nike, and Nestlé, each of which has addressed various health impacts of global capitalism.[14] Such strategies have often implicitly or explicitly centered consumer power—that is, threatening sales through boycotts or other disruptions anchored in potential buyers' repulsion for corporate abuses.[15] By placing the "monitoring ... of wrongdoing" as the responsibility of "concerned groups," these movements in effect suggest a grassroots or consumer assumption of the central regulatory roles of states that may either be too beholden to corporate power or too weakened to carry out that role.

However, other movements to redress or control health hazards have continued to turn to the state to fulfill its regulatory and social welfare roles, such as Chernobyl survivors' demands of compensation from Ukrainian governments, studied by anthropologist Adriana Petryna.[16] Historian Michelle Murphy notes that state-directed health social movements, "tend to conjure a hopeful relation to the state—an optimism about the possibilities of pollution regulation, or about the state's commitment to health, product testing, safe food, and so on."[17] Murphy finds that that optimism runs counter to historical processes of *de*regulation under globalization. Others, however, have suggested that the state might still be "useful" to social movements in new ways in the context of globalization.[18] This is consistent with the recent history of Latin American popular movements, which over the past decades have elected leftist governments (in places like Brazil, Venezuela, Bolivia, and Ecuador) and—more importantly in the eyes of many movement scholars—worked to hold them accountable through democratic engagement that goes beyond electoral politics. State-oriented politics may seek to influence policy-making through protest or other means, or, as in the case of DBCP, uses lawsuits to leverage the legal apparatus to secure tangible personal rewards for harmed individuals, while also punishing past wrongdoing and (hopefully) discouraging future depredations.

Toxic Injustice shows how state institutions were important sites of influence, negotiation, and contestation over DBCP use before, during, and after the 1970s, the decade that saw "an emphatic turn toward neoliberalism in political-economic practices and thinking."[19] State actions cut both—indeed, many—ways. In the history of DBCP, the decisions and action of regulators, officials, judges, and other state actors had impacts and implications that extended well beyond the national borders that formally bounded their power. Regulatory decisions—often at the behest of what might today be called corporate "stakeholders"—were fundamental in creating uneven protections within and across national borders. In exploring these unequal exposures, I further Angus Wright's project of complicating the "circle of poison" formulation, looking at how regulations addressing pesticide manufacture, export, and residues produced uneven levels of exposure between different groups of workers and consumers.[20] However, the state's role in the history of DBCP does not end there. I also look at how state actions served as a focus of resistance and ultimately brought an end to DBCP use nationally and internationally. Finally, I explore the place of state institutions and actors in worker efforts to hold fruit and chemical companies accountable for harms caused by DBCP. Banana workers pursuing litigation in the United States faced exclusion from legal forums there, as defendants mostly successfully argued that court rooms should not accept these "foreign" claims. In response, *afectados* turned to their own states, asking lawmakers, courts, and other institution to either provide compensation or assist in the legal process. In both Nicaragua and Costa Rica, *afectados* made the state both a target for their demands and a tool to address inequalities on a transnational scale.

Another central aim of this book is to engage with a growing body of historical and social scientific work on environmental and occupational health that considers the distribution of and meanings attributed to exposure-linked diseases. Some of this literature has roots in the environmental justice or labor movements and has often focused on the heightened risk faced by poor people and people of color.[21] Recent studies on what sociologist Phil Brown calls "contested illnesses" have emphasized how conflicting scientific claims lie at the center of struggles over defining the causes, treatment, and compensation for disease—particularly "new" diseases or those linked to occupational or environmental exposure.[22] While some of the scholarly literature on these struggles is transnational in scope, the majority of works consider the problem at the local or national scale.[23] By projecting the concerns of this scholarship into a transnational frame, *Toxic Injustice* expands and complicates our understanding of the nature and terms of scientific debate.

Scientific claims regarding DBCP span a wide spectrum. At one end of that spectrum lie early denials from Dow and Shell that worrying animal tests foretold danger for humans, and Dole's ongoing insistence that DBCP cannot cause problems in agricultural workers. At the other end lie Nicaraguan and Costa Rican former banana workers' attribution of a wide range of health problems to DBCP, including skin, liver, bone, kidney, and reproductive problems. The wildly differing definitions of DBCP damage were shaped by transnational forces including the political economy and labor history of banana production, the historical relationships between the United States and Central American nations, the influence of transnational corporations on governments in the United States and Central America, and the power of citizens to influence scientific definitions of harms at various geographic scales. The story of DBCP shows us what debate over disease and science can look like when it takes place across different political and democratic traditions, with sometimes-conflicting understandings of causation and justice.

Just as work by and about the environmental justice movement in the United States has taught much about the effects and critiques of racism, looking at scientific struggle over DBCP in a transnational context helps us understand and critique the workings of the "informal" imperialism that has long structured relations between the United States and Central America. During the decades of DBCP use, corporate and state determinations of what constituted acceptable DBCP use in Central America diverged from protections required at home: while regulators declined to require U.S.-based corporations to protect their workers abroad, corporations largely failed to extend even basic protections from DBCP to their workers, even after human health risks were publicly confirmed. In the decades of the struggle for accountability, U.S. actors' exclusionary visions of justice, deeply held prejudices against Central American political culture, and willingness to intervene in Central American affairs all limited *afectados'* quest for accountability, showing the material and ideological persistence of U.S. dominance in the region.

Historically, the banana trade has been central to the informal empire of the United States in Central America since the late nineteenth century, when U.S. economic expansion and burgeoning geopolitical aspirations brought Central America into the northern nation's growing sphere of international influence, and the isthmus became a site for shoring up U.S. power and extracting profit through trade and investment.[24] U.S.-based corporations United Fruit (now Chiquita) and Standard Fruit (now Dole) grew the fruit in Costa Rica, Panama, Belize, Honduras, Nicaragua, Guatemala, and Ecuador, as well as operating ports and transnational shipping, and dominating distribution channels within the United States. Famously, Standard and United Fruit could usually count on the support of the U.S. government, whose interventionist policy toward Central America in the first half of the twentieth century included the establishment of U.S. control over the Panama Canal Zone and frequent military interventions or occupations. By the 1930s, growing Central American demands for social reforms and national autonomy challenged the dominance of the United States and the centrality of the banana industry to Central American politics and economy. By midcentury, the banana corporations faced growing challenges from labor and government as the U.S. government continued to veer between military and economic approaches to maintaining its interests.[25] To cite some notable examples, the U.S. government played a central role in the ouster of reformist Guatemalan president Jacobo Árbenz in 1954, in training and arming brutal death squads in El Salvador and Guatemala in the 1960s, 1970s, and 1980s, and in funding and fomenting counterrevolutionary violence in Nicaragua in the 1980s. Today, U.S. support for the administration of Honduran President Porfirio Lobo, after a 2009 coup and despite human-rights violations of his administration, suggests that the United States continues to support Central American regimes based on its own geopolitical interests rather than in support of democracy.[26] At the same time, U.S.–Central American economic relations are structured by CAFTA-DR, a "trade agreement" implemented in 2009 that has unevenly distributed the growth it promised while simultaneously eroding democratic decision-making and health and environmental protections.[27] In the twenty-first century, banana production is less central to U.S.–Central American trade, but remains an important part of the economy in Honduras, Costa Rica, Guatemala, and Panama.

Banana production and trade also remain important topics of historical inquiry and contemporary politics. Recently, a number of new studies have enriched the historiography of the banana industry in Central America and beyond, complicating our understanding of worker life and resistance, differences between banana-producing regions, and banana worker engagement with transnational forces.[28] Important questions about transnational trade have been raised by the recent "banana wars" that centered around the European Union's preferential trade agreements with some formerly colonized banana-producing nations, as well as by various efforts to produce, label, and sell a "fair banana."[29] *Toxic Injustice* is, however, the first monograph to focus on banana worker occupational health history, adding an important new perspective to the literature and addressing concerns fundamental to workers' experience of banana labor. When critical or nuanced accounts of the banana industry have turned to the issue of occupational health, they have usually noted the toll of the work on laborer's bodies wrought by painful repetitive labor, traumatic injuries, heatstroke, infectious and parasitic disease, and—especially in the postwar period—pesticides.[30] While banana corporations positioned themselves as "modernizers" of health care in Central America, for example building hospitals in the banana zone, critical accounts were (and are) more likely to point out the violence—both "slow" and fast—that has long characterized U.S. actions—both military and commercial—in Central America.[31] In the vision of Costa Rican labor activist and novelist Carlos Luis Fallas, for example, disease and injury form part of the texture of workers' daily lives. In his 1940 novel *Mamita Yunai,* Fallas suggests that workers' lives are the ultimate cost of banana production—that they will "leave their bones as fertilizer for the bananas."[32] The macabre image suggests workers' very bodies were sacrificed in order to be transformed into the export commodity, itself meant to be exchanged for profit upon sale in the United States. For Fallas, these are the harsh terms of the transnational banana trade. Such terms also stoke resistance to those circumstances—after all, Fallas himself was famous for leading Costa Rica's 15,000-strong, Communist-led banana worker strike of 1934.

Pesticides added a new dimension to banana worker occupational health and resistance. Pesticides were used on banana plantations as early as 1938 with important consequences for health and labor politics.[33] But it was only after midcentury, with the postwar burgeoning of the chemical industry, that pesticides assumed a central place in Central American banana production.

For Standard and United, as well as newcomer Del Monte, synthetic pesticides represented a new tool in their long struggle against tropical fungi and bacteria that plagued vast plantations of genetically identical—and therefore biologically vulnerable—bananas. In his book *Banana Cultures,* historian John Soluri explains how, "by the early 1970s virtually every phase of production—from plant propagation to boxing operations—involved chemical inputs."[34] From then on, workers have donned backpack sprayers to apply herbicides, insecticides, and fungicides; used blowers to dust plastic bags with diazinon; introduced chemicals into watering systems and carried out chemical-laden irrigation; treated rhizomes with chemicals; dipped hands of bananas in fungicidal baths; sanitized tools; and otherwise come into contact with chemicals.[35] They have smelled unfamiliar scents or watched as planes treating the crop also flew over worker housing, or as pesticides applied to bananas by sprayers or irrigation "drifted" into domestic areas. Workers' perspectives on contacts with chemicals undoubtedly varied from person to person. In some cases, workers may have welcomed the use of chemicals—such as when herbicides reduced the effort needed to weed.[36] But whatever their perspective, workers who came into contact with herbicides, fungicides, and insecticides were exposed to new health risks.

While pesticides' proponents made rosy predictions of the chemicals' ability to free humanity from the scourge of pest problems—in banana cultivation and elsewhere—concern about the health and environmental impact of pesticides was never absent.[37] In the mid-1950s, toxicology had emerged as a new discipline meant to measure the health effects of the increasing number of industrial chemicals. Government regulations on pesticide use depended on toxicological data to create legal guidelines for pesticide use. Together, toxicology and regulation were supposed to bridge the gap between promise and suspicion by ensuring pesticides could be safely used. The first major pesticide regulation in the United States—the Federal Insecticide, Fungicide, and Rodenticide Act (FIFRA)—was enacted in 1947, and the 1960s–1970s saw the proliferation of U.S. laws and agencies using science to regulate potentially dangerous products or processes (workplace exposures, environmental pollutants, food and water contaminants, and so on).[38] Nicaragua, Costa Rica, Panama, and Guatemala all passed pesticide-related laws in the 1960s and 1970s.[39] However, as the history of DBCP shows, neither scientific research nor government regulation constituted a simple solution to the problems of pesticide toxicity. Rather, they served as grounds of debate and contention between government, corporations, workers, and others.

In the story that follows, my aim is to show that debate in all its geographic and historical complexity—that is, how its contours were shaped by the banana trade; by relations between nations; by the limits, failures, and successes of national regulatory systems; by both visionary and flawed plans for achieving justice in the context of stark transnational exclusions and inequality. The sites of this history include research labs, regulatory agencies, and plantations. But they also include courtrooms, activist assemblies, and streets and squares filled with protesting *afectados*. For at the heart of this project is an effort to understand not just inequalities of exposures, but the potential for popular resistance and corporate accountability in the context of neoliberal globalization.

I decided to write what eventually became this book after I had taken a year off from graduate school. Drawing on my American Studies training as well as my work experience in public health, I had taken a job at litigation consulting firm. Consulting on a variety of civil tort cases, the firm had compiled a rich body of documents and used them as the basis of publications on the history of science, corporations, and public health. Through contacts with attorneys, I learned of the DBCP cases and gained access to a set of documents from Shell, Dow, United/Chiquita and Standard/Dole. I also received other primary sources from other scholars who had in turn received them from attorneys. These documents are available only because of the lawsuits brought by DBCP-affected people; the companies were compelled to produce them during the discovery phase of DBCP litigation. So, while this history is in large part *about* the legal conflict between workers and corporations, it is also made possible by that conflict. Without the rich body of primary sources produced during litigation, portions of this book—largely the first three chapters—could never have been written.[40]

In addition to internal corporate documents, I rely on legal deposition testimony, newspaper accounts, trade publications, judges' memoranda and orders, records of regulatory hearings and other government publications, and the writings and publications of activists and activist organizations. Published and unpublished scientific reports have also provided insight into the risks and health effects of DBCP. I also obtained primary sources on U.S. intervention in Nicaraguan DBCP politics from a Freedom of Information Act request (which resulted in many highly redacted documents after nearly two years of waiting) and a Wikileaks search (which resulted in a few uncensored documents after a quick internet search). In addition to consulting archival sources, I have conducted interviews with a variety of people from the United States, Costa Rica, and Nicaragua, including DBCP-affected banana workers and their families, movement leaders, attorneys, former and current corporate personnel, scientists, and others whose personal, political, or professional experience has brought them in contact with DBCP. Although many of these interviews are not directly quoted in my text, each of them has informed my understanding and analysis. All translations of interviews and texts, unless otherwise noted, are my own.

I use a fairly conventional chronology to tell this story, with chapters one through four proceeding for the most part chronologically. Chapters five and six, focusing on movements in Costa Rica and Nicaragua, respectively, overlap substantially in time in order to privilege a geographical focus. Each of the chapters draws on a wide range of sources, mostly located in the United States, Costa Rica, and Nicaragua, as well as on the internet. The first half of this book addresses the development and use of DBCP in the United States and Central America. Chapter one focuses on the development and regulation of DBCP in 1951–1964. I show how, in their initial development and testing of DBCP, Dow and Shell's nominally scientific inquiry into its risks was ultimately concerned with securing both regulatory approval and a profitable market for the new pesticide. In the closely interwoven processes of marketing, toxicological testing, and regulation, science was less a reliable method for determining the dangers of the chemical than a tool for both securing regulatory approval and selling the new product. Dow and Shell scientists and other personnel emphasized or obscured research findings—whether from field trials of branded DBCP products or laboratory tests involving monkeys, rats, and guinea pigs—in accordance with whether they promoted the chemical companies' goal of making money from new products. Dow and Shell promoted the chemical in terms that borrowed from the scientific and economic vernaculars, promising profit through control over nature; at the same time they downplayed the importance of animal studies showing testicular damage in order to win official permission to sell their products. Regulatory approval and market expansion went together, and the companies sought authorization for DBCP use on an expanding number of crops, including one that was grown only outside the United States—the banana.

Focusing on the years 1961–1977, chapter two places expanding DBCP use in the United States and Central America in a transnational context. In the United States, farmworkers, many of whom were immigrants from Mexico or Central America, had significantly fewer mandated health protections than industrial workers, and bore the brunt of agricultural exposures. In Central America, the intersection of processes at various geographical scales led to the uptake of DBCP; these included changing corporate agroecological practices, chemical company marketing strategies, national regulation by U.S. and Central American governments, and conflicts between labor and management. Banana workers, left largely unprotected, developed bodily knowledge of DBCP from workaday contact with the chemical. I look at their remembered experiences and knowledge of DBCP, and consider how DBCP application became a site of struggle over labor practices.

Although DBCP had long been known to be toxic to animals, revelations of its undeniable impact on human health came when in October 1977 a group of male pesticide production workers in California linked their high levels of sterility to their DBCP exposure. Chapter three shows how evolving controls exacerbated inequalities along lines of geography and occupation (a category deeply shaped by race and immigrant status in the United States). Production workers in the United States were well protected, while protection of U.S. farmworkers was slow and uneven, and production workers in Mexico and some banana workers in Central America faced new and continued exposures, respectively. U.S. regulation shaped this new redistribution of DBCP risk in contradictory and surprising ways, including regulatory affirmation of continued use in the Hawaiian Islands and in banana-growing nations. Costa Rican responses to DBCP damage occurred in parallel to the U.S. processes after doctors identified unusually high rates of sterility among banana workers. Quiet and unofficial regulation there dealt with DBCP hazards in a narrow, national frame, increasing inequalities in exposure among Central American nations. An end to DBCP use came in the mid-1980s—only after water contamination at multiple sites in the United States showed that the DBCP threat extended beyond farmworkers. While the United States never banned DBCP export, regulations there had profound repercussions for use on banana plantations.

The second half of the book turns to efforts of Central American banana workers to hold the fruit and chemical corporations and, in some cases, their own governments, accountable for the harm done by DBCP. Chapter four explores the emergence and early history of a novel transnational litigation strategy, as Central Americans worked with U.S. and Central American lawyers to bring cases against the fruit and chemical corporations in various U.S. venues. I argue that the key issues in this litigation were location, representation, and translation. The fight between plaintiffs and defendants centered on place, with defendants mostly successfully arguing that the cases should not be brought in the United States but instead dismissed under *forum non conveniens,* a nonjurisdictional legal doctrine that gave judges discretionary power to dismiss a case when they felt it could more "conveniently" be tried elsewhere. While they were joined in the efforts to bring fruit and chemical companies to justice, the relationship between banana workers and their lawyers was also fraught. The process of legal representation vested more power and control in the lawyers than in their banana worker clients, while differing experiences and understandings of DBCP's harms and the ends of the legal process, accentuated by material failures, left many farmworkers dissatisfied with the process.

Chapters five and six examine in detail the national movements created by workers in Costa Rica and Nicaragua in response to their disappointment with the transnational litigation strategy. In Costa Rica, DBCP-affected workers developed a national movement that looked to the state as a supplement or alternative to litigation by seeking compensation from the state insurer, the National Insurance Institute (INS). Chapter five argues that their movement built on national democratic and protest traditions, combining street protest with partnerships with public health–oriented scientists to successfully expand the definition of DBCP harm and win compensation for those exposed to DBCP, including women, who had been excluded from the lawsuits based on a nexus of medical and legal issues. The Costa Rican movement was successful in building a democratic yet scientifically authoritative definition of DBCP damage, something the transnational litigation had never been able to do. As a result of their efforts, Costa Rican *afectados* won widespread but relatively small compensation from the state. However, the national movement effectively ran parallel to transnational litigation; workers were not able to leverage the assistance of the state in the transnational legal fight, or succeed in penalizing the fruit or chemical companies.

Chapter six turns to Nicaragua, where workers also defined a movement in national terms, mobilizing their power as citizens to win legislative change. Rather than turning to scientists, Nicaraguan *afectados* used graphic representations and public spectacles of bodily suffering to win public support. In doing so, they defined DBCP damage as a national problem and built political pressure on the Nicaraguan state. Unlike those of their Costa Rican counterparts, however, Nicaraguan activists' demands of the state were simultaneously an intervention into the transnational litigation. The centerpiece of their movement was a law that facilitated lawsuits against the fruit and chemical corporations in Nicaragua. Special Law 364 had profound transnational effects because it dissuaded corporations from seeking dismissal of cases from U.S. courts, effectively forcing a trial (or at least a more favorable settlement) in one arena or another. Corporations responded forcefully, employing the U.S. State Department, U.S. judges and the "dispute resolution" systems of neoliberal global economic governance in their fight against the banana workers. Remarkably, Nicaraguan *afectados* overcame some of these challenges; but ultimately the corporations have been successful at reversing their democratically won gains.

Despite losses, I end by arguing that *afectados'* successes in both Costa Rica and Nicaragua—though transitory and partial—suggest the state remains an important site of struggle over health and environment on both the national and transnational scales. It is clear that the DBCP story suggests specific ways in which corporate mobility and global economic governance undermine the state as a site of democratic decision-making. However, taken together with activists' very real achievement, even these points of defeat can serve as a roadmap for efforts to control hazards to both health and democracy in an age of corporate globalization.

ONE

Roots of Optimism and Anxiety

THE TRANSNATIONAL HISTORY OF DIBROMOCHLOROPROPANE (DBCP) began in Hawaii in 1951, among the shallow roots of a popular tropical crop. Researchers were not sure exactly *how* the compound protected pineapples from the wormlike nematodes that threatened the plants from within the soil, but they knew it worked. In its efficacy lay the promise of increased profit, and the chemical soon made its way to the mainland United States, where Dow Chemical Company and Shell Chemical Company worked in concert, transforming the experimental compound into their own marketable and branded products, Fumazone and Nemagon.

That transformation took place through interlocking processes of testing, production, and promotion that generated sometimes complementary, sometimes competing explanations of how the chemical would affect not only the agricultural crops and the nematodes it was meant to control, but also the bodies of humans who were exposed to it. In the 1950s and early 1960s, U.S. regulation of markets for new pesticides was still developing, even as synthetic chemicals increasingly became a part of everyday life. In the context of popular and scientific optimism about these still-novel products, Dow and Shell initially marketed DBCP as a chemical that offered control of a new and unfamiliar pest. Marketing used the language of science not only to teach farmers of the existence of the nearly microscopic nematodes, but also to offer DBCP-based products as the best way to control the newly discovered threat. This science-based marketing was accompanied by an economic argument meant to convince farmers that products containing DBCP would boost their bottom line.

Postwar confidence in the new pesticide technologies was not unbounded, and despite the allure of a scientific fix to the problems of agriculture, the use of chemicals meant to poison pests also raised the specter of human health harms. Government regulation and the science of toxicology seemed to offer a resolution of the tension between the benefits and risks of pesticides. However, the standards developed for warnings and precautions rested on the problematic assumption that scientific research would provide reliable information to prevent dangerous products from reaching the market. The history of DBCP reveals something very different: scientific evidence was not a neutral arbiter of pesticide safety, but was itself subject to various and conflicting interpretations. What emerged was a series of scientific narratives that developed in a process of interaction between companies and regulators and that significantly downplayed alarming animal test results in favor of an interpretation of DBCP as "safe for human use." These narratives supported only one conclusion: the continued production and use of DCBP in the United States, with few protections for those exposed to it.

EATING INTO THE BOTTOM LINE: PINEAPPLES, PESTS, AND PROFITS

The pineapple industry, begun around the turn of the twentieth century when the United States forcibly annexed the Hawaiian Islands, had much in common with commercial fruit production elsewhere in the world, including the burgeoning orchard business in California and Florida and banana plantations in Central America. Established in the late nineteenth century, these sites were similarly characterized by large single-crop plantings that depended on low-paid labor and novel distribution networks. At the turn of the twentieth century, the Hawaiian and Central American sites were also part of the U.S. turn to expand geopolitical and economic spheres of influence beyond the continent. On both the islands and the isthmus, U.S. innovators in industrial, corporate agriculture produced "exotic" fruit for the mainland market, and the growing agricultural industry increasingly required the use of pesticidal agents to allow its produce to reach its intended markets in a condition that would appeal to consumers.[1]

By the mid-twentieth century, Hawaiian pineapple growers—agroindustrialists who managed large-scale cultivation as well as canning of the fruit—dominated the world market. However, planting large stretches of land with a single species of pineapple made their crops more vulnerable to pests. This type of monocropping, or monoculture, reduces plant and animal diversity.

As a result, fewer animal species survive, because habitat and sources of food are severely limited. Those species that do survive are often the pests that thrive on the crop itself, usually damaging it in the process. Not kept in check through competition from other species, their numbers expand, and their effects on the crops earn them the title of "pest."

In Hawaii, one of the organisms that flourished in the pineapple fields was the root-knot nematode,[2] a wormlike, soil-dwelling species belonging to the phylum *Nemata*. These tiny creatures emerged as "the most critical of the soil pests" for pineapple growers on the islands.[3] Growers were worried because root-knot nematodes fed on the root system of the pineapple, impairing its ability to attain nutrients from the soil. Nematode-affected pineapples wilted and sometimes failed altogether,[4] resulting in financial losses for the growers. To rid themselves of the profit-eating worm, growers turned to the chemical industry.

The first chemical widely adopted for nematode control was chloropicrin. The most widely used poison gas of World War I, chloropicrin was produced for battlefield use by the U.S. Chemical Warfare Service, which collaborated with the Federal Entomology Service and other federal agencies to test its potential against insects.[5] Beginning in the 1930s, this erstwhile weapon was applied to soils before the pineapples were planted, assuring a nematode-free field.[6] Chloropicrin was both effective and cheap; the government was selling it as war surplus at rock-bottom prices.[7] But the inexpensive chloropicrin supply was running out by the 1940s, and pineapple cultivation was becoming even more important to agroindustrialists in Hawaii. Pineapples were the Hawaiian Islands' second most lucrative crop, and after the bombing of Pearl Harbor were subsidized by the U.S. government as a wartime necessity. Production boomed—by 1945 three-quarters of pineapples grown worldwide came from Hawaii—and research efforts went into finding new chemical controls for the nematode.[8] By the late 1940s, chloropicrin had been largely replaced with two new soil fumigants—dichloropropene-dichloropropane (DD) and ethylene dibromide (EDB).[9] But those chemicals had their own limitations—they killed not only nematodes but plants as well, and so both DD and EDB had to be applied to soil well before planting. A new chemical was needed to fully secure the uninterrupted commercial production of the lucrative crop.

By 1951, the Pineapple Research Institute (PRI), a research group founded by the Hawaiian Pineapple Packers' Association, had come up with a chemical with promise: DBCP. Like the nematicides before it, it was a fumigant.

The liquid form of the chemical was injected into the soil at a depth of six to eight inches, usually with a mechanized applicator that was pulled behind a tractor, introducing the chemical into the soil through chisel-like injectors.[10] Once injected, the liquid quickly volatilized, dispersing throughout the soil in vapor form. Although it was applied like its predecessors, the new compound had a powerful advantage over other nematicides. As the trade magazine *Farm Chemicals* noted about DBCP: "For the first time, an effective nematicide that can be injected into sites of certain living plants without injury, became available."[11]

The chemical had promise, both as a way to control nematodes and as a moneymaker. Two companies—Dow and Shell—would transform the compound into two branded products—Fumazone and Nemagon—that would dominate the DBCP market for decades to come. Even more significantly, the research and marketing efforts of these two private corporations would author the scientific narratives by which DBCP's impact on human health would come to be understood.

TECHNOLOGIES OF ABUNDANCE AND CONTROL

The expanding market for synthetic pesticides at the mid-twentieth century was only part of a much larger agricultural transformation. Farming in the United States had been undergoing a process of specialization and concentration for several decades. By 1939, Carey McWilliams had documented the struggles of laborers in California's "factories in the field."[12] In the years after World War II, the pace of industrialization picked up, and food production became increasingly specialized and mechanized.[13] As the number of farmers plummeted and the size of individual farms swelled, growers increasingly "measure[d] 'success' in terms of a narrowly defined set of economic and productivity criteria."[14] Enlightenment values of reason, science, and control over nature suffused the transition to factory farming, and pesticide chemicals seemed key to meeting the requirements of this emergent agriculture by promising both agricultural abundance and unprecedented control over nature.

Synthetic pesticides had been developed precisely as a scientific tool for controlling the numerous pests that could threaten crops grown in specialized cultivation. As early as the late nineteenth and early twentieth centuries, arsenical compounds were being used in orchard cultivation and were associated with the development of a modern and research-based industrial agriculture in the United States.[15] Federal funding for agricultural experiment stations in the first half of the twentieth century privileged basic research, and the development of new pesticides was part of an approach that saw science as the driver of a prosperous agricultural sector.[16] Around World War I, the growth of the chemical industry intertwined with federal war efforts, exemplified in the conversion of chloropicrin from weapon to insecticide.[17] By the 1940s, another chemical with roots in federal war efforts came to exemplify the promise of scientific control. Dichloro-diphenyl-trichloroethane, or DDT, was a powerful and persistent broad-spectrum bug killer. This organochloride was effective in low doses, had a long-lasting effect, and killed a wide range of insects. In the multifront theater that was World War II, a little DDT—applied to people rather than plants—went a long way toward controlling the lice that spread typhus and the mosquitoes that brought malaria to troops and civilians in North Africa, Italy, and elsewhere. To many nonscientists, the white powder seemed a scientific miracle on the order of other new disease-fighters: the sulfa drugs and penicillin.[18] Glowing press for DDT in the United States suggested the wartime tool might have civilian applications. The chemical was characterized as "one of the great scientific discoveries of World War II."[19] Together with other "repellents and insecticides," DDT was considered by some to be "the biggest contribution of military medicine to the civilian population after the war—a contribution even greater than blood plasma."[20] In 1945, the War Production Board, on the advice of the DDT makers themselves, authorized the sale of DDT for nonmilitary uses.[21] *Time* magazine predicted that the insecticide would "wipe out the mosquito and malaria . . . liquidate the household fly, cockroach and bedbug, [and] control some of the most damaging insects that prey on the world's crops."[22] The manufacturers of DDT promoted the chemical for household and industrial use as well as for fruit, vegetable, and dairy production.[23]

The rhetoric of pesticide promotion echoed militaristic terms.[24] During and after the war years, U.S. consumers were urged to "fight your insect enemies,"[25] an admonition that equated mastery over household pests with U.S. military victory. Agribusiness responded favorably to these urgings: sales of synthetic pesticides shot up and by 1949 surpassed the use of metal- and plant-based formulas.[26] The postwar years saw similar growth in the chemical industry, as pesticide production rose from just under 125 million pounds in 1947 to over 637 million pounds in 1960.[27] By then, the Korean War had ended in a stalemate and the Soviet threat loomed large. But the United States still hoped to prove its mastery over the lowly bug.

After World War II, as deprivations fell away in the United States and an assortment of new consumer and scientific technologies seemed to offer comfort and plenty, pesticides promised bounty as well as control. A clear motive for chemical control of pests was to ensure a profitable harvest. While individual farmers were likely to base estimates of pesticides' benefits on a case-by-case financial calculus, some pesticide proponents envisioned a utopian "era of plenty."[28] The abundance they envisioned would spread beyond national borders to encompass the impoverished nations of the world, where the United States was now vying with the Soviet Union for military, political, and cultural hegemony. In May 1955, for example, a Public Health Service official, addressing the annual meeting of the National Agricultural Chemicals Association, argued that pesticides would help "swell the food supply" and "curb and eliminate diseases" worldwide.[29] This vision fit neatly with the ideology President Truman had outlined in his 1947 Doctrine that Third World population growth, hunger, and lack of scientific and technical development necessitated the judicious application of U.S. technology and capital.[30] According to the president, "More than half the people of the world are living in conditions approaching misery. Their food is inadequate, they are victims of disease. Their economic life is primitive and stagnant." In this troubling context, Truman asserted that "greater production [was] the key to prosperity and peace. And the key to greater production is a wider and more vigorous application of modern scientific and technical knowledge." In language that echoed Truman's but emphasized the nefarious role of pests in creating human misery, a Shell entomologist wrote in 1954 that "half the world is presently underfed, ill-housed, poorly clothed, and suffering from insect-borne disease."[31] For that scientist, and for many others in the United States of the 1950s, the science of pesticides offered the key to creating the prosperity and abundance that would eliminate hunger and cure disease on a global scale.

MARKETING NEMATODES, SELLING PESTICIDES

Not surprisingly, Shell and Dow would mobilize powerful ideas about pesticides' ability to provide both scientific control and agricultural abundance in their efforts to sell DBCP. However, the companies faced a challenge. Unlike DDT, which controlled familiar pests like the "mosquito . . . household fly, cockroach and bedbug," DBCP targeted a pest that most farmers knew little to nothing about.[32] The nematode was almost microscopic. Nematodes lay hidden in the soil and their effects on plants were not always obvious.[33] In order to sell DBCP, then, Shell and Dow had not only to promote the pesticide, but also to publicize the pest it was meant to control. The companies used the themes of scientific control and agricultural plenty to market both the nematode and the nematicide.

Except for special cases like pineapple growers, most U.S. farmers in the mid-1950s knew nothing about nematodes. As one plant pathologist described the pests: "Nematodes that parasitize plant roots are insidious pests. Being of microscopic size and hidden in the soil, they may easily escape detection. Moreover, the symptoms they cause often are so slight or nondescript that they either do not attract enough attention to cause concern or are attributed to other unfavorable circumstances."[34] As A. L. Taylor, a nematologist with Shell, would explain in a history written in the early 1980s, "Many farmers had never heard of nematodes and had certainly never seen any. Some were familiar with root-knot nematode galls, but did not know what caused them and did not associate their presence with reduced growth and yield."[35] Farmers' ignorance of the nematode meant that before Dow and Shell could sell farmers a pesticide, they would have to explain why nematicides were necessary and how to fit them into their daily practice of farming.[36] As Taylor would recall decades later, the promotion of nematicides in the chemicals' infancy involved a number of steps:

To sell nematicides, farmers must be persuaded to believe that:

1. There are little worms called nematodes which attack crop plants. Nematodes, which are too small to recognize without a microscope, exist in enormous numbers in farm fields.
2. Nematodes damage roots and are a primary cause of reduced plant growth and crop yields.
3. Nematodes can be controlled by application of nematicides.
4. Control results in increased crop yields.
5. With crops of moderate to high value per acre, the selling price of the yield increase is 4 to 5 times the cost of the nematicide and its application.
6. Nematicides are therefore a good investment.[37]

Shell and Dow attempted this "persuasion" through a varied marketing campaign that mobilized the themes of control and abundance using the language of science.

Shell's marketing, in particular, linked the new product with the power and authority of science. According to *Shell News* writers, the discovery of nematodes' damage to crops had not only spawned interest on the part of the chemical companies, but "a new science" as well.[38] Another commentator remarked that it was the development of a chemical solution that had provoked such a burgeoning interest in the field of nematology.[39] In any case, the marketing for DBCP was intimately linked with the expanding research field. Marketing efforts emphasized the latest reports in the chemical control of nematodes, including the publication and then reprinting of research.[40]

Perhaps most notably, beginning in 1957 Shell sponsored a series of "Nematode Workshops" in various locations. Complete with a roster of academic and government experts, as well as corporate employees, the workshops were framed as technical seminars. According to reporters for Shell's newspaper, at the first Nematology Workshop, "for two days the hotel meeting room resembled a large classroom as college teachers, government officials, and other agricultural specialists . . . got expert information on the effects and control of nematodes."[41] Workshop participants from Oregon to Puerto Rico had the opportunity to examine nematodes under the microscope, look at fumigant application equipment, and ask questions of leading authorities.[42] The workshops presented a wide scope of information on nematodes, including how they affected regional crops. The educational-scientific format of the program, as well as the presence of non-Shell experts, helped create an atmosphere of impartial inquiry. At the same time, the workshops gave Shell an opportunity to build the market for DBCP and its other nematicide, DD. Not all information promoted Shell's pesticides, but DBCP was positioned as the latest and a very effective way to control the underground pests. Participants, who were likely to have influence through their teaching or policy-making roles, were urged to "disseminate . . . the many aids Shell ha[d] prepared for spreading the nematode story."[43]

Borrowing a technique from university-based agricultural experiment stations,[44] Dow and Shell also brought the science of DBCP from the workshop "classroom" to the field. A. L. Taylor recounts his work developing a nematicide market for Shell:

> Eventually a marketing technique based on field demonstrations was developed. With the generous help of Extension Service personnel, arrangements would be made to cooperate with leading farmers in a community. Equipment and chemical was furnished to fumigate areas in fields where tobacco was to be planted. The rest was done by the farmers, who planted, fertilized and

cultivated the whole field as usual. Early in the growing season, results began to be visible. Tobacco plants in the treated area were larger and more uniform than those in the adjacent untreated part of the field. The improvement continued all through the season. At any time, it was possible to dig roots for comparison of stunted, knotted or rotted roots from untreated soil with extensive, healthy roots from treated soil. The growth difference was plainly visible in the field and far more convincing than any amount of talking and advertising.[45]

The field demonstrations took on all the prestige of scientific experimentation, marshaling scientific evidence that farmers could see with their own eyes.

What the farmers were not told is that only positive field results made it into the promotional materials, presenting a skewed picture of the utility of the chemical. Promotional materials did not include caveats on the extent to which the efficacy of nematicides depended on application technique, weather, and the degree of nematode infestation. For example, one study, summarizing nematicide fieldwork from 1955 to 1960, showed that the use of nematicides (of unspecified type) could actually *decrease* harvests in fields with slight or moderate nematode infestations and provided widely varying levels of improvement in crops with severe infestation.[46] Despite the variance in crop results, field tests were used by Shell to claim that DBCP—now branded as Nemagon—had often proved a "startling success."[47] Field test results might be published in trade journals by corporate personnel or even university professors, giving them additional scientific authority.[48] Published articles were sometimes reprinted as promotional pieces, further erasing the distinction between scientific and marketing discourses.[49]

The corporate and scientific promotion of nematodes embodied a view of the farm environment in which threats to productivity were not necessarily visible and could not be accommodated within farmers' traditional repertoire of remedies. Part of the broader midcentury transformation of farming, pesticide technologies such as DBCP required a displacement of traditional ways of understanding the land. From this new view of the farm, familiar agricultural ailments could take on new meanings. As one article explained, "until the 1940s, crop blight stemming from nematode appetite was blamed on other factors, such as lack of water, insufficient fertilizer or 'worn-out' soil."[50] However, as the nematode workshops emphasized, these symptoms might actually be attributed to an invisible and unfamiliar pest. The nematode's tiny size meant it required special technology to identify the true cause of the problem. As one workshop expert put it, the mere presence of symptoms like stunted growth "is not proof of nematode injury . . . for many other pathogens can cause injuries similar in appearance. An examination of soil and roots by a specialist is the only reliable way to be sure of a nematode infestation."[51] This was a farming that called for more than farmers' knowledge—it required recourse to the technical skills of scientific experts. Shell stood ready to offer that expertise, promoting its research facilities as "portals of progress in pesticide research."[52] Underlying the scientific promotion of nematicides, then, was a new vision of the farm environment as potentially besieged by tiny pests, their symptoms unreadable by the average farmer, and their control requiring the intervention of an expert and the modern technology of chemical control.

DBCP marketing was not conducted entirely in the language of science, but made economic appeals as well. For example, a recurring advertising motif personified nematodes as surreptitious enemies who targeted farmers' financial well-being. Shell produced a film called "Thief in the Soil," which one journalist described as a "ten-minute, full-color drama describing the damages inflicted by nematodes.[53] The worms figured not as a natural part of the soil life, but as "thieves" who would steal the farmers' profit, linking the threat of the hidden nematode with the economic damage that the pest could cause. Such a motif used fear to convince farmers that nematicides, despite the fact that they were more expensive than many other pesticides, would ultimately improve their bottom line. Similarly, the 1958 Dow "Soil Fumigation Handbook" described nematode damage as "plunder underground." The booklet told farmers that "from Maine to California, these microscopic, wormlike pests are tapping the farmer's till for a large percentage of all he grows."[54] The play on the word till—which means both "to cultivate soil" and "money box"—effectively reinforces the message that nematodes threaten profit on the farm. With "the new soil fumigants," Dow maintained, farmers would "no longer have to watch their prize acreage sapped of its profit by nematodes and other soil-borne pests. They can fight back . . . and win astonishing profit rewards."

The twin promises of control and abundance were brought together in a frequently repeated image. Taken from the scientific vernacular, the image is of two plants shown side by side; one is several times the other's size. The larger plant may or may not be labeled "treated," but it is understood that the comparison is in fact an illustration of the effects of the chemical. The message is clear: the larger plant is the best symbol of increased control and profit.

FIGURE 1. A comparison of treated and untreated grapes, meant to illustrate the crop-enhancing powers of Dow's Fumazone brand of DBCP. From L. E. Warren, "Response of Established Tokay Grapes to Soil Fumigants," *Down to Earth* Spring 1960, 13–16.

REGULATING HAZARDS, ADDRESSING ANXIETIES

If pesticides at the mid-twentieth century were widely understood to offer scientific control of the environment and the key to agricultural abundance, they also provoked anxieties about the potential negative effects on people. It stood to reason that chemicals developed precisely for their harmful properties on bugs—chemicals also observed to kill birds and cattle—might also be able to hurt humans. Anxieties about pesticides' dangers interrupted optimistic assessments of pesticides' contributions, focusing instead on their effects on human bodies. Addressing these anxieties required a method for understanding chemicals' effects on the body, the delineation of authority to define what damage was acceptable, and mechanisms to control unacceptable harm. While military institutions exercised control over use of chemicals like DDT in wartime, after World War II other agencies of the federal government emerged as the primary sites to address the tensions between the perceived benefits and dangers of the new technologies.[55] However, regulators' position of authority was shaped by their relationship to the pesticide industry. The corporations that made up that industry had a clear material interest in the success of their products. They managed to both influence and limit regulators' efforts, while simultaneously working to protect themselves from potential lawsuits and more generally to reinforce perceptions of the necessity and safety of their products. To resolve the tensions between pesticides' dangers and benefits, both industry and regulators looked to the science of toxicology. This still-emergent discipline seemed to offer an impartial assessment of chemical dangers that could be used to set acceptable exposure levels and develop effective user warnings.

Although U.S. government concern with pesticides' health risks has often been dated to the early 1960s, when worries about DDT intensified and Rachel Carson's *Silent Spring* was published,[56] fears of pesticide toxicity in fact began much earlier. When DBCP first came on the market in 1955,[57] the potential hazards of pesticides had already emerged as a field of contention both in specialized forums and in the media more generally,[58] with debates about the safety of DDT and the effects of insect eradication campaigns on livestock and the environment taking place among scientists and regulators as well as among farmers and wildlife enthusiasts.

Midcentury regulatory developments made the federal government responsible for protecting people living in the United States against the hazards of pesticides. The Federal Insecticide, Fungicide, and Rodenticide Act (FIFRA), which became law in 1947, required manufacturers to register the "economic poisons" named in its title before they could be sold in the United States.[59] (Nematicides like DBCP were not covered in the original act, and would not be incorporated into FIFRA until the 1960s.) Although FIFRA was primarily concerned with assuring the purity and efficacy of pesticides, it did contain some safety provisions. For example, the law required pesticides to be accompanied by a label that adequately warned users of risks and provided instructions that allowed users to avoid those risks; the most dangerous pesticides were required to be labeled with a skull and crossbones, as well as the word "POISON" in red. Some state governments also passed additional laws for the regulations of pesticides within their borders. By the end of 1951, nineteen states had adopted laws that paralleled the federal statute, and the same number had laws that imposed unique requirements that the chemical companies prepare various labels to meet state-level obligations.[60]

Despite individual state legislative efforts, FIFRA remained the dominant force in setting pesticide policy nationally. As the federal government assumed the responsibility for protecting bodies, it also defined the terms by which harm would be understood. To be sure, bodily harm was minimally defined in the 1947 legislation. While the sale of "injurious" chemicals without warning the consumer was forbidden, the definition of "injurious" was not spelled out. Perhaps most importantly, given the transnational scope of this inquiry, it is critical to note that, from the beginning, FIFRA had an explicitly national frame. The legislation read "no article shall be deemed in violation of this Act when intended solely for export to any foreign country," meaning that anytime a company shipped chemicals abroad, it could effectively ignore the new law's provisions.[61] Therefore, although authorities and industry representatives alike had touted pesticides' promise in helping feed the developing world, people outside of the borders of the United States were denied protection from those very same pesticides.

While FIFRA was primarily concerned with protecting people who purchased or applied pesticides, the federal government also assumed the obligation to safeguard those who encountered pesticides in their foods. In 1949, Paul Dunbar, the U.S. Commissioner of Food and Drugs, told the members of the pesticide industry group NACA (National Agricultural Chemicals Association), that his agency had "an obligation to protect the consumer" from pesticide residues.[62] Dunbar traced that obligation to the passage of the 1938 "Pure Food Law," formally known as the Food, Drug and Cosmetic Act (FDCA), which required the Food and Drug Administration (FDA) to set legal allowable levels, known as tolerances, for "unavoidable poisonous substances" on foods.[63] The FDCA contained an export exemption similar to FIFRA's but was slightly more protective in that it required the exported item to comply with the laws of the country to which it was exported. Dunbar noted that the subsequent proliferation of pesticides during the war years had meant his agency had been unable to effectively carry out the terms of the law even within the confines of the United States.[64] In 1954, the Miller Amendment to the Food, Drug, and Cosmetic Act formalized the procedure for regulating pesticide residues on "raw agricultural commodities."[65] The new legislation more thoroughly elaborated the means by which the federal government could assess the harm caused by a pesticide, requiring "full reports of investigations made with respect to the safety of the pesticide chemical," as well as documentation of the residues remaining on food and methods for their removal.

While legislation such as FIFRA and the Miller Amendment signaled the U.S. government's role as a site for promoting social welfare and protecting citizens' health, limits on that role reflected the power of industry interests. For example, FIFRA ceded much authority to the companies it was meant to regulate—if the Secretary of Agriculture refused to register a chemical, a manufacturer could always achieve a registration "under protest" and legally sell its product. This provision for "registration under protest," which would not be eliminated until 1964,[66] provided a built-in check to the power of the

USDA (United States Department of Agriculture) over chemical producers. By allowing companies to register even a chemical found dangerous by the agency, protest registrations left regulators to shoulder the scientific, legal, and economic burden of proving a chemical's harm.

The influence of industry manifested in other, more hidden ways as well. The USDA of the post–World War II era has been cited as a classic example of a "captured" regulatory agency, in which the regulated industry actually controls the workings of the regulators behind the scenes.[67] For example, within the USDA, pesticide policy was handled by the Agricultural Research Service (ARS). This official regulatory body, however, shared personnel with pesticide manufacturers and industry associations, and also carried out large-scale pesticide application campaigns in an attempt to eradicate pests like the fire ant.[68] Industry and regulatory personnel were connected by strong and regular channels of communication, as well as a shared faith in the ability of pesticide to provide the control and abundance promised in DBCP marketing. The USDA at midcentury largely worked with industry interests to deny or minimize public health or environmental threats posed by pesticides.[69] While individual doubts sometimes emerged within the ARS, by and large the culture of the USDA shared in industry optimism. The USDA was not the only regulatory body to play a part in the nation's pesticide regulation, and interactions with other agencies could at times counter trends within the Department of Agriculture. However, while officials from the Public Health Service and Fish and Wildlife Service were able to question certain USDA campaigns,[70] the Department's dominance in pesticide regulation remained secure. Ultimately, this meant that pesticide regulation usually favored the goals of the pesticide industry.[71]

If regulators faced conflicting demands to protect public health and serve the interests of the chemical industry, pesticide producers themselves could not ignore public anxieties about the dangers of the chemicals. Not only were Dow, Shell, and their contemporaries subject to regulatory demands, but they also feared losing their market or facing damaging lawsuits.[72] As early as 1952, NACA signaled this concern by coauthoring a "Manual of chemical products liability; an analysis of the law concerning liability arising from the manufacture and sale of chemical products" jointly with the leading chemists' professional organization, then known as the Manufacturing Chemists' Association.[73] The agricultural industry also reacted to concerns about pesticide health effects by producing materials and public relations strategies that trumpeted the benefits and safety of pesticides.[74]

In 1958, NACA published a booklet, "Open Door to Plenty," that epitomized the ideology of control and abundance, assuring readers that "today, scientific control of insects, plant diseases, weeds and rodents is also used to make home gardening more fun, to keep highway roadsides beautiful and safe at low cost, to protect our forests from pest damage, and to make pest-free outdoor living enjoyable for millions of people."[75] The pamphlet built a grand narrative of "man's struggle against pests" that began with the biblical locust plagues and ended with the promise of "progress through research."[76] In this document, NACA sought to rhetorically position its industry as responsive to the needs of the small, self-sufficient farmer who had "called on chemical companies" for help.[77] The pamphlet acknowledged that "questions were raised" about the safety of pesticides, but negated the seriousness of any such questions by identifying the askers as "city people."[78] NACA also sidestepped concerns about health and the environment by using the existence of government regulation to reassure readers. "The company has to prove to USDA scientists that the new product will be useful in agriculture and will be safe to use. The company has to prove to the Federal Food and Drug Administration scientists that it can be used effectively on food crops without any hazard to consumers."[79] With this claim, NACA used the existence of regulation as an imprimatur of legitimacy and safety for their industry products at the same time the organization exerted powerful behind-the-scenes influence on the development of the regulations themselves.

In addition to public relations and legal strategies, the tools and language of science played a central role in regulatory and industry efforts to resolve the contradictions between the ideology of abundance and control and the dangers of pesticides. Both FIFRA and the Miller Amendment required companies to submit not only residue data, but also "full reports of investigations made with respect to the safety of the pesticide chemical."[80] These investigations were made by scientists working in toxicology, the study of adverse effects of chemical compounds and the methods to control them.[81] As Christopher Sellers has argued, beginning in the 1920s, toxicological approaches to industrial hazards increasingly removed the consideration of threats to laborers' bodies away from the workplace itself, avoiding contentious labor politics and focusing on the effects of individual chemicals, rather than of job sites considered more holistically.[82] Toxicology rested on an assumption that animal bodies and human bodies shared basic vulnerabilities to environmental exposures, and further, that these vulnerabilities could be determined by the tools of laboratory research. By looking for adverse effects

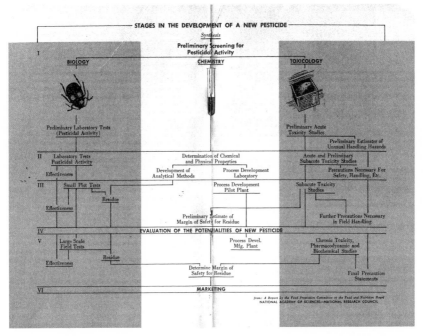

FIGURE 2. Created by the National Academy of Sciences and reprinted in a booklet produced by the National Agricultural Chemicals Association, this chart illustrates the steps, agreed upon by industry and regulators, for bringing a pesticide to market. In violation of the flow shown here, DBCP was marketed before testing was complete. From *Open Door to Plenty* (Washington, DC: National Agricultural Chemicals Associations, 1958).

in animals exposed to chemicals, laboratory research resulted in information that was used to set a "safe concentration" or "threshold level" for individual chemicals. Promising an impartial view into the health effects of chemicals, the methods and language of toxicology became a shared argot as industry representatives, federal regulators, and university researchers grappled with setting rules for pesticide use.[83]

The normative process for toxicological testing of pesticides was graphically laid out by the National Academy of Science's National Research Council in a flow chart showing the biological, chemical, and toxicological tests that a pesticide was supposed to undergo before marketing.[84] Illustrating a process ranging from animal trials to studies on residues, this clear hierarchy assured readers that nothing came to market without undergoing a battery of tests. Pesticides were to be subject to the careful controls of science before they could come in contact with the human body. NACA reprinted the flow chart in the "Open Door to Plenty" booklet, assuring readers: "No products in use in the world today—not even pharmaceuticals—are more thoroughly tested before they are sold or used." Leaving behind any scientific uncertainty (for example, that today's test could not predict tomorrow's risks), NACA made the expansive claim that a strict testing regime meant: "Nothing is left to chance."[85]

If toxicology was a language shared by company researchers and regulators, its specialized language was not accessible to most pesticide users. A key endpoint of the toxicological testing process, then, was the development of warnings for inclusion on pesticide labels. By translating toxicological results into nonscientific language, warnings were meant to briefly and clearly communicate the salient points of a chemical's dangers to a potential user without special training. A rational individual thus warned of a hazard could make an individual informed decision about exposure to the chemical.[86] Before World War II, warning labels were used on dangerous products including poisons (some of them early pesticides), hair dyes, and some pharmaceuticals.[87] In 1945, the Manufacturing Chemists Association issued its *Manual L-1: A Guide for the Preparation of Warning Labels for Hazardous Chemicals,* which specified general principles of warnings for hazardous chemicals.[88] Two years later, FIFRA set enforceable guidelines for the creation of pesticide labels, which stipulated that all products must bear a label with instructions for safe use and required that "highly toxic" chemicals be labeled with a skull and crossbones with the word "poison" printed in red ink.[89]

While warning statements obviously do not lessen a chemical's toxicity, at their best they do enable at least some chemical users to better protect themselves. One problem standing in the way of protection, however, was the question of how factors such as literacy, design, and vocabulary affected label interpretation and efficacy.[90] But the use of these labels created another, more insidious, problem: by including warning labels on their products, manufacturers could claim to have resolved the problem of pesticide exposure. And they did so not by making safe pesticides, but rather by shifting the responsibility for safety to the user. This shift was illustrated in claims such as that of one Public Health Service official—who was also an unapologetic industry booster—that when "*properly used* [pesticides] . . . do not cause any disease or increased susceptibility to disease in either man or animals" (emphasis added).[91] Accusations of failure to "properly use" pesticides functioned as a blanket response to critiques of the chemicals' health effects[92] and, significantly, kept the blame with the user and not with the pesticide.

At midcentury, then, regulators and industry addressed public concern about the health effects of pesticides by using toxicology as proof of the products' safety. But the convoluted alliance between the regulators and those they were entrusted to regulate used science in a way that served to protect business rather than health. For both industry and regulators, toxicological research and the warnings it produced seemed to resolve the contradictions between the ideology of agricultural plenty and control and the frightening dangers posed by pesticides. If the health harms of pesticides were both knowable and preventable, they did not pose a real threat to the growing use of these chemicals. But this was no resolution. Rather, both the science and the precautionary labeling language expanded the ideology of agricultural plenty and control by asserting that the risks of pesticides were themselves controlled—indeed, that these risks *had already been controlled* by the time the product hit the market. Toxicology would function to make risks evident, and good communication practices would mean that the user could protect herself from bodily harm. Disease, like pests, could be obliterated by the tools of science.

CONCURRENT MARKETING AND RESEARCH: "NO SPECIAL HAZARDS"?

The history of DBCP reveals the fault lines in the idea that pesticide harms could be contained by toxicological research and state regulation. As graphically conveyed in the flowchart produced by the National Academy of Science and reproduced by NACA, industry and regulators asserted, together, that toxicological testing was conducted *before* pesticide compounds were marketed, and, implicitly, that chemicals that did not pass a strict testing procedure would not make it to market. When it came to DBCP, however, this was not to be the case. First, the chemical was marketed while toxicological testing was incomplete, and therefore before risks to workers or consumers were known. Second, while toxicology seemed to promise a clear-cut answer to the questions of a chemical's hazard, testing results were in fact open to subjective interpretation. In the case of DBCP, company personnel downplayed worrisome results, instead adopting a sanguine understanding of evidence that allowed them to move forward with their new product.

Shell was the first of the two companies to market its DBCP formulation, with Nemagon appearing on the market in 1955.[93] The corporation sold its "technical grade" product to formulators, who used it to manufacture products for end users. By 1956, a Shell "agricultural bulletin" explained to readers that "various formulations of Nemagon are prepared and sold by many insecticide manufacturers under their own labels and are available at many garden-supply stores and insecticide dealers."[94] Solutions, emulsifiable concentrates, and granules were marketed for use on crops including cotton, vegetables, fruit trees, grapes, berries, ornamental plants, and turf grass.[95]

With the first sales of Nemagon under way, both companies continued toxicological testing. At Dow, Ted Torkelson led an in-house research team; while Shell hired Charles Hine, a physician and toxicologist at the University of California Medical Center, to conduct their research. Both groups submitted the chemical (or perhaps more accurately, a small army of laboratory animals) to a battery of tests to determine what effects it might have on humans who touched, ingested, or inhaled it. In January 1955, Hine's team concluded from vapor and 24-hour skin exposure studies that while the chemical was "moderately toxic following single respiratory exposure, no special hazards [were] anticipated in the handling" of the compound.[96]

The next year, however, research at Dow suggested that DBCP could cause various health and developmental problems in test animals—including fatty livers, retardation of growth, and shrunken testes.[97] Shell, Dow, and University of California personnel carried on a correspondence discussing the discrepancies between the two labs' findings, and both research teams rechecked their results, each at first reaffirming its own findings.[98] In September, Dow toxicologist V. K. Rowe wrote to Hine reporting that Dow's Biochemical Research Laboratory had determined that the dose required to kill half of exposed animals (a toxicological standard referred to as the "LD50," for "lethal dose 50%") was lower than the corresponding figure settled on by Hine's team at Shell. Rowe asked if Hine would like to cooperate on research to find out why DBCP was apparently "more toxic in Midland than it is in San Francisco."[99] Rowe also sent tissue samples from Dow to Hine, and the University of California pathology laboratory confirmed that they contained testicular and tubular abnormalities. Reviewing his own tissue slides, Hine reported to Rowe in March 1957 that he now found "incidental readings of testicular damage with atrophy and aspermatogenesis," but not with the frequency detected by Dow.[100]

Despite the clearly unfinished nature of the toxicological testing, in 1957 representatives of eight departments at Dow signed off on a "release to sales" document for a DBCP formulation that would compete with Nemagon. The signers included Rowe, who added a handwritten statement next to his signature that is illegible in the archival copy but appears to place a condition on the release.[101] Whatever the condition, it was apparently met: by the end of the year, Dow proudly claimed Fumazone nematicide as one of the latest products in its 50-year history of producing agricultural chemicals.[102]

Selling DBCP before testing was completed meant that warning labels would necessarily be based on incomplete information. Labels developed in 1956 and 1957 ignored the existing data on testicular pathology under discussion between the two companies' researchers. Indeed, early DBCP warnings used stock language, concentrated on potential harm to skin, and offered vague warnings about vapors. For example, the warning from Shell's 1957 "Label Bulletin" for Nemagon read:

> WARNING: VAPOR HARMFUL. ABSORBED THROUGH SKIN.
> Avoid breathing vapor. Avoid contact with skin, eyes, and clothing. Do not take internally. In case of accidental spillage on person or clothing, immediately remove clothing and flush skin or eyes with plenty of water; for eyes, get medical attention. Wash clothing and air shoes thoroughly before reuse. Keep away from heat and open flame.[103]

Shell's warning on the final draft of a label designed for export shipments bore the same warning, in English only, with instructions to contact the local Shell subsidiary for more information.[104] Whereas Shell told users that DBCP vapor was "harmful," a label developed by Dow in 1956 gave the slightly longer but no more alarming warning: "Harmful liquid and vapor causes skin irritation and blisters on prolonged contact."[105]

Rather than fully reflecting toxicological results, these statements used standard language codified in the Manufacturing Chemists Associations' *Manual L-1*.[106] The language was likely to have seemed familiar (and perhaps subsequently unalarming) to those who regularly handled chemicals. Neither company's warning statements even gestured toward the possibility of serious long-term health effects. And neither noted on its label that the health effects of the compound were still undiscovered, even though the *Manual L-1* included a provision for experimental samples to be labeled with a cautionary statement explaining that the hazards were unknown. The early labels for DBCP projected a false image of DBCP's safety. Meanwhile, the full toxicological picture of DBCP was still emerging in the corporate and university laboratories where scientists studied the chemical.

In June 1958, a paper bearing the mundane title "Dibromochloropropane: 50 Vapor Exposures and Ancillary Blood Studies" detailed some of the most disturbing DBCP results yet. Hine's research team, it reported, had found that many of the animals exposed to DBCP vapors had abnormally small testicles and could not produce sperm.[107] The report unambiguously spelled out the effect of the chemical on the testicles. Four groups of 15 rats had been exposed to increasing levels of DBCP; a fifth group was unexposed. Half of the rats exposed to the lowest concentration of the chemical—5 parts per million (ppm)—showed testicular damage that in some animals reached "complete degeneration with complete azoospermia [absence of sperm in the semen]."[108] Rats exposed to 10, 20, and 40 ppm of DBCP did even worse. The document reported that "all of the testicular sections in the 20-ppm group showed complete azoospermia with necrosis of the epithelium lining the tubules."[109] In other words, DBCP was somehow preventing animals from generating sperm and killing the cells on the inside of the seminiferous tubules, the site of spermatozoa germination, maturation, and transport in healthy animals. "At 40 ppm," continued Hine, "the effects were similar and more extensive." None of the 40-ppm rats survived the full length of the experiment.

Hine's results were supported by research at Dow, where a July 1958 report concluded that laboratory animals exposed to Fumazone "were seriously affected by only a few repeated seven-hour exposures."[110] The report continued: "All of the male animals had small testicles at autopsy. . . . Effects upon the spermatozoa are characterized by reduction in number to complete absence of sperm cells or the development of abnormal forms."[111] In addition, the seminiferous tubules had degenerated. Shell chemist Louis Lykken, manager of the Technical Services Department within the company's Agricultural Chemical Division, reported to his colleagues that Dow's researchers were "very upset by the effect noted on the testes."[112]

Toxicologists at both companies knew that the disturbing animal results meant that the chemical might have serious health effects in humans as well. For example, Dow's report concluded that because of the "extensive effects" in animals, levels to which humans were exposed should be much lower. In addition, the report cautioned: "Personnel handling this material should be made aware of the hazards involved."[113] Alerted to the results, Mitchell Zavon, consulting medical examiner for Shell's Agricultural Chemical Division, recognized that the results made it "more urgent that we pay additional attention to Nemagon and its effect, if any, on those people exposed to the material." Zavon was a physician who had worked for the Public Health Service, taught at the Kettering Laboratory at the University of

Cincinnati, and served as director of the occupational health services at the Cincinnati Health Department.[114] He suggested that sampling be carried out during production, transportation, and use of the chemical.[115]

In the summer and fall of 1958, the industrial hygienist at the Denver plant where Nemagon was manufactured took measurements on "atmospheric contamination" in the plant.[116] Plots of three separate measurements in June and October showed concentrations mostly between 6 and 20 ppm, with spikes reaching almost to 60 ppm. Measurements from November showed that the installation of a new hood had prevented the most egregious exposures, but exposures still hovered between 1 and 4 ppm. For its part, Dow conducted a study on exposures faced by applicators and found that applying Fumazone through a sprinkler irrigation system could cause concentrations at about half the levels "known to cause serious injury to laboratory animals."[117] Researchers concluded that "this may or may not present a real hazard," but admitted that "this technique of application does offer a ready opportunity for excessive skin contact with the material."[118]

At Dow, a July 1959 internal document listing "toxicological information on Fumazone suitable for use in a bulletin" suggested that the chemical should be handled in a closed system (one engineered so that no vapor could escape) if at all possible.[119] Otherwise, the document advised that exposures be monitored and kept to 1 ppm and that skin and eye protection should be worn by workers. However, the company seemed reluctant to make these recommendations public. A September 1959 Dow "Bulletin" introducing formulators to a new Fumazone product contained milder guidelines.[120] No mention was made of the importance of maintaining a 1 ppm standard; readers were told that "single short exposure to small amounts should cause no more than mild irritation," and were counseled to avoid repeated skin contact or inhalation of the "fairly toxic" vapors. Finally, no information was given on testicular risk.

In the midst of managing worrisome toxicological results, Dow and Shell also faced a new hurdle. In August 1959, legislation passed that would bring nematicides and a few other classes of pesticides under the terms of FIFRA as early as March 1960.[121] For the first time, DBCP would be subject to USDA approval of product labels, as well as FDA establishment of tolerances, the levels of residues that were allowable on or in agricultural commodities treated with the chemical. The USDA oversight in particular would cast a spotlight on toxicology results, as that agency was charged with assuring that labels contained adequate user warnings and instructions for use.

Dow and Shell joined forces, sharing fees and data to complete the process of registering their products in accordance with FIFRA.[122] In January 1960, they asked the FDA to push back by one year the deadline for DBCP registration.[123] By April, the FDA's affirmative response was published in the Federal Register, giving the green light for continued use of DBCP without registration.[124] The reprieve gave the companies time to further address the potential health problems of the compound.

One matter to be settled was the impact of exposure on agriculture workers. DBCP could be applied using a "chisel" that injected the fluid into the soil, through the broadcast application of granules, or by the addition of a liquid formulation into irrigation water. Various methods posed different risks to workers, with the irrigation method raising the most red flags with toxicologists. A June 1960 toxicological report on the Fumazone formulation used in irrigation water bluntly stated that short-term inhalation exposure and dermal exposure from "relatively small amounts" could cause "serious systemic injury, even death."[125] In September of the same year, V. K. Rowe wrote that he was "somewhat concerned" about the application of Fumazone with a "spray technique," noting:

> Persons involved in this operation should certainly stay away from the spray so that they do not become wet with it. This could cause serious adverse effects.... There is likely to be a rather high vapor concentration in the treated area. It would seem to me that this could be quite hazardous if persons entered it and attempted to work in it. Another hazard would be if a gentle breeze were blowing the vapors from such a treated area into a populated area.[126]

Faced with these alarming conclusions, Dow advised Florida officials against the overhead-sprinkler application of DBCP.[127]

For its part, Shell undertook a study in citrus groves of that state to "show whether or not applicators and others are exposed to the concentrations of Nemagon which Dow considers hazardous."[128] Shell's results showed exposures during application ranging from 0.83 ppm to 3.31 ppm, with most application exposures in the 1–2 ppm range.[129] Although many of the concentration studies were above the 1 ppm limit previously recommended by Dow[130]—and although neither company had evidence on what constituted a safe exposure limit—Zavon wrote to the Florida State Plant Board that the exposure levels found were "comparatively low," and that irrigation could continue. He nevertheless recommended that "anyone who is exposed or potentially exposed wash thoroughly and change to clean clothing after exposure. If it appears that the exposure will be more than a brief matter, a respirator should be worn with a filter suitable for organic vapors."[131]

By October 1960, Dow also seemed to back down on its concerns regarding irrigation application, at least publicly: an information sheet on that technique recommended protective clothing and eye protection, but counseled respirator use only in poorly ventilated areas—otherwise, users were simply advised to "avoid breathing vapor or fumes."[132] The warning provided was substantially similar to those developed in 1956 and 1957.

A HISTORY OF SAFE USE?

Although the matter of DBCP's health effects on manufacturing and agricultural workers remained unsettled, Dow and Shell continued joint efforts to register Fumazone and Nemagon. In early 1961, Dow and Shell sent letters to the FDA and the USDA in hopes of meeting the extended deadline for nematicide registration under FIFRA.

From the FDA, the companies asked that tolerances be set for a number of crops to which DBCP might be applied, including bananas, a variety of vegetable and fruit crops, dry and succulent beans, and corn for consumption as ears and popcorn as well as in animal fodder. From the USDA, they asked for a "certification of usefulness," verifying that DBCP was effective on each crop for which they requested tolerances.[133] After certification was received and tolerances were set, the companies could move ahead to request final registration from the USDA.

The multistep registration process was not always a smooth one, and involved compromises by all parties: Dow, Shell, the USDA, and the FDA. For example, at various points, the regulators delayed action on the required certifications, asking for more information on certain crops.[134] While official letters suggested that these regulatory protocols proceeded as part of a formal and professional relationship, internal company communications hinted at a more friendly alliance between the companies and the regulators. The two sides met informally, and at least twice the USDA provided Shell with drafts of government evaluations of the companies' petition as a "means of simplifying the problem of clarification."[135] In early April 1961 the head of the Chemistry Section at the USDA's Pesticides Regulation Branch sent draft opinions, telling Shell's C. C. Compton: "As soon as you have completed your review of the enclosures, we would appreciate a call from you."[136] When the USDA finally issued its decision that same month, it certified most of the crops included by the companies, and the registration process moved forward on these. But Shell would still be disappointed. Indeed, the omission of some fruit and vegetable crops left Lykken complaining that "no value came from the informal discussions with USDA in April. Their original views persisted!"[137] Evidently, the relationship with federal regulators had not been collegial enough for Shell to get everything it expected.

While the companies waited for the FDA to issue tolerances, they prepared for the next big hurdle—USDA label approval and registration. The head of the Pesticide Regulation Branch of the Agricultural Research Service told Shell and Dow that the companies would need to provide additional data in order to achieve final registration of the compound, including more information on uses in certain crops and for particular pests.[138] It soon became clear that the Pesticide Regulation Branch's Toxicology Section also harbored concerns about the chemical.[139] In a list of comments "handed to" Shell and Dow, Toxicology Section personnel suggested a much stronger precautionary label for the chemical. Shell and Dow's own labels began with a nonspecific warning that DBCP vapors were "harmful"; by contrast, the USDA's text began, "WARNING: may be fatal if swallowed, inhaled or absorbed through skin."[140] It went on to list precautionary measures that were in some cases much stronger than those suggested by the companies, including an admonition to "wear polyethylene gloves and goggles when handling material."[141] The USDA scientists also had concerns that the product might be "considerably more toxic by skin contact than might be expected," and asked for data on that route of exposure.[142] The testicular data drew special attention, and led the scientists to request that, "in view of the testicular atrophy demonstrated to occur in experimental animals," the corporations provide information regarding health records of those individuals who have been employed for an extended period in the manufacture or formulation of products containing 1,2-dibromo-3-chloropropane."

Not surprisingly, the regulators' concerns met with resistance at Shell. According to Lykken, "the consensus" at his company was that the USDA's proposed label was "overcautious."[143] Shell's resistance was not based on a scientific assessment of the level of caution merited by the evidence, but rather by the fear of financial losses. Lykken affirmed that "the precautionary statements proposed could have an adverse effect on the sale of this product."[144] A strong warning would endanger the profitability of a chemical that Dow and

Shell had already spent a great deal of time, money, and effort developing and testing. To protect their investment from the threat of a stronger warning, Lykken wrote that Shell representatives should meet jointly with Dow toxicologists and the USDA officials to "settle this issue."[145]

Personnel from the USDA agreed to meet with company representatives. Dow and Shell left the meeting with the regulators with what they had wanted: a chance to refute the concerns of government toxicologists. The USDA, according to Shell's Mitchell Sprinkle, was "willing to relax their labeling requirements if we can provide them with a history of safe use experience in the field and in the manufacturing plant."[146]

Given this opening, Shell set about interpreting DBCP data to emphasize the so-called "history of safe use" of the chemical. The logic of safe use held that, since there had been no adverse reports of human health problems with DBCP, the health effects noted in experimental animals were not generalizable to humans. This approach flew in the face of the fundamental premise of toxicology that informed the companies' own earlier reactions to DBCP animal findings: animal studies provided the best proxy for humans, who were generally protected from experimentation by scientific ethics.[147] In 1949, Commissioner of Food and Drugs Paul Dunbar had told the National Agricultural Chemicals Association that scientists depended on animal studies to infer cumulative or chronic effects in humans. Discussing feeding studies of DDT in rats, he said:

> Now we cannot perform similar experiments on babies. We cannot exactly translate the results of rat experiments into terms of human effects. Certainly in any study of possible toxicity, if we know that one species of animal is affected, there is only one course to follow and that is to play safe and assume that the same results would follow if we could use human animals as experimental subjects.[148]

As Dunbar explained, animal studies were especially important to test chronic effects—those that would appear only after prolonged exposure to a substance. Studies over a short animal lifetime could give insight into analogous exposures in humans, who have much longer life spans and in whom chronic disease caused by extended exposures or surface contact might become apparent only after a latent period of several years or even decades.

Faith in animal testing was reflected in the Dow and Shell scientists' original investigations into DBCP. As the registration process moved forward, researchers from Dow and the University of California jointly submitted a report of their results from those animal studies to the journal *Toxicology and Applied Pharmacology*.[149] When the study was published in May 1961, it recounted the finding of both labs, giving prominence to the abnormal reproductive findings:

> 1,2-dibromo-3-chloropropane caused general manifestations of toxicity, including poor growth, predisposition toward secondary infection, and a specific histological alteration in the testes in male rats receiving 50 repeated 7-hour exposures to 5 ppm. This was the lowest concentration studied. The effect upon testes resulting from exposure to higher concentrations was particularly severe, resulting in atrophy, degenerative changes, reduction of spermatogenesis, and the development of abnormal sperm.[150]

At the same time Dow and Shell staffers were working to *downplay* the effects of DBCP human health effects in the process of registering their products, authors of the *Toxicology and Pharmacology* article extrapolated from their animal findings a number of suggestions for safe handling by humans. For example, although none of the researchers had tested for "no effect level"—the concentration of DBCP animals could be exposed to without developing health problems—they recommended that air concentrations be kept below 1 ppm. They also warned that concentrations would need to be measured in order to maintain that level, since although DBCP had a bad odor at higher concentrations, the smell of the chemical at 1 ppm concentrations was not strong enough to warn people away. In addition, they recommended that if "adequate ventilation" was not available, the chemical should be handled in a "closed system" that did not allow any vapor to escape. Alternately, people exposed to the chemical should use respiratory protection; the authors recommended use of "a full-face gas mask equipped with an organic vapor (black) canister, an air- or oxygen-supplied respirator, or a self-contained breathing apparatus" to protect from exposures remaining after a spill. Finally, they concluded that safety glasses and impermeable clothing be worn when handling the chemical, and that clothes and shoes contaminated with DBCP should "be promptly removed and not worn again until completely free of the material."[151]

In the regulatory realm, however, Dow and Shell moved away from such normative interpretation of their own evidence, and instead developed an explanation of the chemical's toxicity that downplayed the animal results and insisted that human bodies were not subject to the same risks as the monkeys, rats, guinea pigs, and rabbits in Hine's and Torkelson's labs. At the same time, they excised earlier expressions of doubt or precaution on the part of researchers, resulting in an interpretation that insisted that DBCP presented no special health hazards.

These efforts are evident in Shell employees' 1962 edits to a draft report on the toxicology and pharmacology of DBCP. Dow and Shell had rehired Dr. Hine to produce a report that would become part of the documentation submitted to the FDA for residue deliberations. In early 1962, Shell employees Mitchell Zavon and Louis Lykken sent Hine a letter and an edited manuscript of his own earlier report. Certain changes were needed, they stressed. Most importantly, "human use experience must take precedence over animal experiments and the report ought to show this clearly."[152] In addition, they told Hine that his recommendation that DBCP users wear impermeable clothing was "impractical." Where Hine had written "the compound may be classified as moderately to highly toxic following single respiratory exposures and highly toxic on repeated exposure," a reviewer, presumably Zavon, had underlined the second "highly" and written a suggestive "I wonder!" in the margin.[153] And despite the company's focus on human health experience rather than animal data, Hine's reviewers told him to "leave out speculation about possible harmful conditions to man. This is not a treatise on safe use."[154] On the other hand, Zavon *did* want Hine to cite the "considerable human use experience" that showed no harmful effects. "If Dr. Hine does not have it available," he wrote to Lykken, "I am certain that between your files and my own we can either make it available to him or write something for inclusion in the final report."[155] Zavon and Lykken's edits show that Shell not only wanted to direct attention away from animal data, they only wanted human health data if it presented a sanitized picture of the potential risks of DBCP.

Such a picture was presented in the company's "Summary of Use Experiences in Manufacture and Application of NEMAGON Soil Fumigant," finalized in October 1962 for presentation to the USDA.[156] A key part of the "evidence" offered in the report was a short summary of the medical records of workers at Shell's Denver manufacturing plant. Although Zavon had known there was a need to look at production worker exposures as early as 1958, more than three years passed before he obtained the data for inclusion in the "Summary":[157] "Routine examinations are made on operators assigned to the NEMAGON unit. These routine examinations consist of a complete physical checkup including chest X-rays, blood tests, and urine test. As far as could be determined by subjective symptoms, results of physical examinations, and clinical tests, there were no indications of harmful effects due to NEMAGON."[158]

The report also included the air concentrations of Nemagon vapors recorded in the Denver plant in 1958. The concentrations, "normally maintained below 1 ppm," occasionally rose to 10 or even 20 ppm for up to thirty minutes, with occasional spikes as high as 50 ppm. Even at these levels, the report maintained, unremarkable medical records showed that such exposures did not harm workers who wore "coveralls, rubber shoes, and impervious plastic gloves," but normally worked without gas masks or respirators.

Next, the report built on the supposed evidence of safe use within the plant to present a case that field applicators would not be exposed to dangerous concentrations of the chemical. It cited Shell's 1960 study of air in Florida citrus groves showing airborne concentrations ranging from below the level of detection (0.5 ppm) to 3.31 ppm after overhead irrigation.[159] Shell reassured the USDA that these concentrations would not harm workers, based on the fact that they were lower than the levels found in the Denver plant.[160] Because no adverse effects had been reported in production workers, "occasional exposure" at the field levels reported would supposedly present no risk. The report also asserted that workers exposed to Nemagon at levels above 1.5 ppm would be protected because they were likely to "seek vapor-free areas" due to eye and nose irritation caused by the chemical.[161] Like warnings that counseled users to "avoid breathing fumes," this assertion was based on an assumption that users would have both the inclination and the freedom to walk away from a task in process. Ignoring the practical realities of pesticide application, the "Summary" contended that agricultural workers exposed to DBCP through overhead irrigation were safe, and by extension all other application methods could also be considered innocuous: "The application of NEMAGON by overhead irrigation would undoubtedly lead to the highest vapor concentration likely to be encountered from any practical method of application. Therefore, it is inconceivable that injection methods could give rise to more than a small fraction of the concentration in air that occurs from application by overhead irrigation."[162] Finally, the report contended that even workers who did not leave the areas of heaviest overhead irrigation were safe. "Based on plant exposures," the report read, "occasional exposure to even [the heaviest documented field exposures] would not be hazardous to humans."[163]

In short, Shell based factory and field risk assessments on the premise that no damage had been found to the health of exposed plant workers. However, there is no evidence that these exams actually tested for the one effect that the USDA toxicologist had specified as the greatest area of concern: testicular damage. The report itself does not claim that workers were examined for any testicular or reproductive effects, a point taken up many years later by journalist Cathy Trost, who wrote in 1984 that her examination of those medical records showed no results of any sperm count or testicular function tests.[164] In the absence of such testing, Shell managed to parlay incomplete worker health data into a blanket claim of safety for all agricultural uses of DBCP.

Finally, the report addressed USDA's concerns about dermal toxicity. Shell had hired a consulting laboratory to conduct tests similar to the agency-commissioned tests showing rabbit death from skin exposure to Nemagon. Industrial Bio-Test Labs (IBT) conducted the test for Shell, exposing ten rabbits to DBCP "at a dosage of 200 mg per kilogram . . . for 24 hours."[165] None of the rabbits died, and none exhibited any "untoward behavior reactions."[166] Translating rabbit results to humans, the "Summary" maintained that, "it is inconceivable that small amounts of NEMAGON contamination on the skin of humans would have any effect. Similarly, gross amounts spilled on the skin would have no effect, provided it was not allowed to remain on the skin for extended periods." Here, in another remarkable sleight of hand, Shell switched its position on the use of animal testing to substantiate human health effects, or rather, the lack thereof.

While the companies had been preparing their retort to worries about DBCP health harms, they continued to hash out other issues with regulators as well. The FDA had refused to establish tolerances for a number of crops, citing lack of data and dangers that the residues would reach unsafe levels in livestock fodder, leading to meat and milk residues.[167] The FDA also refused to set tolerances for dried fruits; since these were considered processed foods, the companies would need to apply for a "food additive tolerance."[168] The food additive tolerance in particular caused consternation at Shell and Dow, as some at the companies felt it would set a precedent for future onerous regulations.[169] After some conflict between the companies, Dow threatened to involve outside attorneys, and the FDA agreed to treat the existing application as a food-additive petition, saving the companies reams of research and paperwork and revealing the limits of the FDA's willingness to place demands on the companies it was charged with regulating. In 1963, while the debate over a new book—Rachel Carson's *Silent Spring*—was raging, the FDA set tolerances for DBCP residues on crops including lettuce, bananas, carrots, broccoli, cabbage, eggplant, some melons, pineapples, tomatoes, a number of berries, grapes, walnuts, citrus crops, peaches, and a number of other vegetable and fruits. At the same time, the agency granted Shell's petition for allowable residues in some dried fruits and processed tomato products.[170]

In March 1964—after one more deadline extension—the USDA registered Nemagon and Fumazone.[171] Dow and Shell had successfully convinced the regulatory agencies to accept their version of DBCP toxicity. Despite initial concerns, the USDA and FDA had yielded to the companies' characterization of DBCP as safe for human use. Although wording differed slightly between labels, approved warning language was startlingly similar to copy that had been used before toxicological testing results were in, and contained no mention of testicular damage or risk of any other chronic effect. For a Nemagon formulation, the label stated: "WARNING! HARMFUL LIQUID AND VAPOR MAY CAUSE IRRITATION OF SKIN, EYES, NOSE AND THROAT."[172] Although both Dow and Shell had at some points recommended respirator use under certain conditions, the label contained no respirator recommendation, instead advising users to "avoid prolonged breathing of vapors" and to "use in a well ventilated area." The label also offered the commonsense guidance to avoid contact with skin, eyes, and clothing; wash after use; avoid swallowing; and keep away from heat and flame. Finally, the label cautioned: "In case of accidental spillage on person or clothing, immediately remove clothing and flush skin or eyes with plenty of water; for eyes, get medical attention. Wash clothing and air shoes thoroughly before reuse." With USDA approval of this feeble warning language, DBCP—already on the market for over six years—was now officially sanctioned as a tool for fighting the nematodes all but unknown to most farmers just a few decades before.

The early marketing of DBCP, in line with a broader mid-twentieth-century technological optimism, promised easy abundance and control of nature. The emerging science of toxicology was supposed to provide a clear answer to the question of the safety of new compounds, while careful regulation based on that science would prevent dangerous exposures. However, in DBCP's case, the picture was much murkier. Testing and marketing happened at the same time, and test results were subject to interpretation by interested parties, even in variance with accepted tenets of toxicological testing. While federal regulators did at some points question Dow and Shell's explanation of DBCP toxicity, they were apparently convinced by the chemical companies' arguments, and ultimately registered the chemical. Despite the companies' insistence that a "history of safe use" meant production and agricultural workers would be safe, the real health toll of the chemical would only become known after more and more workers were exposed to it. In the meantime, registration under the terms of FIFRA opened the door to increased DBCP use not only in the United States, but also around the world.

TWO

DBCP on the Farm

ONE CROP STOOD OUT IN Shell and Dow's 1961 residue petition to the USDA. Bananas were different because they were not commercially grown in the United States at all, but imported from nations including Honduras, Costa Rica, Panama, Guatemala, and Nicaragua. The inclusion of bananas in the chemical companies' DBCP petitions to the USDA signals that the use of the pesticide must be understood in its transnational as well as its local and national contexts. In fact, tracing the rise and fall of DBCP use from 1961 to 1977 points to the complex interconnection of multiple local, national, and transnational actors and processes.

In the United States, DBCP use increased as new crops and application methods were approved by the federal government and as growers in different regions, growing various crops, found it to be effective. Even exposures within the United States could not be understood as purely "national" phenomena, as a large proportion of farmworkers there were themselves immigrants, and the material realities and politics of farm work were deeply shaped by transnational dynamics, including a U.S. willingness to let Mexicans, Filipinos, and other immigrants go unprotected. The landscape of DBCP use outside the United States was even more complex, as local, national, and transnational sites and processes combined to produce uneven exposures to the chemical. The long history of banana export production in Central America, as well as the changing exigencies of the chemical and fruit markets in the United States, impacted where and how the chemical was used. Corporations based in the United States developed practices that were intimately related to the movement of pests, and even the banana plant itself, in transnational space. Perhaps an even more surprising transnational influence on DBCP use was the impact of U.S. regulations—meant to protect U.S.

consumers—that ended up implicitly authorizing the use of DBCP in Central America, putting banana workers at risk from uncontrolled exposures to the chemical. At the national level in the Central American republics, failures to enforce regulations further shaped DBCP use. Local and company-wide policies on day-to-day interactions between workers and managers also helped determine how DBCP was used, as did regional and local soil and weather conditions. These processes, taking place at various scales, ultimately affected workers at the most local level of the body, as they faced largely uncontrolled exposures to the toxic chemical.

A GROWING MARKET IN THE UNITED STATES

After 1964, Dow and Shell sought federal approval for a growing number of DBCP uses and formulations. Within a few years, federal regulators had approved the use of the chemical on cotton and peanuts, as well as on turf and ornamental plants.[1] Dow and Shell manufactured their own products and also sold "pure" or "technical grade" DBCP to formulators, who used it to create products with varying forms and levels of active ingredient, marketing them as Fumazone or Nemagon, or under names like "X-Cel Nemagon Granules" or "Security Nema-Kill."[2] Occidental Chemical company, or "Oxy," also began synthesizing DBCP under an agreement in which Dow supplied the bromine, epichlorohydrin and allyl chloride used to make DBCP, and Oxy, through its recently purchased subsidiary Best Fertilizers, manufactured Fumazone products at a Lathrop, California plant. Oxy also formulated its own line of DBCP-based nematicides.[3]

From the mid-1960s to the mid-1970s, DBCP use spread quickly across agricultural sectors in the United States. In the 1960s, DBCP producers marketed it for use with various application methods on citrus, cotton, walnut, lettuce, melons, carrots, prunes, grapes, peaches, tomatoes, and cabbage. By 1970, DBCP was among the top three fumigant nematicides and was being used to treat 1.6 million acres in the United States.[4] In 1972, 12 million pounds of the nematicide were purchased in the United States.[5] DBCP was still the only material that could be applied after planting, making it especially valuable in perennial crops such as fruit trees, grapes, and pineapples. Efforts to gain approval for new application methods and crops, including successful petitions for soy and lima beans, continued into the middle of the decade.[6]Increasing DBCP use meant, of course, increasing farmworker contact with the chemical. Farmworkers in the United States were in large part a transnational population. Informal hiring practices as well as formal government programs, like the federal government's "guestworker" or *bracero* program of 1947–1962, had resulted in a farmworker population that consisted in large part of immigrants from Mexico and Central America.[7] Immigrant workers carrying out the difficult, low-paid, and intermittent work of farming also faced language barriers and deeply entrenched racism and nationalism. Immigrants, especially those who lacked official papers, were particularly vulnerable to employer exploitation, such as nonpayment of wages and dangerous working conditions.[8]

Just as immigrants were denied many of the basic rights of citizenship, federal laws denied farmworkers the rights and protections afforded other workers in the United States. In 1935, the National Labor Relations Act excluded farmworkers from its guarantees of union organizing and collective bargaining rights. Agricultural workers were also excluded from the 1935 Federal Unemployment Insurance Law and denied coverage under the 1938 Fair Labor Standards Act, which included minimum wage, overtime, and record-keeping requirements.[9] Some states did enact minimum wage laws for agricultural workers in the 1960s, and some federal legislation gestured toward ensuring better working conditions, but farm work remained outside the mainstream of labor protection.[10]

Lack of protection for farmworkers extended to health and safety issues. By the 1960s, working conditions and protections that had become commonplace in the industrial sectors were still missing in the agricultural fields, including such basics as the provision of drinking water, toilets, and washing facilities. Hazards faced by workers included food poisoning, heatstroke, and operating or working around large machinery that could cause traumatic injury. Cases resulting in worker deaths also included individual and mass poisonings by exposure to highly toxic pesticides. In 1962 alone, such incidents killed 83 farmworkers in the nation's most important agricultural state, California. That same year, there were 2,696 incidents of what the state classified as "non-fatal occupational diseases." That tally did not include chronic diseases and conditions such as those caused by exposure to chemicals including DBCP.[11]

Workers could be exposed to DBCP in a number of ways. Soil injection applications were made with tractors drawing chisel injectors or other types of spreaders behind them. Liquid formulations could be added to water used for various types of irrigation, including by sprinkler or "flood" irrigation, where water is pumped directly onto the field's surface. DBCP could also be applied as a "drench"—that is, it could be mixed with water and applied with a sprinkling can or hose. These uses all presented opportunities for exposures. When DBCP was applied using a tractor and chisel system, the major avenues for contact were in mixing and loading, as well as in handling or repairing machinery. Irrigation workers were exposed as they mixed DBCP with water or repaired broken irrigation systems pumped full of the pesticide or contaminated by residues. In all of these cases, workers could come into skin contact with the chemical or breathe in fumes. Handling granular formulations could also result in dermal exposure and, even when used in accordance with labeling instructions, could lead to inhalation exposure from highly volatile formulas.[12] In addition, workers tasked with cleaning spills could be exposed to skin and inhalation exposure. Anyone near or in a field or orchard during or after treatment could inhale remaining vapors of the chemical.[13] Drenching presented the greatest risk of exposure and was limited to ornamental and turf uses with treatment by professional applicators only.[14]

One place DBCP was used extensively within the United States was on the Hawaiian pineapple plantations for which it was originally developed. There, it was generally children or grandchildren of immigrants from Asia who worked for a handful of fruit companies.[15] In 1963, a short article by Earl Anderson, a plant pathologist at the Pineapple Research Institute (PRI), demonstrated one example of Hawaiian DBCP exposure in a newsletter meant "For [PRI] Members only," a group that included Dole Corporation and just four other growers and packers. Anderson urged pineapple growers to be careful when using DBCP products (which he called BBC). He reported that when PRI, "with the cooperation of the Dow Chemical Company," monitored experimental application of the nematicide, air concentrations ranged between 6.2 and 11.0 ppm, depending on the concentration used, wind speed, timing of sample, and position of the sampler. "The indication," Anderson concluded, was that "toxic levels of BBC vapor in the air were reached. The odor of BBC near the operation was pronounced." He warned pineapple growers to pay "close attention" to DBCP toxicity and protective measures, noting DBCP's capacity to cause "damage to liver, kidneys, and various tissues including sperm cells and seminiferous tubules, dermis, bronchioles, renal collecting tubules, lens and cornea, and alimentary canal" at levels as low as 5 ppm.[16] While Anderson's article showed the kind of information that was available to Hawaiian plantation owners, it is unlikely that workers would have ever had the chance to read or fully understand that information. In Hawaii, as in the continental United States, no regulations required farm supervisors to ensure that everyone in contact with a chemical knew the hazards they worked with.

THE BANANA TRADE IN CENTRAL AMERICA

In Central America, DBCP would find a place within the export-oriented banana industry established in the nineteenth century. The banana trade between the United States and the nations of Honduras, Costa Rica, Panama, Guatemala, and Nicaragua was dominated by two U.S.-based multinational corporations. United Fruit and Standard Fruit famously exerted their influence over both banana workers and Central American governments, using their vast resources and dominance over shipping to squelch competition, gain an upper hand in negotiations with governments, and set the terms in dealings with contract growers—often with the backing of U.S. military and political power.[17] The history of pest control on Central American banana plantations also reached ambitious proportions: in the first five decades of the twentieth century, these efforts included wholly abandoning large and established plantations in favor of new land, flooding vast tracts in hopes of killing pathogens, and investing in massive infrastructure to apply an early inorganic pesticide known as Bordeaux mixture.[18] However, none of these strategies seemed to adequately control pests, which continued to cost Standard and United millions of dollars each year.[19] As in the United States, the workers who labored for the banana companies were a transnational population in their own right, including at different places and times internal migrants from the Central American highlands, Garifuna people from the Central American Atlantic coast, and West Indian immigrants.[20] Workers not only labored directly in the fields, but also built and maintained railroads, cleared and tended plantations, and worked in extensive banana shipping operations.[21]

While the storied influence of the banana companies over Central American governments resulted in the moniker "Banana Republic," the companies always faced some resistance on the part of workers, citizens, and even governments.[22] By midcentury, U.S.-based companies faced growing challenges. Critiques of U.S. imperialism heightened, and the enormous profits reaped by U.S. companies were increasingly questioned. At the same time, pressure for land reform made it difficult for the fruit companies to hold on to the large tracts of land they had accumulated over the first half of the century. Finally, diversification of agricultural exports meant that sugar, cotton, and cattle joined bananas (and, especially in Costa Rica and Nicaragua, coffee) as major export crops, weakening Central American governments' reliance on banana companies.[23] By 1961, several nations established the Central American Common Market, which sought to boost industrial production in the region and spur economic growth in sectors that did not include bananas.[24] Growing labor power was evident when banana workers sparked a three-month general strike in Honduras in 1954,[25] as well as in four major Costa Rican strikes involving banana workers in the 1950s.[26] (Banana workers in Guatemala at the time, in contrast, were suffering under the oppression of the military rulers installed after the U.S.-backed ouster of populist Jacobo Árbenz.[27])

Banana companies responded to the challenges of midcentury conditions by changing their business structure as well as their model for agricultural production. These moves were intended, at least in part, to minimize their own risk while continuing to profit from the banana trade. United and Standard sold off land, reduced the number of workers they hired directly, and increasingly turned to contracts with so-called "independent growers."[28] This arrangement allowed the big companies to retain control over "schedules, methods, and material inputs necessary for proper irrigation, fertilization, pest control, planting, growing and harvesting," as well as research, shipping, and other operations and inputs,[29] while shifting risks to independent growers.

On the farms they continued to operate directly, the big fruit corporations changed their own agricultural practices in hopes of defeating pests, increasing productivity, and minimizing labor needs. For example, Standard and United changed the cultivar, or variety, of banana grown on all their plantations. Switching to the Cavendish cultivar was meant to address problems with a damaging fungal pest, but the change had implications for laborers as well. On the one hand, the adoption of the new cultivar necessitated on-farm packing, a job for which the companies for the first time hired women in large numbers.[30] On the other hand, Cavendish plants could be planted closer together, increasing yields per acre and decreasing the number of workers needed overall.[31] Along with such alterations in production, the companies increasingly looked toward the burgeoning chemical industry for solutions to intractable pest problems as well as to reduce labor costs. For example, herbicides destroyed weeds more efficiently than machetes, saving money for the companies.[32]

These interlocking changes solved some problems for the banana companies, while also creating new ones. One of those new problems was the proliferation of the nematode. By 1959, researchers estimated that the pests were present in substantial percentages of banana plantings in Guatemala, Costa Rica, Honduras, Colombia, and Ecuador.[33] As worries about nematode damage grew, it seemed the pest was poised to assume a key role in the ongoing history of Central American banana production.

NEMATODE PROBLEMS AND SOLUTIONS

Both the nematode problem and the responses to it were shaped by an interplay of local, national, and transnational forces and actors. At the most local level, the nematode problem involved contact between the pests and plants themselves. The nematode most often found feeding on bananas was *Radopholus similis,* or burrowing nematode. A colorless, wormlike parasite measuring less than a millimeter at adulthood, *R. similis* fed on root cells, destroying each cell before moving on to the next. The nematodes' feeding pattern caused rot and lesions that were visible on the outside of the root.[34] Damage rose with population size, and the pests reproduced quickly, with each female laying more than 1,000 eggs over a six-month period.[35] The banana plant that *R. similis* found so nutritious was, despite popular perception, not actually a tree but a large herbaceous perennial. It had a shallow root system, a "pseudostem" of tightly furled leaves, and a crown that grew heavy when fruit developed. The banana propagated either from shoots emerging from the base of the plant or from underground stems call rhizomes, which were cut from existing plants and replanted elsewhere.[36] The meeting of nematode and plant did not bode well for the banana. The pests fed on the roots but could also migrate from the roots into the rhizomes or the corm— the bulbous above-ground structure where roots meet stem. Such an infestation could cause the entire plant to fall, a problem that came to be named "toppling disease." Nematodes could also cause the corm to rot and turn black; in those cases, the infestation was referred to as "blackhead disease."[37]

The relationship between nematode and banana plant was structured by the transnational history of banana production. The nematode *R. similis* was an import to Central America. Indigenous to Australia and New Zealand, by the early 1960s *R. similis* had made its way to Guatemala, Ecuador,

Colombia, Costa Rica, and Honduras on the roots of infected plants, likely including bananas.[38] Banana company practices continued to spread the nematode; as Shell nematologist C. W. McBeth described it, "the planting piece is taken from old diseased planting sites and transferred, pests and all, to the new planting."[39] Once transported to a new area, local or regional environmental processes could facilitate the nematodes' spread. As one nematologist put it, "once such natural obstacles as oceans and land masses are overcome, the nematode can be disseminated locally through plants, soil, water, or other organisms."[40] Local and transitory environmental conditions could also exacerbate nematode damage, as when high winds blew over plants whose root systems had been weakened by the pests.

For its part, the banana originated in Southeast Asia, where multiple varieties were bred and cultivated for thousands of years. The exact circumstances of the banana's travels to the Western Hemisphere are uncertain; estimates of the time of its arrival vary from 500 to more than 2,000 years ago.[41] A number of varieties were cultivated locally in the Caribbean, some by slaves who grew it on sugar plantation grounds long before it entered the transnational trade.[42] In the place of the multiple varieties of bananas traditionally grown, the fruit companies settled on a single cultivar that they felt best met the needs of production and transport. The companies then spread that cultivar across the Central American isthmus. The plantation model—thousands of genetically identical plants in close proximity to each other—meant additional vulnerability as pests thrived among a plentiful food source.

The nematode problem was intimately linked to the fruit companies' agricultural practices, which were developed on the model of industrial plantation agriculture that hoped to maximize yields—both of bananas and profits—and to produce what was considered to be the most attractive banana. As environmental historian John Soluri explains it:

> Routine pruning and fertilizer applications intended to boost yields tended to exacerbate the degree of damage caused by nematode infestations. Pruning ... removed most of the young suckers from banana plants so that nutrients would be concentrated in one or two stems. This technique helped to produce full, long-fingered bananas, but it deprived plants of the structural stability provided by the lateral shoots. Heavy fruit bunches in turn placed great stress upon the stems and roots, leaving fruit-bearing plants vulnerable to uprooting. In other words, the impact of *R. similis* on export bananas was not an entirely "natural" phenomenon; a singular emphasis on high yields helped to create the problem.[43]

The switch to the Cavendish cultivar also exacerbated the nematode problem.[44] An article in *Nature* in 1958 explained that side-by-side plantings of old and new banana cultivars in Jamaica exhibited a stark difference in infection, with the Cavendish plants "riddled with" nematode symptoms.[45] The Cavendish connection firmly linked new nematode infestations to the broad regional political and economic forces that pushed the banana industry to make fundamental changes after 1950, further showing how the pest problem was the outcome of processes on different scales.

The solutions proffered to control nematode damage also were also shaped by phenomena at scales ranging from the local to the transnational. Just as Shell had waged an educational-*cum*-promotional campaign on its nematicides in the United States, as early as 1959 the company produced a Spanish-language booklet on the banana's "pests, diseases, and weeds" that promoted Nemagon and its other products and was apparently geared toward local producers. With the same educational tone that characterized much of the company's U.S. materials on nematicides, the booklet emphasized the nematode problem. The first four pages were devoted to images and text on "nematode control," more than twice the space devoted to each of the better-known and more destructive diseases dubbed "the big three" by United Fruit— Sigatoka, Panama Disease, and Moko. Shell offered information on nematode types, habits, and infection symptoms, along with the impressive increase in yield that it claimed could be had by applying its nematicide products, DD, EDB, and DBCP. In Nemagon's case, Shell claimed that yields could be increased from 6.7 to 25.6 tons per acre with the use of its product.[46]

Increasing yields was an attractive prospect for United Fruit managers, who in 1961 reported "losses in production and quality of fruit in some five thousand acres of bananas due to deterioration of the plantings as a result of a high infestation of nematodes."[47] While this acreage was only a small percentage of total banana lands, it was enough to merit the attention of the company. United Fruit looked for a variety of solutions to the nematode problem, drawing on a transnational research apparatus with tropical headquarters in La Lima, Honduras; research stations in Coto, Costa Rica, and Almirante, Panama; and laboratories in New York and Norton, Massachusetts.[48] Researchers came up with a range of nematode remedies. Planting material could be trimmed of material that might harbor nematodes, then dipped in hot water to kill any remaining creatures and planted in noninfested soil.[49] Fallowing nematode-infested land and rotating with other crops were also thought to get rid of the wormy pests.[50] In addition, research into resistant cultivars was considered as a solution to the nematode problem.[51]

Both United and Standard were also considering DBCP to end their nematode problem, and were in communication with Shell about its still-novel product in the late 1950s. United added DBCP tests to its research roster in 1958, and within two years the company's Associate Director of Research concluded:

> This nematicide has provided beneficial results by increasing the number of stems of fruit produced per acre, increasing the weights of stems produced and decreasing the loss of stems of fruit from plant tip-over, due to loss of root system just prior to the fruit reaching maturity. The weight of fruit harvested per acre has increased from twenty to forty percent following the control of the nematodes.[52]

Standard Fruit apparently did not conduct tests of its own, but a Standard entomologist located in La Ceiba, Honduras did contact Shell in early 1960 with a request for information on Nemagon. Researchers at the chemical company happily sent along a summary of research and recommendations for the use of its DBCP product.[53]

More than research was needed to assure DBCP application on bananas, however. The impact of U.S. regulation on agricultural practice in Central America was indirect but undeniable: widespread use of DBCP on banana plantations would not be feasible without the approval of U.S. regulators. In 1961, United Fruit's Associate Director of Research, Norwood Thornton, wrote to M.J. Sloan, a manager in the Sales Development Department at Shell Chemical Company, noting that experiments at United's tropical locations had found DBCP "effective" for controlling nematodes. Attributing serious losses to nematodes, Thornton wrote that United "would appreciate [Shell's] immediate and concerted effort in progressing clearance with the FDA for use of Nemagon in banana production."[54] Nominally, United Fruit's Central American banana farms were outside the scope of U.S. law precisely because they were not located within the geographical boundaries of the United States. Indeed, the most important piece of regulatory legislation, FIFRA, explicitly exempted chemicals sold outside of the United States from its provisions, meaning warning labels and other requirements would not apply to DBCP used on bananas.[55] However, residue tolerances set by the FDA applied to foods sold in the United States, regardless of where the foods were grown. Therefore, U.S. approval of DBCP residues on bananas would be fundamental to the adoption of the chemical in Central America.

Shell and Dow, undoubtedly hoping to secure the banana DBCP market, included the tropical fruit in their application for residue tolerances in 1961.[56] United Fruit supported the application by providing data on the residue level in samples of treated bananas.[57] Although theoretically meant to protect U.S. consumers, the tolerances requested for bananas seemed designed to meet the needs of the transnational fruit growers. The original application, submitted in February, asked for a tolerance of 100 ppm of bromide—DBCP's breakdown product—in bananas. The chemical companies maintained that the levels they proposed were "well below those associated with bromide intoxication."[58] Although they used a man weighing only 50 kilograms (110 pounds) as an example of an average U.S. consumer—an assumption that would skew toward a more protective standard than assuming a larger banana-eating adult—they failed to account for impacts on children in various stages of development.[59] Furthermore, although bananas were consumed in greater quantities in the United States than citrus, grapes, or berries, the banana tolerance requested (100 ppm) was one of the highest proposed. Only lettuce, endive, and feed corn carried the higher figure of 130 ppm.[60]

The chemical companies' applications for residue tolerances were stalled by FDA requests for more data. Meanwhile, United Fruit told Shell in 1962 that, " the present petition . . . will need to be amended to request a higher tolerance of 175 ppm" in or on the banana, based not on health considerations, but rather "on data obtained by the United Fruit Company on samples taken from plots where dibromo chloropropane was used at rates which would be practical." In a phone call to George Lynn at Dow's Agricultural Chemicals Department, Shell's M.J. Sloan, "indicated that United Fruit would use dibromo chloropropane on considerable acreage if proper tolerances were established." Although the two chemical companies would later compete for that business, Dow personnel agreed to split the $750 fee for amending the petition.[61] Meanwhile, however, decision-makers at United's Boston headquarters had withdrawn "experimental approval" for DBCP use on its Central American plantations. The withdrawal was precisely due to worries that a DBCP-treated crop "would be subject to confiscation if the tolerance for bananas is not granted." The fruit company's decision to play it safe when it came to the possibility of violating U.S. rules signaled the importance of those regulations for companies operating transnationally but selling to a U.S. market.

Regardless of United's hesitance, the chemical companies moved ahead with efforts to secure an official residue limit for bananas, even as the FDA required some changes to the tolerance petition. By August of 1962, the manager of Shell's Agricultural Chemicals Division would resignedly note that yet another tolerance proposal—75 ppm in the pulp of the fruit alone—seemed to be the "highest level . . . the FDA will accept in this food."[62] The FDA also asked that the interval between applications be increased and that a statement on maximum yearly dosage be included in labeling to avoid bromide buildup in soil where more than one crop was grown.[63] In May 1963, the FDA established final DBCP tolerances, allowing 125 ppm of bromide in the whole banana fruit (including the peel), with a maximum of 75 ppm in the pulp.[64] When the USDA registered Fumazone and Nemagon in 1964, bananas were among the officially allowed uses.

DBCP GOES BANANAS

A U.S. tolerance was a necessary but not sufficient condition for DBCP use on bananas. The chemical took off in Central America only after 1967, when nematode populations were resisting other control methods and impacting banana profits. Bananas in Costa Rica and Honduras were not growing well, yields were down, and fruits were sometimes misshapen. Root damage meant that stormy weather could knock down up to one-quarter of the plants on an affected plantation, creating an urgent need for some kind of nematode control.[65] Standard Fruit researchers, who had been trying to control burgeoning nematode populations with fallowing and rhizome treatments, were stumped. Looking to see if DBCP would do the job, Standard ordered 28,500 gallons of concentrated Nemagon for use on its Honduras plantations.[66] This liquid formulation was added to the existing overhead irrigation system and applied to 17,000 acres at a rate of four gallons an acre.[67] By April of 1968, the nematicide seemed to be doing the trick.[68] According to Shell's D. J. Miner, a sales manager who handled many of the fruit company's orders, "the Standard Fruit people all feel that they are seeing a marked improvement in the plantations" and were "most enthusiastic about the results of the project."[69] He continued, "They intend to make a second NEMAGON application in the same areas this year after a seven-month interval and thereafter, one application per year." Standard estimated they would need an additional 40,000 to 60,000 gallons by June. Although Standard also got a competitive bid from Dow, the fruit company signed on with Shell, who would supply 85% to 100% of Standard's DBCP requirements, an amount that totaled over 132,000 gallons for 1968, which at the contracted price of $5.75 a gallon totaled over three-quarters of a million dollars.[70] This was just one event in a big year for Standard: the company was purchased by Castle & Cooke, a food and freight corporation that had merged with Dole seven years before.[71] While "Standard Fruit" continued as a division of Castle & Cooke,[72] "Dole" labels became ubiquitous on the company bananas.

Standard continued to increase banana production, a process that would involve adapting DBCP to local environmental conditions. After learning of Nemagon's success on the company's Honduras plantations, Standard's plantation manager in Costa Rica wanted "to make the same treatments in his area."[73] However, in Costa Rica, regular and predictable rainfall made fixed irrigation systems unnecessary, presenting the question of how DBCP would be applied. In 1967, Standard had tested a granular formulation of the nematicide that was applied without the benefit of an irrigation system. The volatile chemical, however, had evaporated before it could be washed into the soil.[74] Faced with this dilemma, Shell's D. J. Miner wondered whether Shell could provide a special formulation of DBCP "which can be applied to the soil surface around the banana plants and allowed to remain undisturbed until the rain washes the NEMAGON into the soil."[75] Standard's adoption of DBCP would be significant: its Costa Rica operation encompassed 5,000 acres of land farmed by the company and another 15,000–20,000 acres cultivated by independent growers.[76] Miner also thought that Standard might "pay substantially more" for such a formulation. But the promise of a new Nemagon formulation was greater than the mutually beneficial arrangement between Standard and Shell alone: "Other banana plantations in Costa Rica, as well as banana plantations in other areas of the world, where irrigation equipment is not available, are potential markets for a new NEMAGON formulation."[77]

The fruit company planned to test the new granular or encapsulated forms of DBCP developed by Shell,[78] but by June of 1970, Standard's Director of Research, Jack DeMent, notified Shell that he and his colleagues were discontinuing its tests on granulated and encapsulated material. Instead, Standard had discovered a different means of applying DBCP. DeMent wrote: "A hand gun injection of Nemagon was not only feasible, but very economical. . . . This discovery obviated the necessity of obtaining a material that could be applied to the surface of the soil."[79] The "hand gun" in question looked like a cross between a syringe and a pogo stick—a worker pushed the tip of the injector into the soil, then depressed the plunger to release the chemical. Hand injectors had been suggested as a means to apply DBCP to bananas as early as 1959,[80] but by 1970 the method seemed poised to take off commercially. By the end of that year, Standard's operations in Costa Rica alone used about 12,500 gallons a month of DBCP. They kept twice this amount—on consignment from Dow—on hand at their warehouses.[81] In 1971, they expanded use of the chemical to other plantations, using a total of 128,300 gallons in Costa Rica.[82] The following year, the company was planning DBCP applications in Nicaragua as well.[83] Standard had found an efficient and cost-effective application option that worked with local environments; by 1972, it had fully embraced the use of DBCP.

United Fruit—which would become known as United Brands after a merger in 1970[84]—was slower to adopt DBCP. The company had been using a rhizome sterilization process to fight nematodes, but, faced with nematode infestation levels reaching 25% on some farms, restarted their DBCP trials in 1970.[85] By midyear, however, researchers voiced concerns on cost and efficacy, as well as worries that bromide levels remaining in the bananas were simply too high to meet the U.S. tolerances.[86] United's annual research report for that year showed DBCP-treated fruit with residues well under the FDA limit, but reported mixed results with regard to yield and other outcomes.[87] Research on nematode control continued, with more mixed results. In Armuelles, Panama, researchers felt that, "Despite . . . large yield losses, indications are that we can live with the form of *R. Similis* in Armuelles for an as yet undetermined period of time if other farming practices such as drainage, pruning, cleaning, population control etcetera are performed efficiently."[88] United shifted its position in 1973, when company researchers reported that "sufficient data from nematicide trials became available . . . to recommend commercial use of DBCP." That year, injection of a Dow Fumazone formula was instituted in Costa Rica and Panama.[89] Despite ongoing concerns about bromine toxicity, and continuing tests on other nematicides, DBCP use continued to expand.[90] By 1977, United's Honduras plantations were also using DBCP and applying it through existing irrigation stations.[91]

The chemical was also adopted by a new player in the banana field, Del Monte Fresh Produce Company, which had bought plantation lands United was forced to sell after an antitrust investigation in the United States.[92] Del Monte bought its DBCP from an Israeli outfit known as Dead Sea Bromine Company/Bromine Compounds Ltd. and applied it with injectors.

According to representatives of Dead Sea, on one farm Del Monte had boosted banana production from 2,000 to 3,150 boxes per hectare after using their DBCP product, Nemabrom, for two years.[93]

As the Central American market for DBCP grew, Dow and Shell continued to compete for fruit company business.[94] In addition, the Israeli producers courted the banana growers, and even Occidental attempted to enter the Central American market.[95] By 1977, Castle and Cooke estimated that 600,000 gallons of DBCP were used on Latin American banana plantations per year.[96] DBCP had become a key part of doing business for both the chemical and the fruit companies.

EXPOSURE, CONTROL, AND RESISTANCE ON THE BANANA PLANTATION

While decisions about purchasing and application methods were made by United and Standard managers in the United States and Central America, it was workers who carried out the labor of applying DBCP to the banana plants, which took place within a larger context of uncontrolled pesticide exposures. Nicaragua, Costa Rica, Panama, and Guatemala all passed pesticide-related laws in the 1960s and 1970s,[97] but by and large these governments lacked the resources and political commitment to follow through on regulatory reforms and ensure safety for workers.[98] For their part, companies regularly failed to take measures that would protect workers, with occasional exceptions.[99]

Although workers received little protection from DBCP, and companies failed to educate them about its health risks, workers accumulated bodily experiences of DBCP's acute effects and developed locally derived knowledge of the pesticide from working so closely with the chemical. Workers' specific knowledge about DBCP mixed uneasily with their sense of severely constrained power at work, as both corporate documents and workers' testimony would reveal years later. In a series of depositions taken in the 1990s, a number of banana workers retrospectively explained their own knowledge of and experience with DBCP.

Jorge Colindres Cárcamo worked as a mechanic for most of the 25 years he was employed by Standard Fruit in Coyoles Central, Honduras.[100] His responsibilities included maintaining and repairing the pumps and engines essential to applying DBCP. The chemical, he explained in 1994, would be hand-pumped from a barrel to a larger tank and then added to a reservoir holding water diverted from a canal. A mechanical pump distributed the water to turbines, which pumped the water to the top of towers, where devices known as *mariposas*, or butterflies, sprayed it over the banana plants below.[101] Colindres came into contact with DBCP as he cleaned and fixed turbines and other equipment that would become sticky after contact with the chemical. Another Standard laborer, Germán Muñoz Moncada, who worked in Chinandega, Nicaragua, beginning in 1978, described a similar process for applying DBCP.[102] As an irrigator, his job included climbing the towers to fix what he called not "butterflies" but "guns" that shot out water onto the plants below: "Sometimes the gun gets stuck or it goes crazy. And then one has to be paying attention to go and control it . . . one has to climb up to correct it so it can continue." Fixing the guns often meant becoming entirely soaked in the toxic material.[103]

Other workers would get wet with a mixture of DBCP and water as they moved through a newly treated section.[104] Irrigation application of DBCP was often carried out at night so lower temperatures would mean less loss of the chemical to evaporation. Men and women arriving to work the next morning could encounter still-dripping leaves and wet groundcover. Manuel Antonio Valderramos Erazo recalled his work bagging bunches of developing fruit at Standard's Naranjo C plantation in Campo Bálsamo, Honduras:

> The work that I performed required that I start work at 6 o'clock in the morning. And because of that, Nemagon had just been applied. . . one hour before. . . . What you have to do is you have to get a ladder and get on top of the ladder and put a bag around the plant. And when you do that, the . . . leaves of the banana plant are impregnated with Nemagon. So, you're practically drenched from head to toes. . . . And, on the other hand, on the ground there are puddles of Nemagon. The method that you use to support fruit is to dig a hole in the ground and stick a pole to support the plant. And when that plant is harvested, you take the pole away and use it on another plant. So, after you take the pole away, the hole is left in the ground. . . . And these holes get filled with Nemagon and sometimes there would be water covering the ground so that you would step into this hole up to your knees and you'd be covered in Nemagon all the up—up to your knee.[105]

Workers who injected DBCP by hand had a different experience of the chemical. The first step was to mix DBCP with water. The mixing method would depend on the plantation and the formulation used. On Standard Fruit's Río Frío plantation in Costa Rica, for example, two or three workers would get a barrel of the chemical from the warehouse, then lift it onto a sawhorse and place a spigot in a small valve on the barrel. Then, an injection crew of about ten, called a *cuadrilla,* would fill large plastic jugs called *pichingas* with a mixture of water and DBCP from the barrel.[106] The workers would carry the *pichingas* on their shoulders to the transport cable, where they would hook them and pull them to the part of the plantation being treated.[107] Older *pinchingas* sometimes leaked or split, dripping or spilling DBCP onto workers as they handled the jugs.[108] Hermenegildo Salazar Marroquín recalled that, on Del Monte's Bandegua farm, he was in charge of mixing the chemical and water; the formula used would be mixed in a one-to-one ratio with water by emptying out half of the barrel and then filling it again with water. From there, barrels were brought to the application site by tractor or on cables.[109] No matter how DBCP and water were mixed, the process inevitably exposed workers' skin directly to the chemical.

Injection workers filled their applicator tanks with the DBCP mixture once they were at the section of the farm to be treated. This had to be done with care, as one worker, Ernesto Garbanzo García, noted, "because it makes a lot of foam. If you spill it, then they . . . would scold [you]."[110] After applicators were filled, teams of workers injected the DBCP around the base of each banana plant, making sure to tamp down displaced earth into the holes made by the injectors to keep the volatile chemical from escaping.[111] Soil conditions affected the ease of work as well as the extent of exposure to the chemical. In wet conditions, injectors could clog with mud, causing pressure to build and liquid to spray back onto the workers.[112] Alternately, as Garbanzo explained, "when the soil is very hard, you hit it down, and it'll hit on a root, and it will jump up at you, or it'll . . . hit on a rock on the side of the roots and every time it will jump. It'll get on your pants and—the pants and your legs, all of it will get wet where the product jumps at you."[113] Splashes on skin and even in eyes could be a daily occurrence.[114] And at the end of a work day, Standard Fruit recommended that workers clean injectors "thoroughly with water . . . and subsequently [rinse] with a 0.5 litre mixture of lubricating oil and kerosene," a process that also presented a risk of dermal exposure to DBCP.[115]

Banana workers and their families were also exposed indirectly. At sites where the nematicide was applied through irrigation, the chemical-water mixture could drift over plantation worker housing during application, exposing people at home. Workers also sometimes used DBCP to kill fish in streams, bringing the fish home to be eaten.[116] Many women and children worked to assist the men in their families, making them vulnerable to DBCP exposure as well. *Almuerceros,* children charged with bringing their fathers lunch while they worked in the fields, could be exposed upon entering application sites; anyone washing DBCP-impregnated clothing would also have been exposed to the chemical.[117] Sometimes the harm caused by DBCP was intentional: in 1970, a 14-year-old girl—unnamed in a company memo reporting her death—committed suicide by drinking Nemagon concentrate; her death was followed shortly by another DBCP suicide.[118]

Workers and family members were not provided protective equipment or clothing. In the words of Jairo Ramírez Contreras, who worked at Valle de la Estrella in Costa Rica, workers wore "work clothes, any clothes."[119] He usually wore long pants, a short-sleeve shirt, and rubber boots; he was never provided a mask, gloves, or other protective gear. Ernesto Garbanzo recalled that while some workers wore boots, they bought them themselves.[120] While some remembered that protective equipment was given "later"[121] or for other pesticide-related jobs,[122] in interviews, depositions, and written accounts, workers overwhelmingly reported that no protective equipment was given for work with DBCP. In 1976, a consulting nematologist from Cornell University admonished United Fruit because safety precautions for DBCP were not being observed during his visit to the Bocas, Armuelles, and Golfito Divisions. "Specifically," he wrote:

> soap and water were not readily available to wash off DBCP spilled on parts of the body, particularly areas such as eyes. Also it was evident that DBCP was being spilled on clothing and it appeared likely that these clothes were not changed at the end of the day's work. It is suggested that each person be supplied rubber boots and coveralls and that these coveralls be washed by the company at the end of each day.[123]

The recommendation for protective gear points to inherent conflicts—between the environment, workers' bodies, and proposed protective regimes—that characterized not only DBCP use, but pesticide use in hot, humid environments more generally. Although respirators, goggles, gloves, shoes, and long pants and sleeves could help protect a pesticide-exposed worker from harm, the equipment was poorly suited to the hot and humid context of the plantations, and high temperatures made using safety equipment unbearable. Banana companies knew this: in 1974, a Standard Fruit nematologist working in Costa Rica wrote to Jack DeMent, the company's Director of Research, explaining, "Even if we recommend use of rubber gloves and supply them, most workers are going to ignore them because of the heat."[124] This reasoning was often echoed by managers[125] and used as an excuse to avoid the expense and bother of purchasing and using protective gear. Instead of acknowledging the paradox that protective equipment use and hot weather were incompatible, or developing climate-appropriate protective equipment, the companies displaced the dilemma onto the workers, freighting them with the responsibility for ignoring or refusing protective equipment. That is, instead of signaling a problem with the system of production, or the very real limitations posed by the climate, the failure of protective gear was interpreted as a failure of the worker. So understood, companies could mobilize the workers' "refusal" of goggles, masks, and gloves as an excuse to dispense with protective measures.

In addition, as environmental historian Linda Nash has argued, the very idea of protective equipment is based on a concept that human bodies can be separated from the environment through barriers. This concept ignores the multiple, locally specific, and sometimes unpredictable ways that pesticides can "escape" the work context—as drift or runoff, for example—and the unanticipated ways that workers may come into bodily contact with chemicals, such as in food, in their homes near farms, in residues attaching to clothing, and so on. Prescriptions for the use of protective equipment ignored these daily exposures, turning a blind eye to local conditions that impact the potency and persistence of pesticides and workers' relationship to the environment.[126] In Central America, workers often lived near or on banana plantations, and often ate, bathed, fished, and carried out other daily activities on or near plantations. Protective equipment could not prevent exposures in these contexts.

As with their U.S. counterparts, information on the health effects of DBCP was also withheld from workers in Central America. While some DBCP barrels did arrive in Central America with a "warning,"[127] similar to the labels used in the United States, Spanish-language labels did not mention sterility, testicular damage, or cancer. For example, Shell submitted the following label to the EPA (in Spanish) in 1973:

> Can be fatal if swallowed. Can burn the skin. Causes eye, nose and throat irritation. Can be absorbed by the skin. Harmful if inhaled. Keep away from heat and flames. Avoid contact with skin or clothing. Do not use internally.[128]

These notifications were followed by instructions on what to do in case of spills or skin contact and a list of potential harms to wildlife. While this particular label was approved by U.S. regulators, it is unclear whether or where it was used. What is clear is the label's omission of DBCP's chronic health effects. In addition, it appears that the product often made it to Central America without any Spanish-language label at all. National regulation differed on label language requirements: Panama had required Spanish-language labels beginning in 1967, and Guatemala in 1974.[129] In 1978, amendments to FIFRA for the first time imposed minimal requirements on exported pesticides, including labels in the importing nation's language.[130] However, the presence of labels alone could not guarantee that workers would see or hear about the chemical's dangers. Workers might never see a barrel close enough to read it, might be illiterate, or might never be provided enough time in a busy workday to read the label or have it read to them.

Workers' recollection of their exposures to DBCP suggest a complicated push and pull between awareness of the noxious properties of DBCP, accrued through bodily experience, a lack of full information about the health effects, and little control over their working conditions. Workers' experience with the chemical gave them firsthand knowledge of its toxic effects. They could recognize DBCP's smell and described it in varying terms as "strong,"[131] "penetrating," and "offensive."[132] Some remembered feeling dizzy or nauseated on smelling it.[133] The feeling of DBCP on skin was often described by workers as either a cold or burning sensation. As Guatemalan Hermenegildo Salazar described dermal contact, "if it was on the skin on any part of the body, when it would land it would feel cool. But as the minutes were passing you would go feeling a heat until one would feel it was something more like burning."[134] Other workers said it caused burning in their nose and eyes. Ernesto Garbanzo described a more serious reaction to an unusual DBCP spill: "On one occasion I was carrying a *pichinga* on my shoulder and ... [it] split and it got me all wet all on this side."[135] While the "burn" of getting small amounts of DBCP on his skin was a daily experience for Garbanzo, this larger spill caused him "great pain"—he dove into an irrigation canal in search of relief.

Despite citing these symptoms, banana workers recounting the DBCP days also said that they had no idea the chemicals they worked with were harmful. For example, Gerardo Dennis Patrickson said in an interview that he had been "innocently unaware" of the hazards presented by DBCP.[136] Asked in 1994 whether he had thought the chemicals he worked with were dangerous, Ernesto Garbanzo, replied, "No. Who would?"[137] Of course, this "innocence" was undergirded by chemical companies' successful efforts to deny DBCP's risks to humans. In addition, some of the "innocence" may have been attributed to their former selves by now-injured people in an attempt to avoid (their own or others') perceptions of their complicity in their own injury. This may have been especially true in the context of litigation, where a plaintiff's knowledge of harm could jeopardize his or her chance of compensation.

Even taking these points into consideration, workers' coexisting memories of "innocence" along with their bodily experience of the toxicity of DBCP also suggests the constraints workers faced in influencing their own working conditions. Dennis's statement that "you would never imagine that it was going to be harmful. . . . It was part of your work, you had to do it,"[138] has two parts that do not fit together very easily. On the one hand is his negatively expressed claim of faith in chemical safety and on the other, a rationale for that faith which is really no reason at all, but rather a statement of the constraints placed on him. His and other workers' comments suggest that they worked in an environment where they had little say over which chemicals were used or how. Jairo Ramírez Contreras, asked if he reported DBCP-leaking *pichingas* to supervisors, responded that to do so would endanger his daily wage: "By the time you told the supervisors that the *pichinga* had a hole in it, you would already have had the liquid on you. And then you would be replaced."[139] Similarly, Ernesto Garbanzo saw that his livelihood was clearly linked to his willingness to carry out pesticide application, even if he knew it was dangerous, saying, "If they order me to, I would have to apply it because otherwise they would fire me."[140] Garbanzo also suggested that reporting exposures was seen as useless in effecting change. Even when DBCP burned his skin, he recalled, "it was the same to tell or to not tell somebody." Workers' assessment of DBCP dangers, then, took place within a system that gave them little control or voice in the conditions of its use, and at times may have left them no alternative but to believe that the chemicals they worked with were safe.

If many workers felt they had little or no power to set the conditions of DBCP use, company memos suggest that United and Standard personnel saw DBCP application as a site where control over labor practices, independent farmers, and nematodes were inextricably linked. The companies wanted to control nematodes effectively at the lowest cost possible, which required heightened efficiency in both chemical usage and workers' performance. It also required ensuring that planters producing for the companies followed banana company instructions and made sure that their own employees did as well.

Injection application in particular seemed to elicit banana company concern about proper application techniques and worker compliance. Companies adopted various patterns for the injection around the base of the banana plant—sometimes after considerable controversy and conflict among researchers—and wanted to ensure that the small groups of men applying the chemical in the field followed exact instructions.[141] In 1970, Standard Fruit's application instructions, penned by nematologist Johannes Klink, stressed that "close supervision" or, better yet, "excellent supervision" was indispensable. This required a small team of workers—no more than five—with one "reliable *capataz*" or foreman.[142]

In 1973, Paige Taylor, director of the "Experimental Department" at United Fruit's Chiriquí Land Company's Bocas Division in Panama, combined recommendations on chemical formulation, application technique, and labor efficiency to achieve what he thought would be the ideal DBCP injection practice. Taylor suggested using an undiluted DBCP formulation in injection "guns" in order to "avoid the necessity of mixing, transport of diluents, and more frequent filling of injectors."[143] He proposed a variety of injection patterns of varying complexity in hopes of most effectively targeting the nematodes feasting on the banana roots. Along with these technical aspects, Taylor included the productivity of application workers, noting that "the number of acres/day/man varies from 0.1 to 1.0." Taylor's goal was to reach a rate of 0.6 acres per worker per day, but he still worried that this level was too low. The same year, careful calculations were made to estimate the cost of DBCP application at Chiriquí to determine the "variable cost/box/acre," taking into account both Fumazone prices and the effect of a recently signed contract with the union.[144]

By 1974, Standard Fruit had developed a new application pattern that called for the injection of the chemical in a "compact circle" pattern. Robert Dunn at Standard Fruit observed a work crew applying DBCP according to the new plan, and wrote, "it was apparent in a very short time that the injection crew was not familiar with the new system and that even the *capataz* was unsure of many things about the system."[145] The crew was applying five or six more liters of DBCP than necessary, leading Dunn to conclude that: "lack of definition in the field was the main reason for the excessive expenditure of DBCP." Stemming that loss would require "constant [managerial] emphasis" on workers' "strict adherence to the pattern" as well as "close supervision" of injection crews.

Banana companies were not only worried about workers following their instructions, they also wanted independent planters to toe the DBCP line.

By the end of 1969, Standard's superintendent of research in Costa Rica suggested informing associate growers in Costa Rica that, "Nemagon application is the only recommended Nematode Control System, having the approval of the Standard Fruit Research organization. You are urged not to use any other materials pending response data, residue studies and a recommendation from Standard Fruit."[146] Standard also provided independent growers with detailed instructions on how to properly apply Nemagon via hand injection.[147] Monitoring rates of DBCP injection on "associated growers" farms in 1971, Klink found levels were less than ideal, which he blamed on "a number of factors among them plant population, plant distribution and one very important one the human factor." Workers on independent farms may simply have done a "bad job."[148]

Despite managerial efforts to control pests and the laborers hired to eliminate them, associate growers and workers did have some room to challenge unfavorable working conditions. In the 1960s and 1970s, extreme political repression prevailed in many nations, including Guatemala, where violent right-wing military governments targeted civilian organizations, and Nicaragua, where in response to the failings of the Somoza regime, a broad opposition was laying the groundwork for revolution.[149] Nevertheless, these decades also saw a resurgent worker movement in Central America.[150] By 1977, the four largest banana unions in the region had 30,000 members, about half of the total banana workforce.[151] At least once, DBCP application procedures were addressed in a union contract. In 1976, a United researcher's report on conditions in Costa Rica's Golfito division noted: "With the signing of the new labor contract, there was no provision for improved application. For some reason, it was written in that the second cycle was to be injected at twelve inches from the [plant]."[152] While it is unclear why the contract would have addressed the minutiae of DBCP application—or whether the motivating concerns were exposure-related—DBCP's presence in the contract indicates that workers, management, or both saw DBCP as an issue both important and contentious enough to include in formal negotiations.

Worker resistance to conditions of DBCP application took place not only through strikes or contract negotiations, but also through quotidian actions. Managerial complaints indicate that workers' everyday acts of resistance included sparing themselves effort—and perhaps exposure—by making fewer injections than required.[153] By contrast, where workers were paid by volume of DBCP injected, resistance could take the form of over-injecting, as when a researcher at Standard Fruit noted that he had "many times observed that workers hit the plunger 3 to 4 times at the same injection point obviously to get rid of liters fast."[154] Another option was simply to leave. After a visit to United farms (probably in Costa Rica) in 1974, a Dow representative noted that "many of the workers do not return the second day."[155] Leaving a job was most feasible for workers at times like this, when there was "a scarcity of labor and rising labor costs." The Dow visitor reported that the banana growers' turnover problem had been solved elsewhere by changing how DBCP applicators were paid. United could, he suggested, turn DBCP injection into "one of the higher paying jobs," or pay according to the amount of work accomplished. But if worker resistance could win more-favorable conditions, it could also do the reverse—along with the carrot of higher pay, the Dow specialist suggested the familiar stick of "proper supervision" to ensure the success of the injection program. It remains unclear from the historical record whether, or to what extent, labor conflicts around DBCP were based in concerns about health, the tediousness of application, or other factors.

PROBLEMS AND ALTERNATIVES TO DBCP

At various points in the 1970s, local and transnational concerns made both Standard and United consider decreasing or discontinuing DBCP use. In 1974, Standard investigated DBCP substitutes with hopes of keeping down labor costs. The same year, United made the discovery that DBCP—long vaunted because it was safe to apply to living plants—was in fact toxic to banana plants on the company's Higuerón farm in Panama, causing symptoms such as lesions, rot, and malformations, and raising questions about the future of DBCP.[156]

The alternatives to DBCP—Mocap, Furadan, and Nemacur—were acutely toxic chemicals that raised concerns not only among workers, but also among managers. Nemacur (fenamiphos) and Mocap (ethoprop) could cause the poisoning symptoms typical of the organophosphate family to which they belonged: inhibition of the body's production of the nervous system enzyme cholinesterase, which could lead to incontinence, respiratory depression, seizures, loss of consciousness, and even death.[157] Furadan (carbofuran) belonged to the carbamate class, and could also cause depression of the central nervous system, as well as headaches, vomiting, pain, difficulty breathing, and pulmonary edema. When Standard was considering replacing DBCP with either Furadan or Nemacur in 1974, Robert Dunn averred that a "major factor influencing this choice has been safety."[158] Dunn went on to express that Furadan was preferable because it had a much lower dermal toxicity than Nemacur and poisonings were more easily reversed.

At times, worker or regulator reaction forced banana growers to stop using certain chemicals, usually replacing them with another. Del Monte worker Hermenegildo Salazar remembered that after the Bandegua farm switched from DBCP to Mocap, the union brought a lawsuit against the company after one season "because it was a nematicide that was highly toxic . . . and they cancelled Mocap immediately. . . . [Then] Nemacur replaced Mocap."[159] Mocap presented problems elsewhere as well. In 1975, after 311 acres were treated at United's Higuerón farm in Panama, "17 workers became ill. Symptoms were vomiting, nausea and headaches."[160] While the company blamed the problem on a faulty supply, workers and officials were not silenced by the excuse: "[Five] men resigned rather than continue to apply this material. Strong complaints were received from the Labor Inspector and Sindicato [union]."[161] In this particular case, decision-makers at United bowed to local pressures and ceased use of the Mocap formulation in question. Their decision however, seemed based more on the fear of the costs of protection than any genuine concern for workers' health. According to the company's Annual Report:

> The situation reached the point where we were about to be ordered to stop using this material or adopt uneconomical application procedures, such as working shorter hours while paying for a full day's work, furnish laundered clothing daily, install hot water showers for the workers to bathe after working and so on. We feared that these requisites would ramify to all chemical applications such as herbicides, insecticides and fungicides if initiated.[162]

In this case, United appeased local demands primarily as a way to forestall further, more systemic changes in their operations. However, the experience apparently did not cause system-wide precautions in the use of Mocap—in early 1976, a case of Mocap poisoning at another farm prompted the observation that, "greater care is required in use of protective clothing and safety hygiene during and following the use of Mocap."[163]

Worker responses to Mocap and DBCP differed, and this was at least in part because of the nature of the toxic effects of each chemical as well as the impact of individual poisoning incidents. Occasionally, a DBCP-related event called attention to the chemical—as in 1976, when 105 "Fumazone Poisonings" took place within four months in United's Armuelles Division.[164] But more often workers tolerated the daily irritations of contact with DBCP.

The more chronic dangers posed by DBCP were unknown to workers and unknown or disregarded by management. The immediate, self-evident symptoms of Mocap made the toxicity of that chemical more "knowable" than that of DBCP. The composition of each chemical and its effects on workers' bodies shaped patterns of recognition and response by both workers and management.

Despite the toxicity of alternatives, by 1976 the use of DBCP on banana plantations seemed in question. Once again, the fate of the nematicide was determined at the intersection of the local and transnational. United Fruit found bromide toxicity in banana plants and residues in fruit that sometimes exceeded and sometimes merely "approach[ed] tolerance" levels set by the U.S. government, leading the company to discontinue DBCP use in at least one of its Central American locations.[165] Other developments in the United States threatened to terminate use of the chemical in Central America. Regulators had begun to raise questions about the safety of the compound, citing new studies not on its reproductive effects but on its potential to cause cancer.

Standard Fruit, however, was not deterred. In April 1977, Jack DeMent mounted a defense of DBCP, hoping to use the Central American experience as evidence for continued U.S. approval of the chemical. DeMent painted a rosy picture, maintaining that DBCP improved banana quality and boosted yields up to 25%.[166] In contrast to United, where substitutions with other nematicides had taken place, DeMent contended: "DBCP is the only type of soil fumigant nematicide available to us." He also claimed that, "there is no question in my mind that DBCP can be safely handled. In fact, it has been so handled on some 90,000 acres of bananas for more than five years."

DeMent's proffering of DBCP's "safe" use in Central America as evidence for continued U.S. approval of the chemical was ironic, considering that U.S. rules did not cover the safety of workers outside its borders—even those who worked for U.S.-based transnationals like Standard Fruit and its parent company, Dole. It also, of course, showed DeMent's limited understanding— willful, wishful, or earnest—of banana workers' bodily experiences with DBCP. Within just a few months, DeMent and his colleagues in the fruit and chemical industries would be starkly reminded of the health risks posed by the chemical. As the true impacts of DBCP emerged, DeMent's conviction that DBCP could be "safely handled" in Central America would have sobering ramifications at the local, national, and transnational levels.

THREE

Unequal Exposures

IN THE SUMMER OF 1977, pesticide production workers at the Occidental Chemical plant in Lathrop, California, were worried. One "Oxy" worker recalled: "It was rumored [that] anybody that worked in that department for more than two years couldn't produce children. And I haven't."[1] Soon the rumors gave way to evidence, as testing revealed that many workers on the line did have abnormally low or even zero sperm counts. Their sterility was eventually linked to their exposure to DBCP, confirming what Shell and Dow scientists had first suspected 20 years earlier—that the testicular effects of DBCP seen in laboratory animals had a human analogy.

The clear evidence of DBCP's danger to human health did not bring an immediate end to exposures in either the United States or Central America, however. From 1977 to 1985, farm and production workers, environmental advocacy groups, fruit and chemical companies, and scientists and regulators in the United States and Costa Rica engaged in public and private debates over the fate of the chemical. Over that period, regulatory controls over the chemical strengthened, but newly instituted protections did not cover all people equally. In fact, as new rules and practices were hashed out, the *inequalities* in DBCP exposure grew even starker, transnationally and within the United States.

Critics of pesticide use argued that developing-world workers and U.S. consumers shared interests in defeating a "circle of poison," as exported pesticides ended up on U.S. tables in the form of potentially dangerous residues. However, uneven national regulations interacted with transnational markets in ways that complicated the terms in which these interests were shared, as disparities in exposure intensified both within and across borders. Banana workers in Central America and chemical production workers in Mexico faced continued or new exposures. In the United States, responses to the Occidental workers' revelations soon controlled production worker exposures, but farm-workers faced piecemeal and drawn-out responses to dangers in the fields. At the same time, risk shifted within Central America, as Costa Rica's unofficial "regulation" of DBCP left other regional governments uninformed and even led to Standard Fruit's quiet export of Costa Rican DBCP stores to Honduras. Concurrently, U.S.-based transnational corporations used regulations and regulators within their home country to bolster ongoing use of the chemical elsewhere, accentuating the inequalities inherent in the export of DBCP haz-ards. By 1985, however, it was a U.S. national regulation that finally made DBCP use impractical for banana culture and ultimately led to the obsoles-cence of DBCP internationally.

The story of DBCP's slow and uneven decline between 1977 and 1985 reveals how the multiple and contradictory intersections of national regula-tion, transnational markets, public debate, and corporate influence created an uneven geography of exposure to the toxic nematicide. This history high-lights the importance of transnational thinking about pesticide exposures, even as it necessitates "rethinking the circle of poison," as Angus Wright put it.[2] Ultimately, DBCP politics in this period showed that national regula-tions not only created unequal conditions domestically but also had an impact well beyond a country's borders—with sometimes dangerous, sometimes protective results for the health of those beyond political boundaries.

THE "SAFE USE" ARGUMENT UNRAVELS

Production workers at Occidental Chemical's Lathrop, California, plant began noticing problems with sterility as early as 1973.[3] In 1976, chemical production worker Wesley Jones discovered his sperm count was near zero and filed a worker's compensation claim against Oxy. The company sent him to be examined by Charles Hine—the University of California physician and toxicologist that Shell had contracted to conduct their DBCP toxicology research in the 1950s. Inexplicably, Hine—who had played such a large role in uncovering DBCP's testicular effect on animals—maintained, "I found nothing in the medical literature or my files, that would indicate aspermia to be induced by chemicals in the absence of profound intoxication. In fact, there is nothing to suggest that this is likely to occur except in rare instances in man, even with physically overt intoxication."[4] Hine told Jones that his lack of sperm was not caused by any chemical. In fact, he said, it was not work-related at all. The doctor apparently continued to believe in the "safe use" argument he had helped construct to secure regulatory approval for DBCP more than a decade earlier.

Jones and his fellow workers on the "Ag-Chem" line at Oxy, however, had come to believe their exposure to chemicals caused health problems including headaches, nausea, skin problems, loss of smell, heart attacks, and hemorrhaging. A worker named Sam articulated the connection between exposures and symptoms: "It's like out at Ag-Chem, you work in some of those poisons and things, and you go home; when you get home you just feel nauseated you wake up the next day with headaches and, you know, a slight case of poisoning."[5] Another worker, Pat, recounted: "We got a guy down there now, he was loading bulk for almost two years, and every night when he'd go home, he'd blow his nose, and he'd get a bloody nose, every night."[6] Workers likewise understood their growing incidence of sterility as an outcome of chemical exposures. As one man recalled: "I can't think of the last time that anyone, their wives, had a baby. Who's the last one? Ted? And that's been three years, four years ago? Three or four years ago, and, uh, the majority of guys working out there are all under 40, a lot of them are twenty or so, in their early twenties. So, uh, you take it from there."[7] These workers clearly made the link that Hine, even with his experience researching DBCP's effect in animals, denied. Although the Occidental workers could not connect particular chemicals to particular health effects, they were clear that workplace exposure was endangering their health.

The workers' understanding of the health effects of the pesticides they were exposed to were part of a growing concern about the risks of chemicals in the United States. In the 1960s and into the 1970s, scientists, emergent environmentalist groups, and some unions had drawn attention to the dangers of pesticides and other toxins.[8] By 1970, growing popular concern with pollution and other environmental threats coalesced into the first Earth Day, widely understood as the "birth of the modern environmental movement."[9] Large shifts were underway in the federal regulation of pesticides as well. Also in 1970, President Nixon established the Environmental Protection Agency (EPA), which took over the pesticide registration duties formerly carried out by the USDA as part of its "broad responsibility for research, standard-setting, monitoring and enforcement with regard to five environmental hazards; air and water pollution, solid waste disposal, radiation, and pesticides."[10] The new agency's mandate was a break from the federal government's historical role in developing and using pesticides, as well as from the USDA's Agricultural Research Service's ongoing use of chemicals to control fire ants and the U.S. military's more recent (1961–1971) herbicidal warfare campaign in Vietnam.[11] The EPA would not use or develop pesticides, but was charged with evaluating their environmental impact and protecting agricultural workers.[12] Other workers' chemical exposures were covered by another new body, the Occupational Safety and Health Administration (OSHA), which was created in 1971. In 1972, environmental activists achieved a ban on the notorious pesticide DDT, and FIFRA was overhauled with the aims of strengthening controls on pesticides. The new FIFRA for the first time established some—albeit minimal—jurisdiction over exported chemicals, requiring the EPA to notify foreign governments when a pesticide's registration was canceled.[13] These new agencies and new laws provided new protections, opportunities, and targets for activists and workers, but did not quell popular concern with chemical exposure. Indeed, significant gaps remained in controlling farm and factory exposures. Farmworkers pressed for better protections,[14] and among production workers, the Oil Chemical and Atomic Workers (OCAW) union emerged as a leader in prioritizing health and safety issues, including the organization of a 1972 strike and boycott against Shell Oil.[15]

The Oxy workers, confronted with chemical-linked sterility at mid-decade, were therefore not alone in their fight against chemical health hazards. For one thing, they belonged to OCAW, so when Jones and others became suspicious that they might be sterile because of workplace exposures, they were able to draw on the union's experience. OCAW's industrial hygienist advised the Oxy workers to ask management for a list of chemicals they came into contact with. The company refused to provide the information— or the fertility tests the union requested for seven workers who suspected they were sterile.[16] Despite this setback, the workers were on the track of the dangerous chemical.

Oxy management *was* worried enough to bring in outside consultants. The first one, hired in the summer of 1976, recommended the company do a thorough safety review, get sperm counts from workers, or run their own laboratory tests.[17] The next spring Oxy brought in two more industrial hygienists, who suggested DBCP might be the cause of the sterility and took measurements of air levels of the chemical, but stopped short of performing the medical tests workers had requested. While DBCP seemed to be at a safe air concentration the day they were there, the hygienists were concerned that it would be dangerous under other conditions.[18]

While management failed to take action, the workers decided to address the problem by turning to an established OCAW strategy—demanding health-protective agreements as a part of their upcoming contract negotiations.[19] In the spring of 1977, while preparing for contract talks, the union was approached by a pair of young filmmakers. Josh Hanig and David Davis had received a Public Broadcasting Service grant to make a film on safety and health at work. After talking to workers about their sterility concerns, the filmmakers decided to arrange and pay for fertility testing for the seven possibly infertile workers.[20] When the results were mailed to the union doctor, he thought there must have been some mistake—all seven men were sterile. Further investigation showed that one of the seven men had undergone vasectomy; but when 5 of the remaining 6 were retested, the disturbing results were unchanged.[21]

Alarmed by worker reports, corporate officials sought information on the potential causes of sterility even while deciding how to respond to "employees and to possible leaks to the press."[22] Once again DBCP emerged as the likely culprit. On July 20, Oxy product manager Jack Horner spoke with Dow's Product Steward for DBCP, toxicologist Frances O'Melia, about "the tox data of DBCP—especially as it relates to the male human reproductive system."[23] The next day, O'Melia sent Horner a reprint of the 1961 *Toxicology and Applied Pharmacology* article reporting Shell and Dow researchers' findings of reproductive damage to experimental animals.[24] As coincidence would have it, Mitchell Zavon, the former Shell medical examiner who had also been a player in the 1950s toxicology studies, had recently begun working as medical director for Occidental Chemical's parent company, Occidental Petroleum. Consulted on the Lathrop crisis, Zavon wrote on July 22 that he was "skeptical" that "the chemical in question" caused the Oxy workers' sterility.[25] He was, however, concerned with defending the industry from what he considered "the inevitable media sensationalism, attempts to generate legal action, and the almost certain political opportunism."[26]

In fact, in the coming weeks, media attention did project the DBCP issue beyond the realm of worker–employer conflict at Oxy, and other chemical companies and California and federal regulators began to stake out their ground in what would evolve into a lengthy struggle over the fate of the chemical. A San Francisco television station on July 30 and 31 was the first to break the news connecting DBCP to sterility, and the story was quickly picked up by the national TV networks.[27] By August 1, 23 of the 25 men working in Oxy's ag-chem line had been examined. Of the 15 who had not had vasectomies, 10 had abnormal sperm counts. Zavon confirmed, "with one exception, anyone working in the [ag-chem] area for more than 1 year is sterile."[28] By August 2, the state Occupational Safety and Health Administration (CalOSHA) had asked Oxy to cease production, citing a "strong suspicion" that DBCP was the responsible chemical.[29] Two days later, an Associated Press story landed on the front page of newspapers across the nation: "Chemical Workers Found Sterile."[30] By August 11, initial results from Dow showed that 12 out of 14 Magnolia, Arkansas, plant workers with a history of exposure had sterility. In response, Dow halted DBCP production, suspended sales, and urged the return of product from formulators, distributors, dealers, and farmers.[31] Shell, not producing DBCP that summer for unrelated reasons, also stopped sales and began a recall program; its tests of workers also found that 16 out of 21 exposed workers had abnormally low sperm levels.[32] On August 12, the California Department of Food and Agriculture suspended all use of DBCP in the state.[33] Other state regulators took no action on agriculture use, so farmers elsewhere could use the chemical until the federal government stepped in. New national regulations seemed imminent, however, as OSHA, the EPA, and the National Institute for Occupational Safety and Health (NIOSH) began addressing the question of how to respond to the question of DBCP toxicity.[34]

BANANA COMPANIES RESPOND

Even before U.S. regulators took their first steps toward controlling DBCP harms, the two big banana transnationals staked out their positions in the growing skirmish over DBCP. United, already concerned that DBCP was producing excessive residue levels in their bananas, responded to the summer's news with the decision to replace all DBCP with Mocap and Nemacur. Apparently company officials were not concerned about DBCP's health effect on other companies' workers: a terse August 25 memo authorized the sale of existing stocks "at whatever price is obtainable."[35] Standard Fruit, on the other hand, planned to continue the use of DBCP, justifying that decision in the language of both law and science. On August 18, Henry Cassity, materials manager at Standard Fruit's U.S.-based parent company, Castle & Cooke, sent a telex to Dow:

CASTLE AND COOKE WILL CONTINUE TO USE DBCP AS LONG AS (A) ITS USE IS NOT FORBIDDEN BY U.S. GOV'T AGENCIES OR THE PROVINCIAL OR STATE GOV'TS IN A SPECIFIC LOCATION IN WHICH WE OPERATE OR (B) THERE IS NO EVIDENCE THE PEOPLE WHO APPLY THE CHEMICAL HAVE BEEN RENDERED STERILE OR HAVE BEEN HARMED IN ANY OTHER WAYS.[36]

The telex warned that Dow's failure to sell DBCP as agreed would be considered a "breach of contract," a clear intimation of legal action. In addition to its contractual claims, Castle & Cooke spelled out a logic that combined appeals to both regulatory and scientific authority, casting itself as both a good citizen who obeyed relevant regulations and a careful user of chemical technologies who acted according to scientific evidence.

Castle & Cooke personnel implicitly and explicitly affirmed their commitment to following U.S. laws, even in non-U.S. operations, while at the same time expressing anxiety about the effects of such regulation. In an August 16 internal memo, Standard Fruit's Jack DeMent worried about an eventual DBCP ban, because other nematicides were only about "65% as effective as DBCP." He was also afraid that a DBCP ban would lead to a sort of regulatory domino effect: "My overriding concern at this moment is that the adverse publicity over DBCP will lead to cancellation of other soil fumigants. Should this happen, we could lose the DD type fumigants which are the backbone of our current nematode control program in pineapples. This could be a real disaster."[37] It is not clear here if DeMent thought his company would "lose" banned chemicals because they would no longer be manufactured (as appeared to be the case with DBCP) or because the company had a blanket policy that precluded use of banned chemicals. DeMent's fear of adverse publicity was likely linked to continued environmentalist concern with pesticide dangers, including opposition to the export of the most dangerous pesticides from the United States. In 1975, export of pesticides had come to regulators' and the public's attention when four environmental agencies filed a lawsuit against the United States Agency for International Development (USAID) for financing the export of banned, unregistered, and highly toxic pesticides, including DDT.[38] While FIFRA still exempted exported pesticides from its labeling and other regulations, DeMent clearly saw that activist attention could modify U.S. policy, which in turn would influence the companies' pesticide choices, even in locations outside the United States.

In the meantime, Castle & Cooke officials continued their use of DBCP. As they surely suspected, there would be an extensive regulatory process before the EPA came close to "forbidding" the use of the chemical, and there were many policy options short of an outright ban. With Dow "hold[ing] firm on not shipping any product until controversy is settled,"[39] Castle & Cooke turned to the international market for other sources of DBCP in Israel and Mexico, revealing how global supply could undercut the protective actions taken in one nation.[40] The dispersed corporate structure of the company provided another way around what U.S. headquarters saw as "potential legal problems" if the fruit company bought DBCP from the United States. As Henry Cassity warned Standard Fruit personnel in Honduras, Nicaragua, Costa Rica, Ecuador, and the Philippines: they "may have to generate P.O. [purchase order] and payment from your end."[41] Castle & Cooke's claim of respect for bans imposed in the "specific locations in which we operate" was a rhetorical nod to national pesticide regulations that obscured how daily operations in Central America could often violate national safety regulations that fell short of a total ban. In Central America, even in nations where pesticide regulation was relatively well developed, inadequate resources meant enforcement was infrequent, and companies often did not bother to follow the rules.[42]

In the scientific vernacular, Castle & Cooke justified the continued use of DBCP by resuscitating Dow and Shell's early "history of safe use" argument. This time, it had a new twist. The company claimed that no scientific evidence existed that "people who apply the chemical have been rendered sterile or have been harmed in any other ways."[43] That is, instead of claiming—as Dow and Shell had—that human production workers showed no evidence of the harm DBCP caused to animals, it compared farmworkers to production workers. Just because men laboring in U.S. factories had been harmed by the chemical, maintained Castle & Cooke, there was no evidence that banana farmworkers would be sterilized by their exposure to DBCP. In fact, there existed ample evidence to suggest that agricultural exposures were of concern. Both Dow's and Shell's research in the late 1950s and early 1960s had shown levels of exposure among irrigation applicators that at times exceeded the 1-ppm level that Dow was now recommending as the maximum acceptable exposure.[44] Plantation managers on Standard Fruit farms also had everyday evidence of excessive dermal and inhalation exposures, ranging from the smell of DBCP during application to the all-too-common spills and splashes experienced by workers carrying out both injection and irrigation application.

Castle & Cooke's claims to good citizenship and good science at first received no positive response from Dow. Both companies would continue to pay attention to the ongoing and unfinished regulatory process in the United States, however. As the initial media attention to the chemical's dangers gave way to a long and uneven process of control, U.S. laws would soon exert a powerful if indirect influence on the two companies' negotiations.

PUBLIC REGULATIONS, PRIVATE AGREEMENTS, AND UNEVEN CONTROL

In the closing months of 1977 and into 1978, citizen groups, corporations, and regulators weighed in on the future of DBCP in the United States. Over the course of the decade, manufacturers had been successful at expanding the national market for the nematicide, and by 1977 pesticide manufacturers of varying size and reach produced about a hundred unique DBCP-containing products for use on soy, cotton, peanuts, and fruit, nut, and vegetable crops, as well as on turf and ornamentals.[45] The consumption and importance of DBCP across U.S.-grown crops was uneven. Some crops were important sources of sales for chemical producers, with the most DBCP used on soy beans, where it was applied to half a million acres annually.[46] Other crops, such as pineapples, peaches, and grapes, were responsible for far fewer DBCP sales, but were considered highly "dependent" on the chemical for their "long term future existence," as Dow's Harold Lembright put it.[47] Major or minor, "dependent" or not, all uses presented workers with some exposure. For example, the 10,000 workers who applied the chemical to soy fields used a highly mechanized mode of application—an injector pulled behind a tractor—but nevertheless may have had dangerous exposures while mixing, loading and troubleshooting equipment. Thousands more workers were exposed while working on other crops, including an estimated 4,000 involved in irrigation applications used on fruit trees, grapes, and other crops.[48] All uses of DBCP were initially questioned as controls over the chemical began to tighten. However, as regulation of DBCP grew stricter, it was uneven. Paradoxically, inequalities of exposure actually increased as some groups received greater protections than others. These disparities in protection were evident both within the United States and between nations, and were undergirded by both public regulation of and private agreements on the sale and use of DBCP.

On August 23, 1977, two groups filed petitions in federal court seeking the regulations they felt were necessary to control DBCP risks. On one hand,

OCAW asked OSHA to set a factory exposure limit for production workers of one part per billion (ppb)—1,000 times more stringent than Dow's recommended standard, and the strictest standard allowed by law. On the other, the Health Research Group, a coalition that was part of Public Citizen (a consumer advocacy group led by Ralph Nader) had asked EPA to ban all uses of DBCP, an action that would prevent farmworker and public exposure to the chemical in the United States.[49] On September 9, the agencies announced temporary rules that fell short of those suggested by either group. OSHA set an emergency temporary standard of 10 ppb, a level 100 times more stringent than Dow's recommended standard, but 10 times more lenient than the standard proposed by OCAW.[50] The standard effectively halted DBCP production in the United States, at least temporarily. If companies retooled to meet the more stringent requirements, they could legally resume production.[51]

In a weaker ruling, the EPA stopped short of banning DBCP entirely. The agency focused on estimated cancer risk posed by DBCP, as research on human reproductive risks was not yet available. Agency head Douglas Costle's 1977 "Notice of intent to suspend DBCP" found that eating foods with DBCP residues could be deadly. Specifically, assuming lifetime exposure to chemical residues in food, "DBCP may be expected to cause anywhere from 28 to 740 cases of cancer per one million people exposed."[52] Costle also found that farmworkers faced serious risks. For example, he calculated that lifetime exposure to DBCP "may be expected to cause 5.5 cases of cancer per 1,000 applicators in the applicators' lifetime" (equivalent to 5,500 cases per million, for comparison to consumer risks) in soybean application; 9.7 per 1,000 (9,700 per million) in pineapple application; and 1.2 per 1,000 (1,200 per million) in irrigation application. And although "data for conclusively establishing an ample margin of safety for exposure to DBCP with respect to possible damage to human reproductive capacities" were not yet available, Costle also wrote that reproductive risks to workers were "unacceptable."[53]

Despite the fact that the calculations of worker risk dwarfed consumer risk, EPA suspensions were geared more strongly to protecting consumers. The agency suspended its use on 19 crops known or predicted to have residues of the chemical, because of cancer and reproductive risks posed by contaminated foods. For all uses of DBCP not thought to expose consumers to residues—that is, those uses posing a risk only to workers and not consumers—the agency instituted what it called "conditional suspension." This meant that sales and use could continue if certain conditions for heightened worker protection were met. Manufacturers were required to relabel DBCP as a restricted-use chemical, meaning that only certified applicators, who would be required to use respirators and other protective equipment, would be allowed to apply it. The new restrictions, however, left open a likely avenue for worker exposures by failing to mandate a waiting period—called a reentry interval—to prevent people from coming into treated fields after application.[54]

This two-pronged approach to DBCP regulation showed that the EPA was most concerned with preventing exposure to DBCP from *eating* food, not *growing* it, creating another kind of inequality of exposure. Eliminating the use of a substance—as the agency did for food crops with residues—is quite obviously the most effective way of preventing exposures. If the substance is not used at all, no exposure can take place. On the other hand, allowing continued use, even with protective measures, reduces but does not eliminate exposures, which can still take place when humans err or disregard rules, or when equipment fails. Worker and consumer protections had different places on what occupational hygienists call the "hierarchy of controls." Eliminating the use of a substance is the strongest control measure and sits at the top of the hierarchy. The new EPA rule used this option regarding consumer exposures to DBCP. In contrast, measures from lower on the "hierarchy of controls" (such as reentry intervals, labeling, and protective equipment)[55] were the sorts of protections provided to workers, despite the fact that estimates showed occupational risk to be orders of magnitude greater than consumption risk.

This disparity was magnified by the fact that the bulk of DBCP was used on nonvegetable crops—those on which use would not be suspended. As a Shell sales manager noted in October, "the proposed suspension seemingly affects only 11 percent of our business while 89 percent is represented by those crops [sic] uses which will be conditionally suspended."[56] That 89% was dominated by soybeans (which made up 50% of the total), cotton (19%), grapes (8%), and citrus (3%). For these crops and others outside the 19 for which DBCP was suspended, farmworkers—as well as those working nearby or entering fields after application—could still face exposure to DBCP if employers didn't follow regulations, if protective equipment didn't work, or if protections were in some way not 100% effective.

Inequalities of protection were starkly evident on the transnational as well as the national scale. As a Shell sales manager noted, the DBCP suspension "of course, does not include the sizeable off-shore market on bananas."

Banana workers on Standard Fruit farms would see fewer protections than either agricultural or production workers in the United States. After U.S. regulators passed their first rules, Castle & Cooke continued to "plead" for continued DBCP sales from Dow.[57] Even the president of the fruit company reportedly called Dow's president.[58] Evidently, these overtures were successful and by February 1978, the fruit and chemical companies had reached an agreement on the sale of DBCP, with a sales contract "spelling out the volumes, formulations, packages and unit price" for the equivalent of more than 600,000 gallons of Fumazone.[59]

Records of the companies' negotiations did reveal some unease at the potential for ongoing banana worker exposures. Dow, hesitant about bad publicity, developed a "P.R. release we would plan to make in case of adverse press reactions" and suggested that "Castle & Cooke/Standard Fruit may want to draw up a similar release."[60] Dow also insisted on an agreement with Standard Fruit that stipulated that the banana growers provide basic protections to workers.[61] The specific guidelines for Standard Fruit were developed by the toxicologist Frances O'Melia, Dow's Product Steward for DBCP,[62] and then revised after a Dow team visited Castle & Cooke's Honduras operations and observed DBCP irrigation application.[63] The requirements specified that DBCP applicators use protective gear; there were also guidelines for drum disposal, employee training, first aid, and the scheduling of applications, which were to take place at night in order to ensure the fields were empty of workers.[64] In formulating guidelines, the agreement drew from the strict, but at that point still temporary, OSHA standard, rather than the EPA rules that governed agricultural worker exposure in the United States.

While the Dow requirements seemed to provide for the protection of workers, they contained significant flaws. Given the long-standing difficulties regarding the use of heavy protective gear in the hot weather of banana-producing regions, it must have seemed dubious to people at the plantations whether these requirements would be followed at all. There were other problems with the guidelines as well. For example, they ignored the fact that Dow knew "there is no clothing available that is 'impermeable' to DBCP."[65] Another flaw in the document related to the reentry interval. Rather than basing that interval on safe air-level measurements, Dow ultimately established that workers could return after "all odor of the chemical is gone."[66] However, as Dow's own 1958 research on DBCP had established, the chemical's odor was detectable only when it reached levels that were already extremely toxic.[67]

Even without its flaws, the agreement between Dow and Standard Fruit would have failed to protect workers because it existed outside any structure that could provide oversight, enforcement, or accountability to affected individuals. While it drew on the language of the U.S. regulatory apparatus, it detached that language from the context in which it had been formed and would be enforced, rendering it essentially meaningless. Tellingly, when Dow produced a "comprehensive general safety guide" for the use of Fumazone in injection applications, an internal Dow memo and a letter transmitted to Castle & Cooke along with the safety guide both explicitly stated: "We do not expect to monitor the use of FUMAZONE."[68]

Absent any monitoring by Dow or by regulatory authorities, Castle & Cooke's decision-makers instructed Standard Fruit farm managers outside the United States to disregard the rules. In early March of 1978, Jack DeMent, the Director of Agricultural Research for Castle & Cooke, sent a letter to managers in banana divisions with "a few comments" that explicitly instructed them to ignore several key provision of Dow's "requirements."[69] These included provisions that:

> The user will agree not to apply FUMAZONE 86E unless: (1) the treated area is a safe distance away from worker housing and work areas (such as packing stations) or (2) all people have been evacuated from area to be treated and those surrounding areas which may be exposed to the liquid or vapors. . . . People in areas to be treated will be notified to that effect in the language of the workers involved.

DeMent instructed his managers that "this is not operationally feasible and does not need to be implemented."[70] Regarding the requirement that workers exposed to the chemical wear approved protective gear, DeMent commented that "the divisions should supply the equipment to DBCP applicators." However, as he was familiar with the working conditions in the banana fields, he tempered this recommendation with the addendum that "forcing them to use it may be difficult if not impossible but at least such equipment should be available."[71] DeMent's decisions gutted the most important provisions of the Dow "requirements," leaving workers and their families unprotected from DBCP vapors and skin contact.

The form and failures of the Letter of Agreement revealed the limits of national regulation and corporate self-regulation in the context of transnational business dealings. While Dow used U.S. regulations to give rhetorical authority to the safety conditions accompanying DBCP sales, that corporation did not enforce the requirements of the agreement. Ironically, Castle & Cooke—the company that had earlier averred its obedience to U.S. or other national bans—instructed managers to ignore much of the regulatory language that had been written into the agreement. Because the document was exempt from any enforcement mechanism, the banana producer could continue to expose workers with impunity despite its promise to obey safety rules.

U.S. REGULATION, OMISSIONS, AND EVASIONS

Banana workers would not be the only people with continued DBCP exposure. As national regulations were passed to protect factory workers, production workers outside national borders and agricultural workers within the United States also faced continued exposure to the chemical. Production workers in the United States were protected not only by the OSHA temporary standard, but also by the fact that production had come to a standstill in the face of the initial panic over worker sterility and the new factory exposure rules.[72] By March 1978, OSHA set a permanent DBCP standard of 1 ppb—10 times lower than the agency's emergency temporary standard, and the lowest level considered achievable by OSHA.[73] The standard also included a whole range of other requirements, including provision of carefully specified safety equipment and changing and showering rooms. These provisions did not protect farmworkers, however, since agricultural workers fell outside the scope of OSHA. Instead, they were covered by the EPA rules, which were not as strong as OSHA's. The differing levels of protection mirrored workers' political power. Farmworkers—largely immigrants from Latin America and U.S.-born people of color—faced xenophobia and racism, and had little access to economic or political power within the United States. Production workers, on the other hand, were in their majority white males, with access to more economic and political resources, particularly through unions such as OCAW that had recently shown their strength.

Despite these differences, the interests of production and farmworkers were intertwined. Seemingly, farmworkers would not be exposed if strict OSHA rules made it impossible to produce the chemical in the first place. For a time, this seemed likely. In April 1978, the industry publication *Chemical Week* reported that the standard was in large part responsible for Dow permanently "bow[ing] out" of DBCP production.[74] As for Shell, while the company was planning to sell its existing stock of DBCP due to "intense agricultural demand," it insisted that "no manufacturing plans are contemplated."[75]

However, the confluence of continued market demand and strict OSHA standards prompted one upstart chemical company to look beyond U.S. borders to produce the chemical, creating yet another inequality in DBCP exposure. The California-based company Amvac hoped to profit from what it described as the "vacuum [that now] existed in the marketplace."[76] Amvac's own facilities were not up to OSHA standards, so it began by looking elsewhere for DBCP. First buying and reselling existing stores from Occidental,[77] the company soon contracted with two Mexican plants, where workers were unprotected by U.S. regulations and no national rules had yet been instituted.[78] Not surprisingly, DBCP had the same effects on workers in Mexico as it did in the United States. By September 1978, a Mexican health official concerned with the sterility effects conducted testing among production workers at the plants and found all but two of 23 workers had low or zero sperm counts.[79] Although Mexican production was eventually halted by authorities there, Amvac's approach illustrates the conflicts between the national regulatory reach and the national and transnational market for the chemical. While regulations stopped at the border, corporations and chemicals crossed it easily. At the same time, the EPA's weaker regulation was not strong enough to obliterate demand for DBCP, maintaining a market for the chemical wherever and however it was produced.

An EPA action in September 1978 reinforced national inequalities of exposure while also enshrining transnational inequalities. The agency's "Final Position Document," which would actually be far from their last word on the chemical, was essentially a risk/benefit analysis that synthesized research and input collected over the preceding year.[80] The agency's risk analysis was based on accumulated evidence from toxicological tests in laboratory animals, expert opinion on the carcinogenicity of DBCP, and the history of adverse reproductive effects in lab animals and humans. The human data included sterility results from chemical producers, as well as findings from its own epidemiological studies of people exposed in agricultural contexts. These data showed DBCP damage was not limited to factory workers but that "exposure to DBCP in agricultural use situations poses a risk of testicular toxicity (i.e., depressed sperm counts) to exposed individuals, and that the risk increases with increased exposure."[81] Based on the human and animal studies, the report deemed DBCP to be both a "strong carcinogen" and a reproductive toxin. For agricultural work, the report raised the most concern about irrigation application and dermal contact, noting that "exposure to a *single drop* of DBCP could result in [a measurably] increased risk of cancer."[82]

But the EPA process had also given chemical and agricultural companies a chance to weigh in with their take on the question of whether DBCP should be allowed.[83] Dow sought support for the nematicide from its DBCP customers, asking them for feedback and explaining that "a very essential part of the pesticide's defense can be the 'benefit/risk equation.' Simply stated: Do the benefits of its usage outweigh the risks?"[84] Dow and those customers who responded to its requests felt that the answer to that question was yes. Castle & Cooke's desire to retain U.S. approval of DBCP use on bananas was evident in its contribution to the EPA: an 18-page document making a case for the "Benefits of DBCP Use on Bananas."[85] Although Jack DeMent had as recently as August 1977 affirmed that "the possibility of loss of this compound or any other in our arsenal has always been considered in our research thinking,"[86] the "Benefits" document argued that DBCP was irreplaceable. Invoking a rationale that went far beyond the usual price-and-yield comparisons companies used to determine the nematicide of choice, the document argued that DBCP was essential to the banana trade, which in turn was essential to maintaining the health of people in the United States and developing the economies of banana-producing countries. Without DBCP, it claimed, banana companies would be forced to expand production at great expense and "ecological disruption."[87] Other pesticides to replace DBCP were either "more hazardous (at least insofar as acute toxicity is concerned), difficult to use, ineffective, or dangerous to banana plants or wildlife."[88] In this document, DBCP was portrayed not as a dangerous chemical, but as a bedrock of safety, health, and economic development.

When the EPA released its Final Position Document on September 6, 1978, the agency reproduced the banana company's analysis, finding that:

> There are no direct adverse impacts [of canceling DBCP banana uses] in the United States, since little or no DBCP is used in this country. However, if other banana growing countries were to ban DBCP, the price of imported bananas could increase from 23.5 to 27.0 cents per pound. Based on the U.S. consumption of 4.2 billion pounds of bananas, this price increase would result in a total impact of $150,000,000 to consumers.[89]

These price projections were taken directly from Castle & Cooke's analysis,[90] and reporting them as an aggregate seemed to indicate an onerous burden on U.S. consumers. But in fact, were the price increases considered at the scale of the individual consumer, they were almost negligible.

At the same time, the EPA dismissed the risk side of the analysis out of hand: "Because little or no DBCP is known to be applied to bananas [in the United States]"—that is, because no bananas were commercially grown in the United States— "no risks [to workers] were calculated."[91] By imposing a strictly national and consumer-driven frame for the consideration of risks, the agency ignored the dangers faced by banana workers in Central America and elsewhere, despite the fact that the use of the chemical involved two elements that were noted in the very same document to be of great concern to the EPA: irrigation application and high levels of dermal exposure. The omission of the serious risks faced by banana workers biased the risk/benefit analysis in favor of ongoing approval of DBCP. The result was a lopsided analysis that implicitly authorized the ongoing use of DBCP in Central America.

Ultimately, the FDA recommended that for use on bananas and on 27 other crops, DBCP be classified as a restricted-use chemical. For these 28 crops, the EPA would require full-body protective clothing, boots, gloves, and a respirator during "all phases of all applications by all techniques," as well as a 24-hour reentry interval. Use on homegrown vegetables, 19 commercially grown vegetables, and peanuts would be canceled. Once again, the most protective action—cancellation—was taken only on those crops where the EPA determined that consumers would be exposed to DBCP-derived residues on food. In crops where food exposure was not deemed a problem, farmworkers' exposures would be controlled not by the strongest measure— elimination at the source—but by less robust interventions.[92]

Ironically, the same month the Final Position Document was released, FIFRA was amended with changes strengthening the label and notification requirements for the export of highly regulated pesticides. For the first time, all exported chemicals had to bear labels—in the importing nation's language—that truthfully disclosed active ingredients, producing company, toxicity warnings, and, where applicable, a notice that the pesticide was not registered in the United States. The new law also strengthened the 1972 notification procedures for banned pesticides, requiring purchaser acknowledgment that the chemical was banned in the United States, as well as notification of the appropriate authority in the importing nation.[93] These requirements all depended on information exchange rather than controls on sales, maintaining a lower standard of protection for people outside the United States than for those inside. The new DBCP regulations revealed the inadequacy of notification laws. First, the EPA's explicit affirmation of its use on bananas would rhetorically offset any discouragement of use implied by the notification.[94] In addition, the "foreign purchaser" in DBCP's case was an employer who had demonstrated little concern for worker health, and notifying the governments where the banana companies operated, while a starting point, would not give them the resources needed to adequately control pesticide use.

In the United States the continued availability of DBCP, combined with less-than-ideal real-life application practices, meant that the chemical still posed serious risks to farmworkers. After observing what was supposed to be a tightly monitored test application of DBCP in California in January 1979, Robert Reeves from the State Occupational Health Research and Development Section reported to colleagues in the state Department of Food and Agriculture that he had witnessed "exposure [of workers] to unknown concentrations . . . during the transfer operation, equipment hookup, and actual application," including a leak that was not decontaminated. Workers wore no respirators or protective clothing.[95] Reeves concluded: "The cavalier attitude and utter disdain for minimizing exposure to a known carcinogen by these individuals in quite shocking. . . . If this test was an example of how DBCP is handled under controlled conditions, then the circumstances of uncontrolled day to day application must be devastating." For these farmworkers at least, regulation did not mean protection.

Despite dangers to farmworkers in the United States and Central America, it looked like the "Final Position Document" determinations would preserve a future for DBCP both nationally and internationally. Castle & Cooke was not the only corporation to benefit from the EPA's latest word on DBCP. Growers of crops on which high amounts of DBCP were used—including soybeans, cotton, and grapes—could continue to count on the availability of the nematicide, while the chemical producers could look forward to profiting from a DBCP market revitalized and stabilized as a result of the new regulations.[96] While another round of hearings was still in store, for now it was, according to Shell personnel, the "most favorable decision industry might have anticipated."[97]

BANANA WORKER STERILITY, CORPORATE INFLUENCE, AND UNOFFICIAL REGULATION IN COSTA RICA

While DBCP control was being debated in the United States, and Castle & Cooke continued to use the chemical on their Standard Fruit plantations, the first DBCP-linked sterility cases were diagnosed in Costa Rica. This discovery of a sterility effect set off a closed-door negotiation between the banana company and the Costa Rican government. As in the United States, the "solution" to the DBCP problem in Costa Rica was primarily defined in national terms and left workers outside the country's borders subject to further exposures, deepening inequalities in exposure between the banana-growing nations of Central America.

In late 1977, while Dow and Castle & Cooke were negotiating the terms of their continued trade in DBCP, Dr. Carlos Domínguez Vargas, Chief of Urology at Hospital Rafael Ángel Calderón Guardia, in San José, noticed that a high number of unusually young men were coming into his clinic complaining of their inability to father children.[98] While the capital was at least a two-hour bus ride from the Atlantic banana-growing region, banana workers came to the state-funded hospital to avail themselves of specialty care provided through the public health insurance system. By the end of 1978, Domínguez had seen between 20 and 30 men with sterility problems. Another 52 men in a Río Frío clinic had similar diagnoses.[99] Domínguez, noticing that most of the men were banana workers, began to suspect the damage was pesticide-related. He asked some of the affected workers to copy down the names and label information from the pesticides they used on the banana plantations. Among them, not surprisingly, was DBCP.

Although the doctor sought information from affected workers, the regulation of DBCP in Costa Rica would not unfold with public revelations of worker sterility, as it had in the United States. Instead, concerned elites like Domínguez, well positioned to glean information from national and international sources and with access to national officials, set in motion a quiet process of negotiation between the government and Standard Fruit decision-makers. Domínguez's ability to tap into a transnational network of information about pesticides was exemplified by the fact that his brother was the manager of a Costa Rican pesticide importing business. When the doctor asked his brother what he knew about the health effects of the chemicals named on the labels provided by banana workers, the importer confirmed that DBCP was, in fact, known to cause sterility. Domínguez discussed the problem with the director of the hospital, Dr. Fernando Urbina Salazar.[100] By the middle of 1978, the doctors contacted the EPA and secured information from that agency on the risks of DBCP. The EPA had apparently already sent information on DBCP's new regulatory status to Costa Rican officials at the Ministry of Agriculture, as required by the 1972 amendments to FIFRA, although the director of that office later claimed that he did not remember receiving the telex.[101]

By the end of 1978, the physicians felt that there was enough evidence to presume that workplace exposures to pesticides were causing this unusual upswing in sterility. Dr. Urbina, whose high-ranking position made it likely he would catch the attention of decision-makers, wrote the regional director of the *Caja Costarricense de Seguro Social (Caja)*, the Costa Rican Social Security Administration, in November, arguing that the "cause-effect correlation" was sufficient for the *Caja* to "take the steps necessary to stop the sick men's employer from using this product."[102] The next month he followed up with a letter to the Vice President of Costa Rica, a medical doctor himself, "emphatically recommend[ing] taking measures so that no products whose active ingredient is DBCP are used in Costa Rica."[103] Although they were clearly genuinely concerned with workers' health, the doctors' choice to push for change through addressing officials rather than unions or the media meant that the problem of DBCP remained a secret from most of the banana workers, who continued to be faced with dangerous exposures.

By the time Dr. Urbina penned his letter to the Vice President, Standard Fruit had learned of the sterility cases and also contacted government officials, initiating a series of meetings to press their own perspective on DBCP.[104] Records of these closed-door sessions show how the banana company was able to gain direct access to officials and exercise its influence within the context of a Costa Rican economy and government that had already been shaped by the demands of the transnational banana business. At the same time, the transnational's power was not absolute, and officials concerned with workers' health pushed back against banana company pressure.

On November 30, Standard Fruit's Henry Nanne, Randall Ferris, and Alfonso Muñoz met with Minister of Health Dr. Carmelo Calvosa. Henry Nanne's memo describing the meeting suggests that the Minister of Health carefully tread a path between mollifying banana executives and carrying out the demands of his role as protector of national health. Nanne listed "some of the most important statements from the Minister," some of which were sympathetic to Standard's concerns.[105] "I worked with a Banana Company," Calvosa had sympathized, "and know that publicity on these kind of problems is used by many people for other purposes. . . . I also know that there are all kind of Doctors of which will do anything for money."[106] Calvosa framed Standard's use of DBCP in much the same terms as the company did, noting that it was "a product approved by the Minister of Agriculture." He accepted the company's contention that they had made protective equipment available. But Calvosa also acknowledged that such equipment was difficult to use in the muggy banana zones. More importantly, the Minister accepted that DBCP was indeed the cause of the workers' sterility, and "presume[d] more cases will show up, probably because of the result of the long term usage of the product without any protective device."[107] The Minister's contradictory statements bespoke of dual ideological commitments—on the one hand to the banana companies and the presumed economic benefits of the banana industry to Costa Rica, and on the other hand to his role as protector of national health.

The tension between these commitments was resolved with a compromise: Calvosa agreed to keep the disturbing news quiet, saying, "I will inform the press (if I'm asked) that the case is under study, therefore we will try to keep it as confidential as possible."[108] But, according to Nanne's memo, his "first and constant statement was: I am glad you took the right action, stop using DBCP. This will avoid further problems."[109] Calvosa may have been erring on the side of diplomacy—when the meeting began, Standard Fruit had not in fact agreed to stop using DBCP. But the Minister's tactics apparently met with some success, since on the same day of the meeting, Standard Fruit personnel made a compromise of their own by putting a hold on scheduled imports of the chemical.[110] Standard claimed to have stopped using DBCP that day as well, although a later study by the *Instituto Nacional de Seguros* (INS) suggested that the company probably continued to apply existing stocks of the chemical.[111]

In any case, Standard was far from giving up on DBCP and continued to advocate for continued use of the chemical. A draft memo written at the end of December 1978 built on the risk/benefit analysis contained in the EPA's "Final Position Document" to argue that Costa Rica could achieve "a sophisticated balance between maintaining agricultural production and safeguarding public health."[112] Ongoing negotiations with government decision-makers, however, suggested the banana company's power was limited within the political context of late-1970s Costa Rica. Castle & Cooke personnel from the United States made a trip to Costa Rica in January 1979 to meet with government officials, but the trip did not forestall stronger DBCP regulation, and the national Office of Quarantine and Registration stopped authorizing DBCP imports early in 1979.[113] Trying to keep the door open to continued DBCP use, Castle & Cooke's California staff worked with Standard Fruit personnel in Costa Rica, planning for another encounter with the Minister of Health.[114] They decided to ask for explicit permission to use DBCP for experimental purposes only under the EPA and Dow's guidelines for safe use. While they would have preferred authorization "to resume hand injection applications on a fully operational basis under strict handling controls," asking for limited experimental use seemed like it "would have a much greater probability of acceptance" and would "retain a partial precedent for DBCP use in Costa Rica ... [and provide] a mechanism for obtaining permission for further uses if needed."[115]

By May of that year, the banana growers and the government had reached their final compromise. Standard had not been able to dictate government policy: Costa Rica would continue to ban imports of DBCP.[116] But the government did make some concessions. It would stop short of withdrawing the registration of the chemical, which would not happen until 1988.[117] The government apparently did not inform workers or the press about the sterility problem, or hold the company responsible for the increasing numbers of affected men who sought treatment from the Costa Rican health system.

Despite the evidence of DBCP's sterility effect in agricultural workers, Castle & Cooke did not cease use of DBCP in its subsidiary banana plantations outside of Costa Rica. Indeed, once again national control proved incommensurate with transnational systems of production. In January 1979, Standard Fruit re-exported a 20,000-kilogram shipment of DBCP, newly arrived in Limón, Costa Rica, to its plantations in La Ceiba, Honduras.[118] In total, over 180,000 liters of DBCP were sent from Costa Rica to Honduras between January and August of 1979.[119] Not content with using existing stores, in the spring of that year Castle & Cooke agreed to buy more DBCP from Amvac, which had started manufacturing the chemical in Los Angeles after Mexican production was halted.[120] In May, Castle & Cooke's purchasing manager compiled a list of "High Use Items," including DBCP for "nematode control"—in Honduras, Nicaragua, Ecuador, and the Philippines.[121]

By contrast, for Standard's main competitor, United Fruit, corporate policy on DBCP did not result in as many trade-offs between safety and profit. For United, high bromine residues, "labor problems with injections," and phytotoxicity made further use of the compound "unlikely unless new application techniques are developed ... or costs of [alternatives] become excessive."[122] Both were keeping their options open, but Standard was far more dependent on DBCP.

Castle & Cooke and Standard Fruit's willingness to continue using the chemical shows that the Costa Rican government's unofficial "ban" on the chemical was not enough to discourage the company from continuing to use it elsewhere. Instead, it simply shifted the risk. Like U.S. regulation of the chemical, the deal worked out with the Costa Rican government was strictly national in terms. As with the U.S. regulations, this nationally bounded (and in the case of Costa Rica, closed-door) government action implicitly authorized the corporation to put other nations' workers at risk. The fact that the Costa Rican negotiation with Castle & Cooke took place behind closed doors also allowed the company to more easily continue to use DBCP elsewhere. Official, public regulation of DBCP in Costa Rica would surely have been noticed by banana worker unions as well as other Central American governments, which may have been more likely to protest or control the use of the chemical in their own nations. Castle & Cooke was able to take advantage of the nationally bounded actions of various governments to continue using the dangerous toxin in nations without government controls.

DEBATE, POWER, AND DATA IN ONGOING DBCP REGULATION

In the United States, regulation of DBCP continued to unfold in a process that was public, but nevertheless marked by inequalities and conflicts. October 1978 marked the beginning of a year of intense debates not only over inequality in exposure, but also over access to the regulatory process in the first place. Workers' and citizens' groups mobilized for stricter regulation, while corporations—notably pineapple producers afraid of losing access to the chemical—attempted to secure the least restrictive rules for their own favored uses. At the same time, new scientific data emerged that changed the power calculus of regulation by suggesting that DBCP could affect a much larger group. As opposing viewpoints collided and scientific evidence was interpreted, regulation became stricter even as it continued to prioritize some groups' safety over that of others.

After the publication of the EPA's September 1978 "Final Position Document," a group of farmworkers, citizens' groups, and worker advocates, including the groups California Rural Legal Assistance, Migrant Legal Action Program, and the National Association of Farmworker Organizations, filed a petition with the EPA. In what would become known as the Amaya Petition, they maintained that the restrictions and cancellations proposed by the EPA were inadequate to protect farmworkers from DBCP hazards. They argued that the type of respirator recommended was the wrong type, that protective clothing recommendations were inadequate, and that a 24-hour reentry interval was far too short to ensure that workers were not exposed to lingering air concentrations of DBCP. In addition, they argued that farmworkers should receive training, and that the EPA should do a better job of monitoring and educating workers.[123]

Opponents of the Amaya Petition included Amvac and the Pineapple Growers' Association of Hawaii (PGAH), an industry organization comprised of three corporate members, one of which was Castle & Cooke.[124] Those fruit and chemical interests sought to prevent the farmworkers' concerns from being considered at all, arguing that because the Amaya group asked for a more—rather than less—protective standard than that proposed by the EPA, they did not have legal standing to request a hearing.[125] After conflicting interpretations from EPA authorities, in February 1979 Assistant Administrator Steven Jellinek issued a notice that there would be a hearing on the Amaya objections.[126]

In the ongoing proceedings, representatives of the Amaya group drew attention to the double standard in national DBCP regulations, showing how farmworkers were less protected than their industrial counterparts. In a Scientific Advisory Panel meeting in April, Ralph Lightstone, the lawyer for California Rural Legal Assistance, said that their "starting point" for determining adequate protections was "the OSHA standards."[127] While OSHA had called for zero dermal exposure and a 1 ppb inhalation maximum, the EPA process had not set a firm guideline for exposure levels in agricultural settings, opting for more ambiguous guidelines. Lightstone pointed out that the respirator recommended by the EPA was not as effective as that required by OSHA, and that the protective clothing recommended by the EPA did not appear to be effective. What's more, unlike the OSHA standard, the language of the EPA's recommendations did not make it clear that protective measures were the responsibility of the employer. Lightstone explained:

> The OSHA standard relates: "The employer shall provide the protective clothing at no cost to the employee. The employer shall assure removal of clothing, that the procedure for removal of clothing and disposal are followed. The employer shall ensure," and it goes through all of these particular standards.
>
> The EPA proposal, [specifying what] would go on the label, says, "Wear DBCP resistant clothing." It does not give a clear delineation of responsibility.

If you are going to be registering and using a carcinogen, you need clear delineation of responsibility for each of the safety requirements and for meeting the standard of exposure.[128]

With arguments such as this, Lightstone and his colleagues brought the question of uneven protection to the forefront of the DBCP debate.

However, ultimately it was not the question of inequality between farmworker and industrial worker protection that tipped the scales on DBCP regulation. Rather, as the Amaya Petition was considered, new data on residues demonstrated that DBCP posed a larger-than-suspected threat to a much broader group of people, significantly upping the stakes of DBCP regulations. The data came from California, where the director of the state Department of Food and Agriculture (CDFA) was conducting crop residue tests in order to monitor the necessity of the state DBCP ban.[129] Contrary to their expectations, researchers found DBCP residues in two crops, grapes and peaches, which had been cleared for restricted use on the assumption that the fruit would not be contaminated from applications made months before harvest.[130] The Amaya group had included some initial CDFA residue data in their petitions, and it was this data—rather than farmworker-specific risks—that the EPA's Jellinek emphasized when he decided to grant a hearing of the petition in February 1979:

> *Most importantly,* Amaya's objections refer to new residue data developed by the California Department of Food and Agriculture (CDFA), using a new and more sensitive analytical methodology, which indicate that residues of DBCP have been found in several crops (oranges, lemons, peaches, and grapes) for which the Final Position Document had predicted that no residues would be present. (emphasis added)[131]

Apparently, for Jellinek, new information on residues provoked more concern with EPA decision-makers than the important points Amaya petitioners raised about farmworker exposures, even though some farmworker exposure scenarios presented much greater risks than ingestion of the chemical on food. The new CDFA findings also showed that ambient air concentrations lasted longer and traveled farther from the site of application than previously thought, increasing risk for both farmworkers and bystanders.[132]

In early May, it seemed that despite the new residue data and farmworker objections, EPA and CDFA would approve continued use of DBCP with some restrictions.[133] Agricultural and chemical companies rallying to support the chemical included "most of the commodity groups using DBCP including Castle & Cooke and the pineapple growers [as well as] Química Orgánica [one of the Mexican plants from which Amvac had purchased DBCP], Amvac and Occidental." After attending a Scientific Advisory Panel (SAP) meeting, a representative of Occidental Chemical Company felt confident: "SAP appears to favor continued use of DBCP with restrictions. The EPA, which they advise, appears to be in the same mood. So is the California Department of Food & Agriculture. However, there are legal steps to be accomplished brought on by the California Rural Legal Assistance group."[134] As long as the legal steps turned out as the Oxy rep expected, that company was looking forward to the return of a boom DBCP market, as the chemical "represented an annual pre-tax income of over $4.5 million."[135]

On May 16, however, the CDFA announced a finding that signaled a much broader threat to both public health and chemical company profits. Testing had found DBCP in 36 wells in the San Joaquin Valley and Riverside County. The tainted wells supplied homes, towns, and farms in the region. Over a dozen wells had DBCP concentrations exceeding 1 ppb.[136] But this was just the beginning: eventually, California officials sampled 785 wells, and found DBCP in 180 of them, some with concentrations of 10 ppb or more.[137] While the chemical had been found in well water before, notably near Occidental Chemical's plant,[138] the new contamination was found in wells removed from obvious sources, especially concerning because it showed that the chemical persisted in groundwater. Subsequent EPA-led sampling in various states also found contamination in Arizona and Hawaii. No positive samples were found in wells tested in the southeastern United States.[139]

Well-water contamination with DBCP meant the possibility of high exposures in a population extending well beyond farmworkers. Farmworker exposures, which affected a marginalized population and could nominally be controlled by warnings and restrictions, may have been opposed by some organized farmworkers—like the Amaya group—but otherwise did not generate much public outrage or opposition. Exposure through drinking water was a different story, as it was difficult to control with warnings or other measures. Indeed, access to nontoxic public water was—if not always a reality—an expectation whose breach posed the possibility of broad political backlash. Given the mounting water-contamination data, on June 29, the Scientific Advisory Panel recommended bringing an end to all DBCP use.[140] By mid-July, the EPA had temporarily stopped all use and initiated proceedings for a permanent suspension.[141] While it had seemed the Amaya group's concerns might be disregarded by the EPA, the agency could not ignore the broader dangers that were now apparent.

The fight over DBCP had not ended, however. Revoking the registration would require more hearings, and the chemical's advocates presented their arguments for preservation of their pet DBCP uses or formulations. Amvac, for example, argued that the considerable body of toxicological evidence on DBCP did not apply to its particular products, because they did not contain the same additional ingredients as other formulations.[142] Pleading a similar exceptionalism, PGAH and the State of Hawaii argued that DBCP could be safely used on pineapples.[143] Opponents of DBCP once again included the Amaya petitioners, who were joined by the National Association of Farmworkers Organizations in favor of canceling all registrations.[144] After more than two months of proceedings, the EPA judge presiding over the hearings issued his recommendation: all uses should be suspended pending a cancellation hearing.[145]

When EPA Administrator Douglas Costle issued his decision on DBCP suspension on October 29, he adopted most of the judge's recommendations, with a single exception that produced a new inequality in DBCP exposure within the United States. Apparently adopting the reasoning developed by PGAH and their team of attorneys, Costle allowed continued DBCP use on Hawaiian pineapples.[146] The administrator explained the Hawaiian exception in terms of application techniques, residue tests, and a lengthy discussion of Hawaiian geology, arguing that DBCP could be used there without significant applicator exposures, food residues, or water contamination.[147] The Hawaiian pineapple exception was not necessarily permanent—it was accompanied by an EPA plan to cancel *all* uses, including pineapples, pending public hearings.[148] However, it was clear that the last bastion of DBCP use in the United States would be the place it started—a tropical island with an imperial past, racially and culturally distinct from the mainland, with an economy historically based on the same plantation model that characterized banana cultivation in Central America.

DBCP'S LAST GASPS

In mid-November 1979, in the wake of the EPA's almost total DBCP suspension, Castle & Cooke announced that while it would continue to use DBCP in Hawaii, it would no longer use the chemical in any other agricultural operations inside or outside the United States.[149] This announcement seemed to affirm Castle & Cooke's assertion that company policy was to obey U.S. bans in any location in which they operated. This scenario suggested the seamless relationship between U.S. regulation, U.S.-based corporations, and international operations, in which national rules, followed by companies even in overseas operations, would protect people both inside and outside of the United States. In reality, however, the DBCP debate was not over—either in the United States or in Central America. Instead, the early 1980s saw continued wrangling over DBCP, which had contradictory reverberations beyond national boundaries.

Even after the all-but-pineapple ban and with pineapple use cancellation on the horizon, Amvac and Arizona-based chemical company Gowan, as well as PGAH and (in testament to the power of pineapple interests there) the state of Hawaii, strategized to preserve a future for DBCP. They hoped that the chemical companies would be allowed to voluntarily withdraw their registrations for nonpineapple uses,[150] a move that—unlike an agency-mandated ban—would keep alive the possibility that DBCP could be registered again in the future.[151] In addition, DBCP advocates wanted permission to continue pineapple use under strengthened safety precautions.

Paradoxically, Amvac also hoped to use U.S. regulatory authority to retain a DBCP market in Central America. The company developed its own "Nematocide Crop Guide" for bananas, with much of the same information—from safety requirements to application instructions—as a traditional EPA-approved label would have included.[152] However, in contravention of any traditional regulatory channel, Amvac asked the EPA for an approbatory "letter regarding the . . . Crop Guide." Amvac President Glenn Wintemute hoped that "in lieu of a label," such a letter would "provide some protection as to continued use of DBCP on bananas."[153] In other words, Amvac was looking for an unusual kind of EPA authorization for DBCP sales outside the United States. By the end of the year, the EPA had forwarded a draft of such a letter to Amvac. Wintemute forwarded the draft to Castle & Cooke's Jack DeMent at the end of 1980, explaining that it would "assure foreign governments that EPA has been responsibly involved in the development of this Crop Guide."[154] His request for input from DeMent also suggested that the letter was meant to reassure the banana company, an important potential customer for Amvac. An official EPA statement in support of its "Crop Guide" would be a boon to Amvac, but would again disregard banana worker safety.

Company efforts to convince the EPA to preserve at least some kind of future for DBCP came in the context of a growing critique of toxic exports stemming from both fears of food contamination and the health toll of pesticides worldwide (the World Health Organization had estimated 500,000 poisoning cases annually). In 1979, the question of what to do about pesticide exports occupied an important place in federal policy discussions. The U.S. Government Accountability Office (GAO) authored a concerned report finding "Better Regulation of Pesticide Exports and Pesticide Residues in Imported Food Is Essential," and President Carter appointed an *ad hoc* working group to look into the issue. Even the State Department cosponsored (with the U.S. National Committee for Man and the Biosphere) a "Strategy Conference on Pesticide Management" focusing on the U.S. export of pesticides and agriculture and regulation in the developing world. Official recommendations and remedies from the GAO and State Department alike emphasized information exchange and excluded any real possibility of curtailing the export of banned chemicals.[155] A few years later, President Carter would follow suit, issuing a 1981 executive order strengthening export notification guidelines.[156]

While DBCP was mentioned only in passing in a 1979 report from the GAO and not at all in proceedings of the State Department–sponsored "Strategy Conference," the nematicide would soon garner more public attention. In the November 1979 issue of *Mother Jones* magazine, journalists David Weir (who had observed the State Department conference), Mark Schapiro, and Terry Jacobs skewered Standard Fruit for continuing to use DBCP on banana farms, arguing that the reimportation to the United States of banned pesticides—in the form of residues on the bananas themselves—made the ongoing banana trade a "boomerang crime."[157] The same year, two scientific studies were published that made less of a media splash but further suggested agricultural workers had indeed been affected by exposure to DBCP. One paper found sperm-count depression (although not infertility or azoospermia) among "applicators involved in irrigation setup work and in the calibration of equipment,"[158] while another found lowered sperm counts in formulators, custom applicators, and farmers, and "that greater increments in exposure were associated with a diminution in sperm production."[159] In Costa Rica, the first published report of DBCP's effects on farmworkers there would appear in 1980, finding significantly higher rates of sterility among workers with more hours of contact with DBCP and a statistically significant inverse relationship between duration of DBCP contact and sperm count.[160]

Science and journalism notwithstanding, federal concern with pesticide exports would not survive the 1980 elections. Less than a month after famously antiregulation President Ronald Regan took office, he revoked Carter's executive order in the name of ending "excessive regulations."[161] It must have seemed a good omen to Amvac officials. In fact, behind the scenes EPA attorneys penned a letter to the Agency Administrator,[162] detailing a potential resolution to continued use of the chemical on pineapples and offering a rational for approving Amvac's "Crop Guide" in a two-page appendix.[163] The appendix was a mishmash of regulatory and technical reasoning, based in large part on information provided by Amvac and "informal exchanges" with "knowledgeable personnel from major banana growing companies." The best that the EPA lawyers could say about Amvac's "Crop Guide" was the lukewarm endorsement that the "practices and cautions [it] recommended [would] . . . reduce the risks of DBCP application as far as practicable given the realities of banana culture."[164] The appendix was equally ambivalent on the EPA's authority to make policy on DBCP use on bananas, stating that although it had no "statutory requirement to evaluate use directions or recommendations designed for pesticides in other nations," and "could not engage in risk-benefit balancing" for use outside the United States, evaluation of the Amvac guide was "consistent with the letter and spirit of Section 17 of FIFRA." That section required notification of purchasers and foreign governments of suspension or cancellation of pesticide registrations, a requirement arguably distinct from approving instructions for use outside the United States.

The EPA attorneys "strongly recommend[ed]" that the Administrator not only approve Amvac's guide, but also allow continued pineapple use and let the chemical companies voluntarily withdraw their registrations. On March 5, 1981, Acting EPA Administrator Walter Barber accepted the lawyers' recommendations, agreeing to the "compromise" with corporate interests.[165] Barber's decision was a victory for the fruit and chemical companies alike: by approving the Amvac "Crop Guide," the EPA projected the agency's moral and scientific influence beyond national boundaries. This exceptional action allowed Amvac to invoke U.S. authority well beyond the nation's borders, even while the government recognized DBCP's dangers. The agreement effectively gave chemical corporations the best possible deal on regulation within the continental United States (that is, voluntary rather than enforced cancellation), while giving them even more latitude to continue sales of the pesticide in those far-flung places where the threat of sterility would be not raise powerful opposition—the still-new state of Hawaii and the banana-growing nations of Central America.

Castle & Cooke's response was in some ways the inverse of Amvac's—the banana grower, rather than seeking expanded markets outside the United States, announced that they would in fact follow U.S. rules in all locations where they operated. Castle & Cooke's announcement was consistent with their previously articulated policy of not using any chemicals that had been banned in the United States, showing that U.S. regulation did affect the company's policy in their international operations. The public relations benefits of this approach were also undeniable. For example, when the *Honolulu Advertiser* reported the story—including both the *Mother Jones* critique and Castle & Cooke's "discontinuance of DBCP . . . in all of its agricultural operations in the world, except in Hawaii pineapple fields"—banana company personnel were relieved that "we came out pretty well."[166]

However, reports of continued Castle & Cooke DBCP use surfaced even as they struggled to paint a picture of themselves as responsible pesticide users. According to a Castle & Cooke vice president, "existing inventory was utilized with final application in early 1980."[167] However, memos from Standard Fruit in Nicaragua cited applications in the final months of 1980, albeit with the notation that "next year we expect to apply Furadan."[168] In 1981, accusations of continued Castle & Cooke use of DBCP came from the authors of the *Mother Jones* article, this time as part of a book with the evocative title *Circle of Poison*. In the book, the authors quoted an anonymous source from Amvac as saying that Castle & Cooke still used DBCP, they just "[got] the stuff through importers down there" rather than buying directly from Amvac.[169] In the face of this and related critiques, Castle & Cooke continued to assert that they would not use any chemicals outlawed in the United States.[170]

Despite Castle & Cooke's public assertions, questions regarding DBCP use continued to surface. In 1984, a manager in Honduras wrote to headquarters in the United States, asking Jack DeMent whether DBCP could be applied to banana crops through an irrigation system.[171] DeMent answered that there was a "current corporate injunction against using the chemical in any form." His letter showed the impact of critics' efforts when he wrote, "I am reasonably sure that any use of it in any manner would appear in some 'Mother Jones' type of publication within 30 days. . . . I do not think we could face the P.R. fallout." However, DeMent continued, "we are testing." In something of a corporate sleight-of-hand, the company disavowed DBCP use on farms, but continued to research it on plots in Honduras, where workers applying the chemical would still face exposure. In 1983, researchers started a project to "compare all of the currently approved nematicides available in liquid formulations [including Fumazone] for injection through the drip irrigation system."[172] DeMent himself traveled to Honduras, likely in late 1984, and was so pleased with the results of the experiments that he "was prepared to ask for a review of corporate policy banning the use of this chemical in any manner on corporate farms."[173] In DeMent's corporate calculus of pesticide use, the "substantially" better production on DBCP plots outweighed earlier concerns regarding the risk of "P.R. fallout,"[174] while worker health did not seem to enter the picture at all.

The key to the future of DBCP on bananas, however, ultimately lay with U.S. regulators. In DeMent's words, drip irrigation with DBCP was "an EPA unofficially blessed application procedure as long as FDA had not cancelled the legal DBCP residue" limit.[175] The FDA, in the background for most of the post-1977 negotiations over DBCP's regulatory fate, had taken no action on banana residues. This meant that tainted fruit could still enter the country repercussion-free.[176] By 1985, however, changes in U.S. regulation would once more impact DBCP use on bananas. This time, it seemed that the EPA's previous decision to allow DBCP use in pineapple cultivation was flawed—new analysis had shown that minuscule amounts of DBCP in Hawaiian wells—from 50 to 500 parts per trillion—were likely to cause a lifetime cancer risk as high as 9 chances in 100,000.[177] The agency found this risk to be unacceptable, and in January proposed that the remaining pineapple use be canceled. While Hawaiian pineapple growers on Maui would be allowed "careful" use of DBCP until the end of 1986, after that all use would cease.[178] DeMent, anticipating that the residue tolerance for DBCP would also be revoked, called the cancellation the "death rattle" for the use of DBCP on bananas.[179] By January of 1986, the residue limit had indeed been revoked.

Tightening U.S. restrictions precipitated a worldwide decline in the use of DBCP. Costa Rica officially banned the chemical in 1988, and other countries followed suit during the 1990s.[180] By 1986 DBCP had been included on the World Health Organization's "Obsolete Pesticide" list.[181] The complicated, contradictory, and halting U.S. regulation of DBCP had had varying but profound effects on the production and use of the chemical inside as well as outside the nation. As the effects of this final U.S. regulation rippled outward, the DBCP era came to an end.

Although use of the chemical had dwindled, its effects would continue to be felt over the following decades, as tens of thousands of people in Central America linked their health problems—from sterility to cancer and beyond—to exposure to DBCP. By the 1980s, their stories had begun to reach the press in Central America, the United States, and worldwide. Known as *los afectados* (the affected) or *los damnificados* (the victims), the workers damaged by DBCP would begin to join together to fight for both recognition and compensation. For many of them, the fight would take place at the very source of the chemical itself: the United States.

FOUR

An Inconvenient Forum?

IN MAY 1983, SAÚL MUÑOZ Sibaja and 57 other DBCP-affected Costa Rican men and their women partners or wives filed a suit against Dow and Shell in Florida state court, claiming that DBCP exposure had made the men sterile.[1] They were the first of what would eventually be thousands of plaintiffs from Central America to bring DBCP cases in the United States in an attempt to hold U.S.-based transnational corporations accountable for harms done abroad. The transnational litigation seemed to hold the promise of justice for workers whose health had been affected by DBCP, symbolically reversing the flow of DBCP from the United States into Central America as the plaintiffs sought to hold chemical and fruit corporations accountable on their home turf. However, that symmetry seemed increasingly illusory as defendant corporations argued that the cases brought by DBCP-affected workers from outside the United States simply did not belong in U.S. courts and instead should be tried in the courts of the plaintiffs' home nations. From 1983 through the mid-1990s, these cases constituted a complicated field of struggle and negotiation involving workers, transnational corporations, plaintiff and defense lawyers, corporate personnel, and U.S. judges.

First and foremost the struggle centered on questions of location and relocation. In an effort to dismiss the cases altogether, the fruit and chemical companies turned to a once-seldom-used legal doctrine called *forum non conveniens* (FNC). Taking its name from the Latin for "inconvenient forum," the doctrine gave U.S. judges the power to dismiss a case that fell within their jurisdiction on the basis that the case was more "conveniently" tried somewhere else. From the *Sibaja* case onward, debates over FNC became the central issue of the DBCP litigation. As defendants turned again and again to FNC, judges' responses—the orders that dismissed the cases as well as those that kept them in the United States—articulated competing notions of the place of U.S. courtrooms in a globalizing world.

While DBCP plaintiffs and their lawyers jointly struggled against defendants as to the location of their cases, the relationship between the workers and their attorneys was itself transnational, linking personal injury lawyers in the United States with workers (or former workers) and attorneys in Central America. This linkage held the potential of leveraging the skills of U.S. lawyers in tort litigation, and their access to U.S. courts, to let banana workers confront on equal footing the powerful corporations that had shaped their exposure to DBCP. The transnational relationship between workers and their attorneys was characterized by acts of representation and translation. Representing Central American clients meant that the lawyers would be responsible for bringing their stories to the U.S. legal system and fighting for their interests there. While some of the plaintiffs themselves would travel to the United States to provide deposition testimony, for the most part the plaintiffs' transnational movement was conceptual—their demands were brought and forwarded by the attorneys who functioned in their name. That representation included making choices about where to file suit, how to best move lawsuits forward, and when to bargain with defendants. Representation was accompanied by translation. In order to bring the cases, the U.S. and Central American attorneys translated the workers' experience, not only from Spanish to English, but from locally articulated understandings into the language of U.S. tort law. Although the strategy was meant to bridge the power differential between banana workers and transnational corporations, the partnership between lawyers and their clients did not eliminate problems of inequality. The legal strategy itself meant operating within the standard goals and assumptions of the U.S. legal system, where lawyers and not workers held more knowledge and power. Although documentary evidence on the relationship between lawyers and their DBCP clients is limited, as the years passed and the litigation dragged on, it was punctuated with indications of problems in the processes of representation and translation, especially as some cases reached important turning points or conclusions.[2]

From 1983, when the Sibaja case was filed, until the middle of the next decade, the issues of location, translation, and representation featured prominently in the DBCP litigation. Location loomed largest, as the issue of FNC dominated the cases, generating debate on the legitimacy of foreigners' claims in U.S. courts, the rights of corporations, and the role of the U.S.

court system. At the same time, occasional disputes between DBCP-affected workers and their attorneys pointed to the dilemmas inherent in transnational representation and translation. By the mid-1990s, despite some accomplishments, DBCP-affected workers seemed to be losing their struggles for accountability. The sum of responses to DBCP cases had largely affirmed the corporations' perspective, limiting the rights of Central American plaintiffs to bring suit in the United States.

THE QUESTION OF FORUM AND *SIBAJA V. DOW*

For Saúl Muñoz Sibaja and other plaintiffs, bringing a sterility case in the United States represented an alternative to the meager workers' compensation paid by the Costa Rican National Insurance Institute (*Instituto Nacional de Seguros,* or INS), a state agency that was the monopoly insurer in Costa Rica.[3] The INS department of *riesgos de trabajo* (labor risks) determined the amount of compensation by assessing "the percent of limitation in the function of the affected organ," awarding injured workers compensation corresponding to a percentage of their annual income.[4] Although the Labor Code had not assigned a value for sterility, testicular atrophy was compensated as if it were the loss of both testicles, and assigned proportional compensation of 40–100%, depending on the age of the worker at the time of the injury.[5] Men seeking compensation had to complete a process that involved a year of sperm testing and a testicular biopsy; if they did not recover, they received compensation ranging from a few hundred to a few thousand dollars.[6] Although no one knew exactly how much a lawsuit in the United States would bring, U.S. workers injured by the chemical had received six- and seven-figure awards at trial in 1983.[7] For the group that would become the *Sibaja v. Dow* plaintiffs, then, pursuing litigation in the United States offered a chance at a dramatically higher level of recovery compared to remaining within the Costa Rican workers' compensation framework.

For plaintiff lawyers, the case offered a chance to participate in an emergent wave of transnational litigation in the United States. There are at least three explanations for how the *Sibaja* group and plaintiff lawyers first came together to bring DBCP cases in the United States. One lawyer involved in the litigation reported that Costa Rican physicians put their patients in touch with U.S. plaintiff lawyers, while another account held that U.S. lawyers "offered their collaboration" after reading about the history of DBCP in

Circle of Poison.[8] A third report noted that banana worker organizations contacted Costa Rican lawyers for help, who in turn reached out to their North American colleagues.[9] Whatever the specific circumstances, as attorneys in the highly competitive, male-dominated and, generally politically progressive world of tort law, the lawyers who initiated these cases were likely driven by varying combinations of professional ambition, profit seeking, and political commitment.[10]

The new transnational litigation was part of an intensified phase of globalization, which many scholars date to the 1970s, characterized by the adoption of neoliberal economic policies, the ascendancy of transnational corporations, and increased movement of goods, people and ideas across national boundaries. As the deregulatory policies of the Reagan administration narrowed the scope of action for public interest law in the United States and attention to global problems grew, attorneys interested in effecting social change increasingly engaged in a transnational sphere.[11] The expansion in international claims was given impetus in 1980, when a U.S. judicial decision showed that the Alien Tort Statute (ATS) or Alien Tort Claims Act (ATCA), a law passed in 1789 that granted federal court jurisdiction for cases brought "by an alien for a tort only, committed in violation of the law of nations or a treaty of the United States," had new relevance. The law allowed non-U.S. residents—"aliens"—to bring tort cases, which are civil (as opposed to criminal) suits seeking recovery for an alleged injury or wrongdoing. The 1980 decision found that a "violation of the law of nations" included not only the traditional realm of maritime and territorial law, but also international human rights, meaning ATS could be used to litigate human rights cases in the United States even when the violations were said to have taken place elsewhere.[12] The claims made in the *Sibaja* lawsuit did not fall within the definition of the "law of nations": *Sibaja* was a product liability case, alleging that harm had been caused by a defective product—DBCP. Nevertheless, the attention to ATS and pursuing legal justice for non-U.S. citizens likely helped pique lawyers' interest in the case.

For their part, Costa Ricans were facing an economic crisis. In 1981 Costa Rica became the first Latin American nation to default on its foreign debt, an early event in what would become known as the "lost decade" for Latin America. Costa Rica, like other nations, would respond with the belt-tightening economic and social reforms prescribed by the International Monetary Fund (IMF).[13] The promise of a monetary award undoubtedly carried extra weight for affected workers in such dire circumstances. In the United States,

Central America featured prominently in foreign policy with U.S. backing for the violent "anti-communist" governments that waged bloody counterinsurgency campaigns in El Salvador and Guatemala. And even as the Reagan administration funded the "Contras" in an effort to destabilize the Sandinista government in Nicaragua, the United States also poured money into Costa Rica to help stabilize what it saw as a stable, democratic alternative to the Nicaraguan popular regime.[14] At the same time, progressives in the United States joined movements in solidarity with leftist and people's movements in Central America. In this context, it is not surprising that the attention of U.S. lawyers also turned to the isthmus.

Attorneys also undoubtedly shared with their clients a desire for economic gain. If a case brought by Central American workers was anything like the California cases, the potentially hundreds or thousands of Central American plaintiffs could mean a huge payout for attorneys who worked on contingency and took a portion of any award. Victory—"recovery" in legal parlance—would not only would mean money for banana workers, but also fees and expenses for the attorneys themselves.[15]

The lawyers working with the first Costa Rican DBCP plaintiffs were attorneys from the Dallas-based plaintiff law firm Baron & Budd and their Costa Rican associates. The partnership between Costa Rican plaintiffs and their U.S. and Costa Rican lawyers was a relationship of representation. Attorneys in all cases stand in for their clients, seeking evidence, making arguments, and performing the myriad other actions needed to move a case forward in their names. This representation involves more than simply carrying out the clients' wishes, however, as the attorney's role extends to making decisions on the client's behalf. As one legal scholar described it around the time *Sibaja* was filed,

> the traditional allocation of decision making authority is one in which the client decides the "ends" of the lawsuit while the attorney controls the "means." Thus, the client determines such "ends" as whether to settle a civil case ... and the attorney decides, even contrary to the client's express wishes, what legal and constitutional arguments or defenses to raise.[16]

Ethical guidelines elaborated by the American Bar Association required that clients decide the goals of their work with lawyers and that attorneys "consult" with clients on "the means by which [a lawsuit is] pursued."[17] In practice, these guidelines left attorneys with broad authority to speak and make decisions on behalf of their Central American clients.[18] Attorney consultation was complicated by a number of factors, including the size of the plaintiff group (smaller than some other tort lawsuits, but large enough to complicate any client-centered decision-making process); language differences between the clients, some of their lawyers and the legal argot of the U.S. justice system; and the wide gap between banana workers' and attorneys' life experiences, literacy levels, and education. While banana workers entrusted their attorneys to stand in for them in an effort to obtain some compensation in U.S. courts, these factors raised the potential for a very attenuated sort of representation, with clients having little understanding or real input into the "means" their attorneys might use to achieve the shared "ends" of the case.

Legal representation also involved acts of translation, as attorneys transformed the language of the workers' complaints, not only from Spanish to English, but from the everyday private or locally articulated understanding of harms into the technical language of U.S. tort law. Popular understanding of DBCP disease necessarily differed from the legal claims seeking compensation on the grounds of strict liability in tort, negligence and implied warranty. Furthermore, that legal language would require translation back to the workers to make it understandable to them, or indeed, to almost anyone not trained as lawyer. In simplified terms, the *Sibaja* complaint alleged that the chemical companies were liable for sterility injuries because the companies did not exercise the reasonable care that could have prevented those injuries (negligence), and further, that in selling DBCP, Dow and Shell had implied that it was safe to use (breach of warranty).[19] The complaint also held that, even if the companies had acted without fault—that is, even if they did not know about DBCP toxicity or had done everything in their power to prevent exposures—they were still responsible for the injuries and should compensate workers (strict liability).[20]

The issue of location—*where* the case should be heard—quickly became the critical struggle in *Sibaja*. In the United States, plaintiffs have the choice of filing in any court that has jurisdiction over the case, with jurisdictional rules hinging on a number of factors, including the type of case brought and the place of residence or business activity of one or both of the parties. Plaintiff attorneys chose to bring their case in Florida state court in Dade County, which had jurisdiction over any company qualified to transact business in Florida, including Dow and Shell.[21] The defendants, however, asked for and received removal of that case to federal court. The state and federal court systems in the United States have different but overlapping purposes.[22] While most civil and criminal cases are brought in state courts that enforce both state and federal law, and operate according to state-specific laws, when the parties are from different states or nations (called "diversity" cases), defendants can "remove" the case from state court and have it heard in federal court instead.[23] The difference between the state and federal systems was not trivial as far as the outcome of *Sibaja* was concerned. In federal court, but not in Florida,[24] companies could invoke the dusty legal doctrine of *forum non conveniens* in an attempt to have a case against them dismissed.

A doctrine with roots in Scottish common law, FNC allows dismissal of a case even when a court has jurisdiction.[25] An FNC dismissal is granted not according to the statutory guidelines that determine jurisdiction, but rather at the discretion of a judge after consideration of a three-pronged test that has developed in common law over time. That test was based largely on two important cases: *Gulf Oil v. Gilbert* and *Piper v. Reyno*. The *Gulf Oil* case, decided in 1947, derived from a case brought by a Virginian (Gilbert) against Gulf Oil Corporation, which he blamed for causing a fire in a warehouse he owned in Lynchburg.[26] Gilbert attempted to remove that case to federal court because he feared that a Virginia jury would balk at the amount of damages sought. The Supreme Court ultimately upheld an FNC dismissal in the Gilbert case, finding that, while a plaintiff's choice of forum deserved deference, it should not be used to "vex," "harass," or "oppress" a defendant.[27] While the court "declined to 'catalogue the circumstances'" in which FNC could be properly used,[28] the decision did spell out a three-pronged framework for the analysis of FNC motions. First, to invoke FNC, a judge had to determine if there was another "adequate" forum in which the case could be tried. Second, the judge also had to consider the interests of the parties to the suit. These "private interests" included factors such as the cost and ease of travel, discovery, and access to witnesses. Finally, the judge had to make a determination of "public interests," or considerations relating to the general public, such as the burden of jury duty and the general sense that local issues should be resolved "at home," as well as a weighing of the factors that affect the court *per se,* including issues ranging from potential administrative difficulties to concerns about choice of law.

The other case that determined how FNC could be used in U.S. courts, *Piper v. Reyno,* had been decided by the Supreme Court less than two years before *Sibaja* was filed. The lawsuit had been brought in connection with an airplane crash in Scotland that had killed five people. The crash victims' estate administrator brought the case in the United States. Faced with FNC dismissal, the plaintiffs argued that their case should not be dismissed, because they would likely receive less compensation if the case proceeded in Scotland rather than the United States. The court dismissed their case, but when plaintiffs appealed, the circuit court reversed the initial decision, finding that FNC could not be used in cases where it would result in a move to a forum where the law was less favorable to the plaintiff. However, the case advanced to the Supreme Court, where justices reversed once more, finding that the existence of more-favorable law in the forum chosen by the plaintiff was not adequate grounds for defeating an FNC motion. In fact, the Supreme Court's *Piper* decision further limited the rights of foreign plaintiffs, holding that their choice of forum did not deserve the same level of deference as that of U.S.-based plaintiffs. The justices' decision had significant impact on the future of FNC. As they acknowledged, the opposite decision would have rendered FNC "virtually useless," because when plaintiffs could easily satisfy multiple jurisdiction and venue rules, "plaintiffs will select that forum whose choice-of-law rules are most advantageous. Thus, if the possibility of an unfavorable change in substantive law is given substantial weight in the forum non conveniens inquiry, dismissal would rarely be proper." Instead of weakening FNC, the Supreme Court's decision strengthened the position of defendants who wished to rid themselves of potentially costly international cases.

The precedents did not bode well for *Sibaja*. The *Piper* decision meant that *Sibaja* plaintiffs could not argue that their case should stay in the United States simply because that was the place where they stood the best chances of winning, or of winning a hefty verdict. In the eyes of the federal judge, the three-pronged *Gilbert* test also seemed to work against the Central American plaintiffs. He found that private interests were in favor of dismissal: people, documents, and evidence would have to be brought from Costa Rica to the United States in order to try the case, representing a burden for the parties. In considering public interests, the judge found that the logistical complexity of the case would constitute a burden on the court: 58 foreign plaintiffs meant consideration of various legal questions for each, and the court would likely have to consider Costa Rican law in making a decision on the case. The judge added that Florida citizens should not be required to spend jury duty time on a case that, in his eyes, had "no connection with this community."[29] In addition, he found no reason to find that Costa Rica was not an adequate alternative forum. In sum, the judge decided that, in the case of *Sibaja*, "the convenience of the parties, the witnesses and the court, and the interests of justice, dictated that the case be dismissed."[30] The plaintiffs' appeal was unsuccessful.

Underlying the judge's decisions and the very logic of the FNC doctrine were assumptions about the meanings of "convenience" and "justice." As expressed in *Sibaja,* both convenience and justice were defined in terms of the boundaries of the nation—Costa Rican cases presented a burden and a threat to justice because of the imposition of faraway evidence and foreign concerns. The decision reinforced the boundaries of the nation against the entry of Costa Rican claims, insisting that the problems of Costa Ricans—even those injured by the actions of a corporation now based in Florida—really had nothing to do with the citizens of that state or of anywhere in the United States. Although Dow and Shell both did business in Florida, the Court deemed their international activities to be outside the scope of the court's power. Bananas and chemicals could move between Florida and Costa Rica, but according to the federal judges, lawsuits could not do the same.

Despite the U.S. judges' finding that Costa Rica was an "adequate alternative forum," no *Sibaja* plaintiff or lawyer went on to pursue the case there. Legal and practical considerations made it difficult to bring a case in Costa Rica. Evidentiary laws were far more restrictive than those in the United States, and there was no well-established tradition of tort cases with a large number of plaintiffs.[31] In addition, laws and customs made it harder to finance a case. Plaintiffs could be required to post a bond at the beginning of litigation in order to guarantee payment, and lawyers could work on a contingency basis only in limited circumstances, meaning that some of their clients might have had to pay for legal services as a case advanced—an impossibility for most banana workers. While U.S. lawyers might front all expenses for a case in the hopes of winning a large verdict, such a tradition did not exist in Costa Rica.[32] In fact, if a worker did manage to file and win a case, the verdict would likely be low—in the $1,500 range.[33] So while the U.S. court may have found the Central American nation to be a more convenient forum, litigation there would not be pursued. Instead, workers who had been plaintiffs in *Sibaja* joined other DBCP suits in the United States,[34] and Shell and Dow could congratulate themselves on a *de facto* victory in the first transnational DBCP case. Over the next decades, however, as word of DBCP's effects spread in Central America, more sterilized workers would seek compensation in U.S. courts. The doctrine of FNC and the exclusive vision of the U.S. nation it expressed would remain the central battlefield in DBCP litigation.

In September 1986, one of Costa Rica's leading newspapers, *La Nación,* broke a sensational story. "Victims of a Toxic Substance," read the front-page head-line, "Five Hundred Sterile Workers in the Atlantic Zone."[35] By the time *La Nación* brought DBCP to Costa Ricans' attention, U.S. residue tolerances had been revoked and the chemical had been added to the World Health Organization's list of obsolete chemicals. The fact that the issue of DBCP sterility was indeed news to most Costa Ricans at this late date reveals the power and persistence of the deal struck between Standard Fruit and the Costa Rican government in 1977, the relatively small impact of the U.S. legal process in Costa Rica, and—despite a globalization touted as increasing communication across borders—little information sharing between the two countries.

Over three days of coverage, the paper told a story of how DBCP was used on banana plantations, how Costa Rican doctors and workers had discovered the cause of sterility in the late 1970s, and what effect the sterility had had on affected workers' lives. As this first journalistic coverage explored the meaning and import of the banana workers' sterility, it articulated DBCP damage as threatening to Costa Rican masculinity and family structure. This very personal and localized damage was, however, understood in the context of transnational processes. The paper questioned the role of U.S.-based transnationals in causing the newly uncovered sterility epidemic, but also looked to the United States as the best hope for justice for banana worker victims of DBCP.

At the heart of the stories was a dramatic picture of extensive damage to Costa Rican men. Although the paper provided information on the physical aspects of the disease, including a diagram of the male reproductive system and definitions of the various levels of sperm damage, the impact on men's health was mostly described in social and psychological terms, portraying sterility as a threat to a banana worker masculinity that underpinned the normative family structure. For Costa Rican banana workers, having children "in a common-law marriage, in marriage, or by accident is an unmistakable sign of masculinity," explained *La Nación.*[36] If a man cannot father children, he "is not a man." DBCP's threat to masculinity simultaneously threatened the very structure of the family. Women, unable to bear children with their partners, would leave, rendering the family unstable. "As a result of their infertility," reported the paper, "the workers have suffered abandonment by their wives, divorce and separations, sexual impotence and depression which they try to drown in liquor." Workers and their partners emphasized the harm that DBCP had done to their community, and reported social isolation, suicidal ideation, and increased alcohol use among exposed workers.[37] In a culture where masculinity was highly valued and children were at the center of family life, DBCP-induced sterility was portrayed as nothing less than a "familial holocaust."

The high stakes of DBCP sterility were illustrated in the story of one man, Simón Vallejo. Vallejo had been valued as a good worker by his *compañeros*, but his status changed dramatically when his community discovered that he could not father children.[38] In the newspaper's retelling, Vallejo's life took a downward spiral: nagged by his wife, identified only as "Elisa," about their childlessness, he saw a doctor who confirmed his sterility. He subsequently became an object of ridicule among his coworkers, and his home was nearly "unmade" as he gave Elisa "absolute liberty to do whatever she wished, as she gave him his to drown his sorrows" in alcohol. Depressed and suicidal (he had spent several days "caressing the barrel of a borrowed revolver"), Vallejo suggested to his wife that she conceive with another man and bear a child, whom he could then adopt and to whom he could act as a father. Elisa agreed, but after the baby was born, she married the other man, "bewitched by his virility." In this story, Vallejo's infertility not only robs him of the recognition of the masculinity he earned through hard work, but undermined his masculine authority within his own family unit, notably by limiting his control over his wife. The result is depression, near suicide, and eventually, the loss of his family to a man with greater virility. Ending with Elisa's new husband also being diagnosed with sterility, the story suggests that "mass sterility" caused by DBCP could erode entire communities and a valued way of life.

La Nación situated this familial and community tragedy in a transnational context and questioned the role of transnational corporations, especially Dow and Standard Fruit: "Why, if in August 1977 [people] in the United States were warned of the danger and even suspended DBCP manufacture, was Standard still importing and using it in its banana farms? Did Dow Chemical also have something to do with these importations?"[39] The paper looked beyond national borders for a remedy for suffering workers, specifically invoking the relocation of local complaints to "the North American courts of justice."[40] The paper reported on national workers' compensation, but found that alternative inadequate. The U.S. legal process, on the other hand—despite the defeat in *Sibaja* and a host of unanswered questions—seemed to hold a promise of justice. According to *La Nación,* this U.S. litigation was the "strongest and most optimistic option" for workers seeking justice.[41]

CASTRO ALFARO V. DOW

One of the cases the newspaper reported on was *Domingo Castro Alfaro v. Dow.* The case had been filed against Dow and Shell in 1984 in a Texas state court in Harris County by 81 Costa Rican plaintiffs represented by Baron & Budd.[42] The plaintiffs sought to hold the chemical companies liable for sterility in men, their partners' loss of consortium, and psychological harm caused by DBCP-induced sterility.[43] As in *Sibaja,* legal debate in *Castro Alfaro* centered around FNC and the ability of Central American plaintiffs to gain a hearing in the United States. Arguments, counterarguments, decisions, and appeals on the appropriate forum for *Castro Alfaro* would drag on for almost a decade. The case became a site for the airing of anxieties over the permeability of U.S. legal borders, centering on questions of the mobility of both worker-plaintiffs and company-defendants, while at the same time revealing fault lines in the plaintiff lawyers' representation of their clients and the translation of banana worker demands into the U.S. courts.

The contest over the location of *Castro Alfaro* showed a complex interplay between the scales of the local (in this case, the state of Texas), the national, and the transnational, as the parties fought over the place of foreign claims in Texas courts. Baron & Budd lawyers, working with Costa Rican attorneys including Marlene Chávez and colleagues at the firm of Mena, Chávez y Asociados, filed *Castro Alfaro* in Texas state court. Texas, like Florida, had a law that seemed favorable to avoiding an FNC dismissal. A special provision in Texas law afforded state courts jurisdiction over lawsuits based on events that took place outside of Texas, including beyond U.S. borders. Section 71.031 of the Texas Civil Practice and Remedies Code, titled "Act or Omission Out of State," held that: "An action for damages for the death or personal injury of a citizen of this state, of the United States, or of a foreign country may be enforced in the courts of this state, although the wrongful act, neglect, or default causing the death or injury takes place in a foreign state or country."[44] This short rule seemed to promise that Central American plaintiffs could bring their cases in Texas without threat of being dismissed on FNC grounds.

That promise, however, was soon called into question. Defendants, working with Houston-based law firm Baker & Botts,[45] responded to the litigation with an approach that focused on achieving FNC dismissal. The first salvo came soon after the case was filed: Dow and Shell argued that the Texas court did not have jurisdiction, or, if it did, the case should nonetheless be dismissed under FNC.[46] In an unpublished decision, Harris County District Court Judge Shearn Smith found that the Texas state court system did have jurisdiction under section 71.031.[47] On the heels of his jurisdictional finding, however, Judge Smith dismissed the case on grounds of FNC.

The workers' lawyers appealed the decision of the Harris County Court.[48] The continued debates focused largely on the technicalities of Texas's Section 71.031 law. In their appeal, the plaintiffs' lawyers argued that Section 71.031 not only granted jurisdiction, but guaranteed it. In other words, they claimed that the statute bore a jurisdictional mandate. By contrast, the defendants maintained that the statute was merely permissive: even if the case did come under the jurisdiction of the Texas court, nothing in Section 71.031 prevented it from being dismissed.[49] In a 3–1 decision that focused with great precision on the language of the statute, the court agreed with the plaintiffs, ruling that their case could not be dismissed. With this decision, the court affirmed the right of international plaintiffs to bring a case in a Texas court and secure the commitment of the court to see the case through to its end. By insisting on the foreign plaintiff's ability to bring a suit in Texas, the appeals panel firmly located Texan courts and Texans themselves within a transnational sphere. Unlike the federal court's ruling on *Sibaja,* the Texas decision did not define justice according to national boundaries, instead finding that although a dispute may have its roots in a faraway territory or foreign state, the courts of Texas still had an interest in that dispute's settlement.

The differences between Texas and Florida, and between state and federal courts, demonstrate how complicated the question of gaining—or preventing—access to U.S. courts could be. When non-U.S. plaintiffs were able to gain a court hearing in a particular state, that state retained its character as a unit of the larger nation, while simultaneously standing in for the United States as a whole. In *Castro Alfaro,* for example, some aspects of the conflict concerned the local effects of Texas court policy, or Texas's role vis-à-vis the other 49 states of the United States, and drew on ideas of what it meant to be Texan within the context of the United States. At the same time, the discussion of access to Texas courts could at times be properly understood as a discussion of access to U.S. courts more generally. For example, when the appeals court decision addressed Texas law and the capacity of Texas courts, the root of the legal dispute was the fact that the "foreign" plaintiff came not from another state within the United States, but from another nation entirely. Defining Texas against the *international* foreigner made the deliberation in DBCP litigation inherently a question of defending *national* boundaries.

As DBCP cases filed in California and Florida failed, it became clear that Texas granted unusual access to international plaintiffs. *Aquilar v. Dow* had been dismissed by California state courts in 1986.[50] *Cabalceta v. Standard,* like *Sibaja* before it, was filed in Dade County, Florida, removed to federal court, and then dismissed in 1987.[51] While the dismissed cases were not pursued, some of the plaintiffs would join the *Castro Alfaro* group in order to keep their claims alive.[52] As the corporations' FNC strategy proved effective elsewhere, the stakes were raised in Texas.

Cultural Translation in Castro Alfaro

While the Texas court's decision was a victory for plaintiffs, banana workers and their lawyers were experiencing their own challenges born of the transnational nature of the case. Here, I draw heavily on the work of Emily Yozell, an attorney originally from the United States who relocated to Costa Rica and worked with DBCP-affected workers in a number of cases. Yozell provides thoughtful insight into the challenges of this litigation in a chapter written for a collection of essays on "human rights, labor rights, and international trade." Yozell's account constitutes a set of "practical lessons"—as well as a cogent argument—for a culturally sensitive and self-aware U.S. attorney approach to such litigation.[53] Her chapter on the case, published after its conclusion, points to the places where translation and representation can break down when lawyers from the United States, lawyers from Latin America, and Latin American plaintiffs work together. Yozell relates how language and cultural difference between two cultures led to problems and confusion:

> A lawyer familiar with both US and Costa Rican law was hired to prepare plaintiffs for deposition and trial. As part of her preparation, she reviewed each plaintiff's answers to the interrogatories. In the course of the interview,

she realized that they were fraught with factual errors. Dates were inaccurate or imprecise. One plaintiff would give different answers to questions about marital status at different times. This lawyer realized that in the culture of these workers, it was important to give the answer they believed the interviewer wanted to hear, and impolite to say, "I don't know," and that without a knowledge of local history, critical events could not be dated.[54]

Here, cultural differences in norms of interaction and the perception of time result in a breakdown in communication—a failure of translation—that fails in giving full voice to the workers' experience, leaving lawyers without information they can use or discrediting the witness as less than reliable. According to Yozell, problems with communication extended to the relationship between U.S. and Costa Rican lawyers—when some plaintiffs' cases were dismissed because forms were not signed in time, Costa Rican lawyers were "incredulous" because they were not familiar with or prepared for such a consequence.

The failures of translation Yozell describes are not symmetrical. The language, rules, and cultural expectations governing the litigation were those of the United States, reinforcing the power of U.S.-based attorneys and leaving Costa Rican plaintiffs in unfamiliar or even frightening territory. For example, Yozell recounts Domingo Castro Alfaro's first trip to the United States to testify. Along with a coworker, he travelled from Río Frío to San José to Houston, a trip that involved not only his first airplane ride, but also his first elevator ride:

> These two men had hardly even been in any city, let alone the English-speaking metropolis of Houston. Lodged on the 35th floor of a highrise hotel, neither had ever seen an elevator. Their region had recently survived a major earthquake.... Neither plaintiff had even seen a building with 35 floors, but both knew it was a very dangerous place to be during an earthquake. Intimidated by the electric and electronic gadgets provided for their "comfort," the two could not sleep alone in their separate rooms; instead they huddled together in one for reassurance.[55]

While Yozell is careful not to point fingers, the incidents she reports suggest that attorneys at times failed in their responsibilities to both translate and represent, not only leaving their clients and co-counsel confused or frightened, but presenting the possibility of introducing "procedural and factual errors that [could have proven] fatal to the lawsuit." Indeed, Yozell insists that to avoid situations such as this, U.S. attorneys engaged in transnational litigation should educate themselves about the history, culture, and legal system of their plaintiffs' nation; include a "cultural mediator or translator" on the legal team; and above all adopt an attitude of humility in the face of cultures and traditions unfamiliar to them.

A Victory That Complicates

As problems of representation and translation arose between lawyers and their clients, defendants continued to fight for dismissal of *Castro Alfaro*. By 1990, they had succeeded in advancing their appeal to the Texas Supreme Court. As those on both sides of the controversy knew, a Texas Supreme Court rejection of FNC would clear the way for perhaps thousands more DBCP (and other) claims to be brought in the state, potentially representing hundreds of millions of dollars in compensation to affected banana workers.[56] Despite controversy in the business world—according to the *Los Angeles Times*, "about 40 large corporations" filed *amicus curiae,* or "friend of the court" briefs in favor of FNC dismissal[57]—and violently divided opinions on the Texas Supreme Court bench, the Court allowed the banana workers' cases to stay in Texas. In an opinion that focused on the history of the FNC doctrine and the intent of Section 71.031 drafters, the state Supreme Court went beyond previous decisions that merely upheld the right of the workers in this case to maintain their case in Texas. Indeed, the majority of the state Supreme Court went so far as to assert that the State of Texas had "statutorily abolished the doctrine of *forum non conveniens.*"[58] It was a watershed decision that not only preserved the Central American plaintiffs' access to the courts, but also presented an opportunity for others outside the United States to seek justice against U.S. corporations in Texas courts.

However, the decision was far from a unanimous, ringing endorsement of foreign plaintiffs' court access. The case generated a remarkable series of seven separate opinions: Justice Cread Ray's majority opinion, plus two additional consenting and four dissenting. The opinions, some mildly worded, others downright combative, represented a wide range of thought on the utility and legality of *forum non conveniens.* At the center of the controversy emerged the questions of court access and the transnational mobility of both worker-litigants and the corporations they sought to hold accountable.

Dissenting opinions echoed Appeals Court Judge Duggan's concern that opening Texas courts to non-U.S. litigants would expose the justice system to exploitative claims from foreigners. For example, Justice Hecht wrote: "This advantage to suing in American courts has not escaped international notice. England's Lord Denning, for example, has observed, 'As a moth is drawn to the light, so is a litigant drawn to the United States. If he can only get his case into their courts, he stands to win a fortune.'"[59] In this metaphor, the material allure of the U.S. nation and the litigants' drive to obtain its riches are naturalized by comparison to the biological (if fatal) attraction of insect to flame. Naturalizing the flow of litigation into the United States allows Hecht to avoid discussion of the roots of the litigation in the actions of the chemical company defendants. Justice Hecht's invocation of an image of profit-hungry litigants mirrored contemporary discourse about immigrants, who according to anti-immigration rhetoric, drained U.S. resources by taking advantage of welfare programs, U.S. jobs, and other economic benefits. Such discourse refigures immigration as an external threat to the nation only by disembodying it from the nation's own participation in the transnational processes that help produce the movement of people across borders, such as trade regimes, currency policy, conditional lending, and so on.[60] Like discourses of immigration that deny how a nation may help create conditions for immigration, Hecht's moth metaphor actively ignored the broader historical context of DBCP exportation from the United States and a fruit production system that benefits U.S. companies and consumers. The desire of the foreign plaintiffs for U.S. courts is figured as natural, not because the U.S. had any role in causing their harms, but rather because the U.S. is a beacon for foreigners generally. Freed of context, the flow of DBCP litigants can be conceptualized as a natural, if unfortunate, phenomenon that flows from the nations' superiority, not from its complicity in producing dangerous conditions.

A strongly worded dissent from Justice Cook characterized DBCP litigants not (as Justice Hecht would have them) as natural creatures passively drawn to the wealth promised by the United States, but rather as aggressive fortune hunters. Beginning with the fact that at least some members of the *Castro Alfaro* plaintiff group had also been part of *Sibaja* and other cases, Cook maintained that their search for an accepting forum in the United States constituted an abuse of the U.S.—and Texas—court system. "Like turn-of-the-century wildcatters," he wrote, "the plaintiffs in this case searched all across the nation for a place to make their claims. Through three courts they moved, filing their lawsuits on one coast and then on the other. By each of those courts the plaintiffs were rejected, and so they continued their search for a more willing forum. Their efforts are finally rewarded. Today they hit pay dirt in Texas."[61]

For Justice Cook, it was first the *mobility* of these unauthorized legal subjects that disturbed. They not only crossed U.S. borders, but were able to move "all around the nation," reaching from coast to coast and throughout the court system. This movement is, in Cook's rhetoric, outside the legitimate framework of the law. In a metaphor sure to resonate in a state known for its oil resources, Justice Cook compared the DBCP workers to "turn-of-the-century wildcatters," the independent prospectors who looked for oil in regions that were not known to be productive. Wildcatters were outside the corporate system that already characterized the oil industry by 1900, making them, if not outside the law, outside the norms for conducting legitimate business. With the wildcatter metaphor, Cook attributed some measure of outlawry to the plaintiffs.[62] At the same time, though, wildcatters were capitalists. By comparing the workers to these entrepreneurs, Cook shifted the association of both money and agency to the plaintiffs. The actual wealth of chemical company defendants is naturalized by the metaphor that refigures them as the land defenseless against its own drilling.

The wildcatter and moth metaphors both expressed a concern that allowing foreign plaintiffs in would have disastrous consequences for domestic courts and legitimate residents. A third dissent also articulated a vision of a beleaguered U.S. justice system by referencing another example of international litigation that had caught the nation's attention. In 1984, a gas leak at a methyl-isocyanate plant in Bhopal, India, operated by a subsidiary of Union Carbide, immediately killed an estimated 3,800 people; many thousands more experienced delayed health problems or premature death.[63] By 1985, over 144 cases had been filed in the United States. Despite the presence of the Indian government as a plaintiff, the consolidated cases had been conditionally dismissed under FNC to be tried in India.[64] Justice Gonsalvez worried that without FNC in its state courts, "Texas will become an irresistible forum for all mass disaster lawsuits.... 'Bhopal'-type litigation, with little or no connection to Texas will add to our already crowded dockets, forcing our residents to wait in the corridors of our courthouses while foreign causes of action are tried."[65] Here, the Justice described international plaintiffs as a Malthusian threat to the fabric of the U.S. legal system, overwhelming the resources of the nation and leaving U.S. residents literally on the outside of the courtroom, waiting in the hall while the cases of foreigners are tried. This scenario projected a fear that the movement of cases from outside the United States across national borders would displace U.S. residents' own access to the fruits of the nation's well-developed justice system. In other words, the movement of cases into the United States is figured as fundamentally threatening to the exercise of justice, defined in explicitly national terms. In order to protect U.S. residents' access to their own courtrooms, FNC was needed to keep international plaintiffs out.

The dissenters who articulated grave concerns about the mobility of plaintiffs were mostly silent on the mobility of the chemical company defendants, perhaps indicating that the companies' transnational nature was so accepted that it was undeserving of remark. In seeking FNC dismissal, the corporations—like the plaintiffs—sought to locate the litigation in the forum that seemed most conducive to their own interests. Justice Hecht was the lone dissenter to note this symmetry. After arguing that "the benefit to the plaintiffs in suing in Texas should be obvious: more money," he added a footnote conceding that "it is equally plain to me that defendants want to be sued in Costa Rica rather than Texas because they expect that their exposure will be less there than here." Hecht, however, aligned rather than opposed corporate interests to larger public interest, noting that it "seems plain to me that the Legislature would want to protect the citizens of this state, its constituents, from greater exposure to liability than they would face in the country in which the alleged wrong was committed."[66] Hecht would, then, create a border that was permeable in one direction—corporate "citizens" could cross political boundaries not only to seek international markets and export goods but also to seek justice in the international forum of their choosing. Plaintiffs from abroad, on the other hand, were defined as interlopers with no legitimate claim to U.S. courts.

If dissenters authorized the mobility of the defendant corporations while affirming U.S. and Texas boundaries to keep international plaintiffs out, a concurring opinion by Justice Doggett articulated a radically different vision of the mobility of the parties and the role of the Texas courts. Organized around the presumption that the defendant corporations should "be held responsible at home for harm caused abroad,"[67] Doggett's opinion made a series of claims about the national or local "home" of transnationals, arguing that both the local (Texan) and national identity of the corporations meant that the Texas courts had an interest in hearing the DBCP cases rather than dismissing them. Doggett de-emphasized the foreignness of the plaintiffs, focusing instead on defining the defendants as *U.S.* companies. In doing so, he insisted that the facts of the case took place not in Costa Rica, but in his own country:

The banana plantation workers allegedly injured by DBCP were employed by an American company on American-owned land and grew Dole bananas for export solely to American tables. The chemical allegedly rendering the workers sterile was researched, formulated, tested, manufactured, labeled and shipped by an American company in the United States to another American company. The decision to manufacture DBCP for distribution and use in the third world was made by these two American companies in their corporate offices in the United States. Yet now Shell and Dow argue that the one part of this equation that should not be American is the legal consequences of their actions.[68]

In this analysis, the DBCP cases do not appear foreign at all, but have everything to do with the United States and the justice system of that nation. In fact, in Doggett's analysis, it is not the plaintiffs who are making an unauthorized border crossing with their claims, but the corporations who are making a run for the border to escape the law.

Doggett also emphasized the material interconnections between nations in the context of globalization: travel and communication circuits easily connected various countries. In addition, he maintained,

> The doctrine of *forum non conveniens* is obsolete in a world in which markets are global and in which ecologists have documented the delicate balance of all life on this planet. The parochial perspective ... ignores the reality that actions of our corporations affecting those abroad will also affect Texans. Although DBCP is banned from use within the United States, it and other similarly banned chemicals have been consumed by Texans eating foods imported from Costa Rica and elsewhere.[69]

Adopting the *Circle of Poison* argument (and, indeed, citing the book), Doggett linked the economic aspects of globalization with changing legal, ethical, and practical imperatives for local people. He argued that Texans had, in two senses of the word, an *interest* in the behavior of U.S.-based corporations abroad. On the one hand, as people with a commitment to universal justice for humanity, Texans were concerned with that behavior. On the other, as consumers embedded in an increasingly interrelated global system, they had a real stake in weighing in on the actions of transnational corporations.

The varying dissenting and concurring opinions regarding the fate of *Castro Alfaro* articulated competing visions of the permeability of U.S. borders, the mobility of corporations and plaintiffs, and the responsibilities of the Texas courts. Despite the vociferous opposition of the dissenters, the majority opinion—that FNC was not applicable in Texas—preserved the *Castro Alfaro* plaintiffs' ability to pursue their cases in the United States and held the chemical companies accountable within the Texas court system.

NEW OPPORTUNITIES AND NEW OPPOSITION

The Texas Supreme Court's ruling provided the opportunity for the first-ever trial on Central American DBCP exposure and provoked optimism for other, similar cases. By the same token, it gave defendants a strong incentive to settle *Castro Alfaro* in order to avoid what might be substantial jury verdicts.[70] A number of new DBCP cases were in the works, and with the number of plaintiffs now reaching into the tens of thousands, the litigation threatened to significantly affect the defendants' balance sheets. The *Castro Alfaro* case itself had expanded, and now involved around 1,000 banana workers as well as Standard Fruit Company, Occidental Petroleum, Castle & Cooke, Standard Steamship Company, and Dole Fresh Fruit.[71] For both defendants and plaintiffs, settling could avoid a lengthy and expensive process with an uncertain outcome. By June 1992, the lawyers had reached a settlement: the fruit and chemical companies paid about $20 million combined to about 1,000 DBCP-affected plaintiffs.[72]

On one hand, the settlement represented a remarkable achievement in that, in the face of FNC challenges, plaintiff lawyers had not let the companies escape unscathed. Given the difficulty of this still-emergent strand of transnational litigation, this was a significant accomplishment. On the other hand, however, the settlement provoked some plaintiffs to anger over the amount and terms of compensation, and stoked suspicion and discontent about the transnational process. In other words, the process revealed a breakdown in the attorneys' representation of clients and translation between two languages and cultures, leaving some banana workers feeling that justice had not been served.

A primary complaint among the plaintiffs was that the settlement did not even come close to reaching the level of compensation that had been awarded DBCP-affected U.S. production workers; one observer called it "paltry."[73] After the attorneys had taken their 40% fee and covered their expenses, the $20 million was divided unevenly between workers, according to factors such as their level of impairment and number of children prior to sterility.[74] For example, a St. Louis newspaper recounted the story of Luis Chávez, a

Honduran plaintiff who was awarded $3,500; after lawyers' fees and costs were deducted, Chávez was left with only $1,000.[75]

Critique of the *Castro Alfaro* settlement process suggested that plaintiffs had little sense of control over the settlement process. The negotiation of a settlement falls within the scope of an attorney's role as a representative of her client; the details of such agreements are hammered out between lawyers from each side, then attorneys obtain their clients' approval of the deal. Complaints in *Castro Alfaro* suggested a break in communication around this process. Some complaints were rooted in questions of literal translation, with some plaintiffs protesting that lawyers did not translate or explain the English-language settlement agreement.[76] But the settlement discontent also reflected a failure to translate in a broader sense, as the stakes, customs, and risks of the U.S. legal process remained opaque to Central American plaintiffs. Indeed, after the Castro settlement, some workers were angered by what they saw as low compensation amounts and high lawyers' percentages, expressing suspicion and discontent about the transnational process. Luis Chávez, for example, doubted that his lawyers truly considered his interests and needs in their actions in the United States. Instead of feeling well represented, Chávez felt mobilized in service of the attorneys' own interest, saying they "ma[d]e good business for themselves, but not for me."[77] In addition to a sense that money was unfairly distributed, Emily Yozell found that her clients "rarely understood the economic compensation as the be-all and end-all of their struggles," although she does not provide greater detail about their conception of justice. Lawyers on the other hand, may have a clear interest in settling, as a settlement ensures they receive fees and expenses, while proceeding to trial is always fraught with uncertainty.

Questions of translation and representation relating to the *Castro Alfaro* settlement were complicated by the fact that at least one defendant, Dole, had started to offer extrajudicial settlements to DBCP-affected workers in Costa Rica, Honduras, and perhaps other locations.[78] Banana workers reported being offered as little as $135 and as much as $4,000 dollars by representatives of the company.[79] Direct compensation by the company would remove attorneys from the worker-corporation negotiation, eliminating both the perils and the promise of representation by attorneys. Direct compensation could also come more quickly and in a less complicated and mediated process, but at the same time it left workers without legal advice and eliminated the hope of a higher level of compensation. While the amounts varied, signing an agreement always meant that a worker promised never to pursue a court case in the United States.[80]

Lawyers for DBCP-affected workers, of course, had an interest in stopping extrajudicial settlements, as they interfered with ongoing litigation and narrowed the pool of potential future clients; justice-minded attorneys would also be worried that workers dealing directly with the company would receive less money than available through litigation and that the corporations would get away cheaply. Plaintiff lawyers eventually succeeded in obtaining court orders prohibiting extrajudicial payments involving current clients.[81] Other direct settlement offers—as well as their critics and supporters—would persist, offering at least a rhetorical alternative to DBCP-affected workers who questioned the motives of attorneys or the benefits of lodging a transnational complaint.

OPEN WINDOW IN TEXAS?

While some Central American plaintiffs came to terms with the settlement of *Castro Alfaro,* fallout from the case would come to constrain their ability to bring new cases in Texas. Defendant corporations and other proponents of FNC would use both legislation and litigation tactics to undermine the plaintiffs' victory at the Texas Supreme Court. While lawyers rushed to file new cases, their window of opportunity was soon to be closed. Defense tactics would soon redefine Texas courts along the lines envisioned by dissenting justices in *Castro Alfaro,* that is, as venues with high—perhaps insurmountable—barriers to the claims of those from outside the nation's borders.

Although the state Supreme Court's 1990 decision had seemed to seal the fate of the FNC doctrine there, one of the justices offered another option. Justice Hightower's opinion had gently prodded for a legislative reinstatement of FNC, reminding Texans that "the legislature has the privilege of changing its mind" and FNC-enabling legislation could be passed if lawmakers saw fit.[82] This approach appealed to business groups reluctant to face additional claims by non-U.S. residents in Texas courts, and in the early years of the 1990s a controversial debate took place in the Texas legislature over whether to reinstate FNC. A coalition of defense attorneys and business groups lobbied for the passage of FNC-enabling legislation, while plaintiff lawyers and consumer and environmental organizations fought against the proposal. A 1991 bill to codify FNC did not pass, but two years later a "Forum Non Conveniens Implementation Act of 1993" had been added to the Texas Civil Practice and Remedies Code.[83] As of September 1993, FNC could be invoked in the courts of Texas.

The new FNC law would not be implemented until August 30, 1993, prompting would-be plaintiffs and their lawyers to file new cases in time to be considered under the old rule.[84] As the deadline approached, lawyers filed thousands of suits on behalf of agricultural workers from Costa Rica, Panama, Guatemala, Honduras, and Nicaragua, as well as the Philippines, Ecuador, Côte d'Ivoire, and elsewhere. Some of the suits were class actions, bringing the total number of potential claimants well into the tens of thousands.[85]

Faced with plaintiffs clamoring to make it through the small window of opportunity that had been opened in Texas, fruit and chemical company defendants—working with heavy-hitting defense firms such as Jones, Day, Reavis & Pogue; and Fulbright & Jaworski—used novel strategies preserve their access to FNC by moving the cases to federal rather than state court. The approach that proved successful was to involve a "foreign government" in the litigation, as the presence of such a party conferred federal jurisdiction and therefore removal from the Texas state courts. Although no plaintiff had sued a government, the defendants themselves brought another party into the suits—Dead Sea Bromine, the Israeli company that had sold DBCP to banana companies. Dead Sea, defendants maintained, not only shared liability in the cases, but, because it was owned in major part by the Israeli government, its presence in the case was equivalent to that of a foreign government, providing grounds for removal to federal court.[86] By 1995, the Dead Sea gambit had successfully placed a number of cases in federal court. Several of the suits originally filed in Texas state courts had been consolidated by then, meaning the court considered them as a group for some purposes. Each of the now-consolidated cases included a different number and mix of plaintiffs, from single-plaintiff *Erazo v. Shell* to the class-action *Delgado v. Shell*.[87] Together, they were referred to as the "*Delgado* cases."

The judge in charge of the consolidated DBCP cases was Simeon Timothy "Sim" Lake III, a Reagan appointee who by 1995 had served eight years as a federal judge.[88] In an earlier decision, Lake had already agreed with defendants that Dead Sea Bromine qualified as a "foreign sovereign," although he maintained that it was not a government defendant per se but rather a company that was owned in the majority by the Israeli government. Now, the judge's task was to determine whether the cases had been properly removed to federal court, a decision that hinged on questions of timing and procedure as well as whether Dead Sea's presence would mean all defendants were granted the treatment usually reserved for foreign governments. If the judge found that cases had been improperly removed, they would return to Texas and could not be dismissed under FNC. Although plaintiffs argued that by mid-1995 Israel's ownership share in Dead Sea had dipped below 50%, the judge was "not persuaded" that this worked in favor of remanding the cases, finding instead that what mattered was the composition of ownership at the time the exposure had taken place. Denying the plaintiffs' motion to dismiss or sever Dead Sea from the broader litigation, the judge effectively extended the right to remain in federal court to Dead Sea's co-defendants as well. (In two exceptions to this decision, the *Erazo* and *Rodríguez* cases were remanded to state court due to complicated issues of timing, stipulations, and settlement agreements.) Overall, Judge Lake's receptivity to defendants' claims allowed them the special privileges granted national governments in order to engineer a move to the forum they preferred.[89]

Having established that the cases belonged in federal court, Judge Lake moved on to a consideration of the defendants' FNC dismissal motion. Following the standard three-pronged test, he would first consider the availability of an adequate alternative forum, then consider the private factors, and finally weigh the public factors. Together, his deliberations made evident one of FNC's central contradictions—it embedded a commitment to a universal right to justice while at the same time it drew limits around that right by reinforcing boundaries around the national legal space.

True to precedent, Lake began with the question of an "adequate alternative forum." The primacy of the adequate alternative forum analysis within the FNC decision-making structure exposed a tension between the national mythology of the United States as thoroughly and uniquely committed to the rule of law on the one hand, and to the policing of national boundaries of law on the other.[90] The potential of an FNC dismissal created a crisis because an exclusion without an alternative is ideologically untenable for Americans committed to (and self-defined by their supposed adherence to) the rule of law.[91] The crisis was resolved by predicating an FNC dismissal, first and foremost, on the existence of an "adequate alternative forum." As long as a plaintiff has somewhere (else) to go, the ideological imperative of the rule of law is met, and the U.S. courts can be insulated from the demands of foreigners.

However, Judge Lake, like other judges, had a wide field of discretion for making determinations on whether the alternative forums were indeed available and adequate.[92] Hoping to convince the judge, each side argued for a judicial interpretation that supported its goals. Plaintiff lawyers and the witnesses they called made arguments and presented evidence that the alternative forums under consideration were not, in fact, available or adequate, due to factors including low recovery amounts, the unavailability of representation on contingency, limits on testimony, and lack of jury trials.[93]

The plaintiffs also provided nation-specific arguments as to the availability and adequacy of non-U.S. forums. They held that the cases could not be tried in Nicaragua because of an ongoing dispute between President Violeta Chamorro and the Nicaraguan National Assembly over the structure and functioning of the Supreme Court. Plaintiffs also argued that in Costa Rica, despite the existence of an "abstract right to sue for damages . . . no injured person . . . has ever filed an action based on a defective product against its manufacturers or distributors."[94] Government representatives from various nations added their voices to those of the plaintiffs, arguing that their own systems either could not handle the cases[95] or did not recognize FNC and so would not accept dismissed cases for hearing in their own nations.[96] By bringing in state authorities from the various plaintiffs' nations, attorneys hoped to convince Judge Lake that the alternative forums for the DBCP cases were neither adequate nor available.

Defendants, for their part, offered evidence that the civil and/or labor codes of the plaintiffs' various nations provided legal means of recovery. Defendants also brought in the testimony of current or former government officials in the hope of making an authoritative claim. Defendants and their witnesses pointed to various civil or labor codes that could provide grounds for compensation for DBCP claims. Some provided legal grounds for filing a lawsuit, while others enabled injured workers to receive state-funded workers' compensation.[97]

Perhaps unsurprisingly given his early decisions sympathetic to defendants' interests, Judge Lake chose to follow a minimal standard for the determination of the adequacy of alternative forums. Citing a precedent case, he maintained that "a foreign forum is adequate when the parties will not be deprived of all remedies or treated unfairly even though they may not enjoy the same benefits as they might receive in an American court."[98] Affirming various defense lawyer arguments, the judge found that his minimal standard for adequacy had been met. While formally resolving the tension between the right to trial and the fortification of national boundaries, in practice the minimal adequacy standard consigned plaintiffs to "alternatives" that would in many cases offer little hope of monetary recovery or any other measure of justice.

Embracing the idea that *any* remedy was an *adequate* remedy led the judge to accept either alternatives that fell far short of U.S. standards of jurisprudence, or alternatives that were unlikely to be accessible for real-life plaintiffs, despite their availability in theory.[99] In other words, while the adequacy analysis supposedly ensured that the U.S. imperative for rule of law was met, in practice Judge Lake did not insist that central tenets of U.S. jurisprudence be met in order for a forum to qualify as adequate. For example, although the right to testify on one's behalf and the right to a jury trial are cornerstones of U.S. concepts of justice, the judge refused to measure other nations' adequacy against these standards.

Of course, using U.S. practice as a yardstick for an adequacy test raises uncomfortable questions. To hold that another country's court system is either unavailable or inadequate opens a U.S. court to criticisms of chauvinism. Indeed, proponents of FNC have been quick to agree with the judge overseeing the Bhopal litigation that "to retain the litigation in [the United States], as plaintiffs request, would be yet another example of imperialism, another situation in which an established sovereign inflicted its rules, its standards and values on a developing nation."[100] This analysis, however, was contradicted by the fact that in many cases, not only were foreign individuals seeking justice in U.S. courts, but often their very governments supported those efforts. The motivation of foreign government officials to disavow their ability to deal with local claims undoubtedly had varied and multiple roots, likely ranging from political pressure to fears about cost of workers' compensation. Equally unusual, advocates of transnational corporations—not known for their support of Latin American self-determination—found that an "anti-imperialist" argument supported their goals. The defendants' unusual stance ironically insisted that their own wishes must be followed in order to protect plaintiffs' nations from outside interference. The oddity of the situation was compounded by Judge Lake's rejection of national authorities' disavowal of their own judicial competence, coming as it did on the heels of his willingness to grant Dead Sea Bromine sovereign status. That is, while the judge granted Dead Sea the nation-state's prerogative to choose a preferred forum—trial in a state, or in a federal court—he disregarded actual nation-states' articulated preference for a forum—any forum—within the United States.

Finally, Judge Lake's adequacy analysis ignored the fact that, even when a remedy existed on paper, individuals may still have faced barriers that would prevent them from achieving that remedy. For example, without contingency fees, most banana workers or ex-banana workers could not afford a lawyer; in 1994, for example, banana workers in Costa Rica, the best-off economically of the Central American nations, made about $250 a month, certainly not enough to pay attorney fees; by 1998 that salary had fallen to $187 a month.[101] A bar on noneconomic harm could mean that no money would be awarded for an injury that did not necessarily impair work functions. Limits on evidence could mean that a plaintiff's side of the story would not get a fair hearing. Furthermore, many plaintiffs' witnesses indicated that courts in their countries had never before tried these kinds of cases, making the judicial process and outcome unpredictable and uncertain even if a plaintiff managed to secure a lawyer and properly file a case.

Judge Lake refused to consider these types of issues as part of the primary adequacy analysis, instead shunting them into the secondary "private interests" category. Private interests, as articulated in the Supreme Court's decision in *Gulf Oil,* included

> the relative ease of access to sources of proof; availability of compulsory process for attendance of unwilling, and the cost of obtaining attendance of willing, witnesses; possibility of view of premises, if view would be appropriate to the action; and all other practical problems that make trial of a case easy, expeditious and inexpensive. There may also be questions as to the enforceability of a judgment if one is obtained.[102]

The fundamental questions about "alleged deficiencies in the judicial systems of the plaintiffs' home countries" that the judge had excluded from the adequacy analysis, he instead considered under the less weighty rubric of "all other practical problems." His memorandum and order skimmed quickly over questions including the ability of plaintiffs to pay for lawyers in absence of contingency fees, the ability of foreign courts to handle the cases, and "procedural differences" between U.S. and foreign courts. For example, the judge acknowledged that the lack of contingency arrangements in many other countries would mean that "initial expenses will be much greater in [the plaintiffs'] home countries than in this court," but failed to recognize that this could present an insurmountable barrier, meaning the litigation would not be pursued. Similarly, he noted that "most of plaintiffs' home countries employ fee shifting devices under which the losing party is responsible to reimburse the winner for all of the court costs and attorney's fees paid," but did not add that this fee arrangement might actually discourage would-be plaintiffs who worried that if they lost, they would be responsible for the defendants' legal costs. "In most of the countries an indigent plaintiff can also obtain free counsel," continued the judge, but did not address plaintiffs' fate in countries without this right, or whether available lawyers would be prepared to take on the complex and novel cases this litigation would entail. By relocating these issues from the primary adequacy analysis, moving them to the private factors analysis, and then failing to plumb the implications of barriers to remedies in other forums, the judge in effect demoted these fundamental questions of access and weighed in in favor of an FNC dismissal.

Indeed, even within the private factors rubric, issues of access held little weight for Judge Lake: while he found that the so-called "other practical problems" weighed slightly against dismissal, when he considered the private interests as a whole, he found further reasons to dismiss the case. A U.S. forum for the trial, he found, did not allow convenient access to the human and documentary resources that would be needed to decide the case. Plaintiffs, as well as their "co-workers, family members, neighbors, supervisors, doctors, and employers"[103] would need to come to Texas for the trial; and defendants would also need to look at documents that were located in the plaintiffs' home countries. The plaintiffs had offered to pay their own and some witnesses' expenses for travel to Texas. But, the Judge maintained, the court had no way to guarantee (or pay for) the appearance of other witnesses.[104] The private interests in securing witnesses and evidence, he found, were better met by sending the cases "back" to plaintiffs' nations, where documents and testimony would supposedly be easier to provide. However, Judge Lake ignored limitations on evidence and witnesses that existed in some of the plaintiffs' home nations. For example, in some Latin American countries, testimony was forbidden on the part of "the parties themselves, their spouses, their close family members, their friends and enemies, their trusted employees, or anyone having a direct or an indirect interest in the outcome of the case"; in court systems where this was the case, an FNC dismissal would obviously not provide easier access to witness testimony.[105]

While witness and evidence rules were much more liberal in the United States, the judge's analysis depended heavily on a conception that sharing information across borders presented a formidable challenge. This conception was belied by the long history of the century-old banana industry, with its longstanding transnational movement of people and goods. In addition, the judge ignored the changes of the last third of the twentieth century, including improved travel and communication technologies. Fax machines, the proliferation of air-travel routes, reliable international phone lines and even the still-emergent internet meant that coordinating people and documents was much easier, cheaper, and quicker than when *Gulf Oil* was decided.

Finally, the judge made brief inquiry into the "public factors," beginning by stating unequivocally that "the controversy is not 'localized' in the Southern District of Texas (or in any other location in Texas) under any meaning of the word 'local.'"[106] Although two of the defendants were headquartered in Texas, he wrote, they moved there after the plaintiffs were exposed to DBCP; two plaintiffs in one of the consolidated cases, also Texas residents, had likewise moved there *after* being exposed to the chemical. "Plaintiffs," the judge continued, "have produced no evidence that any defendant took any action in Texas that affected whether plaintiffs would be exposed to DBCP.... [There is a] dearth of evidence linking plaintiffs' claims to Texas."[107] This analysis proposed a definition of "local" much narrower than the jurisdiction of his court, which did encompass *Delgado*. For Lake, the foreignness of the case was evidenced in the administrative and legal hurdles it would create for the court; likewise, he felt that asking Texans to serve as jury members would place an unfair burden on their own time. By constructing engagement with an international trial as both an institutional and personal burden, Lake forwarded an image of U.S. legal and ethical interests as bounded by the borders of the nation.

Transnational cases such as *Delgado* did pose some special challenges to the court. One of these is known as "choice of law": although a case is brought in a particular court in the United States, the law of that forum is not necessarily the law that governs that case. Depending on the types of charges brought, where the damages took place, and the nationalities of the plaintiffs and defendants, the court may apply the law of any state of the United States, federal law, or the law of another country. The *Piper* and *Gulf Oil* precedents allowed consideration of choice of law questions as part of the FNC tests. Judge Lake found that in the case under consideration, those questions had not yet been resolved, but that *Delgado* was likely to require the application of "at least twelve different legal standards" once the laws of various states in the United States and several of the plaintiffs' nations were taken into account.[108] According to the judge, dismissing cases to the plaintiffs' home countries minimized but did not eliminate choice-of-law problems because those "home courts" would likely have to consider (only) their own and U.S. law.

As the movement of Texas-based transnational corporations showed, the state was already implicated in a web of political, economic, social, and cultural relations with the home countries of DBCP plaintiffs—especially those from the Central American region, which had a long shared history with the United States. Dismissing cases to other nations' courts, Judge Lake disavowed the demands of globalization on the Texas courtroom. At the same time, he shifted the burden, assuming that courts in the plaintiffs' home nations would face less complex but still significant choice-of-law questions themselves. The U.S. judge in effect enacted a kind of American exceptionalism that denied the place of the United States in an international system and, failing to carry out the responsibilities of that place, externalized them onto other nations.

Also under the rubric of the public interest, Judge Lake found that a Texas trial would be too burdensome on the citizens who would make up the jury. Claiming that the court had neither an interest in the DBCP cases as a "local controversy," nor a responsibility to find a way to deal with admittedly challenging choice-of-law questions, Lake asserted that asking U.S. citizens to make decisions in the case would place an "unfair burden upon local residents forced to serve as jurors."[109] This opinion extended to jurors (that is, citizens) the exceptionalism and disengagement Lake claimed for the court in his choice-of-law discussion. Citizens of the United States, he implicitly asserted, were not to be bothered with decisions about what might happen to people who live as far away as Papua New Guinea or as close as Guatemala.

Having concluded that an adequate alternative forum existed, and that the private and public interests supported dismissal, Judge Lake granted the defendants' motion for an FNC dismissal of *Delgado,* and forbade plaintiffs from filing or joining any other DBCP-related lawsuits in the United States. A clear victory for the defendants, the dismissal did not come without conditions evidently meant to correct some of the most obvious contradictions created by dismissing the case. Defendants were required to participate in 90 more days of discovery in the United States to allow plaintiffs access to sources of proof not available in their home nations. In addition, the defendants had to agree to participate in the legal cases in the foreign forums. Plaintiffs were told that if the courts of their home countries did not accept jurisdiction over the cases, they could return to Judge Lake's courtroom, and "the court will resume jurisdiction over the action as if the case had never been dismissed for f.n.c."[110] The bar for return was set high, however. Despite the fact that Judge Lake was a district judge and not a Supreme Court justice, he decided that the highest court in a plaintiff's home country would need to hear and dismiss the case for lack of jurisdiction before Judge Lake would take it back.

By setting a minimum standard for alternative forum adequacy, embracing a narrow conception of ease of movement of people and documentary evidence across national boundaries, and defining the public interest in narrowly national terms, Judge Lake consigned plaintiffs to pursuing justice in their own nations, with only a difficult and circuitous path back to the United States. Constraining Central Americans (and others) to nationally bound remedies belied the transnational roots of their complaints. Of course, corporate defendants would have to travel to the plaintiffs' home forums if they did indeed manage to file cases there, a situation that did not seem to raise a concern for the judge. His decision perversely affirmed corporations' transnational mobility, while pinning individuals within their own nations against their expressed will, and in some cases the will of those nations' government officials. A defeat for the DBCP-affected workers and their lawyers, *Delgado* seemed to force either an abandonment of worker claims, or a new legal effort in the Central American courts.

CENTRAL AMERICAN NATIONS RESPOND

While *Delgado* plaintiff lawyers would continue to appeal Judge Lake's decision, for the time being plaintiffs' complaints had effectively been excluded from U.S. courts. Accordingly, some of *Delgado*'s DBCP-affected workers filed cases at home, a course that they hoped would either provide some measure of justice through their own courts and compensation systems, or clear a path back to the United States. As plaintiffs brought suits in Costa Rica, Nicaragua, Panama, Guatemala, and Honduras, the fallout from *Delgado* sparked growing opposition to FNC in Central America as judges and other officials saw Lake's decision as an affront to their judicial traditions and sovereignty.[111]

In rapid reaction to Judge Lake's *Delgado* dismissal, Costa Rican plaintiffs' lawyer Susana Chávez Sell and Costa Rican colleagues filed a case against DBCP producers and users in the Fourth Civil Court of San José, Costa Rica, in early August 1995.[112] Their complaint, known as *Abarca,* contained an unusual element—a request that the judge consider the question of jurisdiction before any other issue. The plaintiffs made it clear that they were seeking compensation from the defendants if and only if the court found that it did have jurisdiction. The primacy of the jurisdictional question showed that for plaintiffs the Costa Rican case was a response to U.S. litigation and, they hoped, an avenue back to Judge Lake's courtroom.

Ricardo González Mora, the first judge to consider the *Abarca* case, met the plaintiffs' hopes by reacting strongly against Judge Lake's terms of dismissal. Judge González Mora's opinion acted as a refusal of U.S. imposition of the DBCP cases on Costa Rica, insisting that the question of his court's jurisdiction over *Abarca* would be decided by consideration of Costa Rican—not U.S.—law. Contrary to the common-law system used in the United States, in which judges are granted wide discretion and law is developed through precedent over time, Costa Rica used a civil law system, based in written codes that must be strictly interpreted by the judiciary.[113] To decide on *Abarca* jurisdiction, Judge González Mora turned to two legal codes, the Costa Rican *Código Procesal Civil* (Civil Code) and the *Código Bustamante,* the international civil code governing private international law for a number of Latin American nations.[114]

Reading *Abarca* in light of these codes, Judge González Mora conceptualized the issue of "localized interest" very differently than Judge Lake had. While Lake said that the case was not of interest to Texans because none of the actions pertinent to the case had happened there, in González Mora's analysis, the most important events had taken place much closer to Houston than to San José. At the heart of the case, wrote the Costa Rican judge, was the matter of "finding liability on the part of the manufacturers and distributors of an allegedly defective product."[115] Because the acts in question involved the production and distribution of the chemical, rather than its use *per se* in Costa Rica, there were not sufficient connecting factors to create jurisdiction under the Costa Rican Civil Code. Jurisdiction, he found, should be governed by the Bustamante Code, which granted jurisdiction to the forum where the defendant was domiciled or resident. Accordingly, Judge González Mora found that the plaintiffs had acted correctly when they filed the case in Texas in the first place.[116]

More than a finding of jurisdiction, the judge's opinion was a rejection of the doctrine of *forum non conveniens* and what he saw as its imposition on Costa Rica. The doctrine, he wrote, was "unknown and therefore cannot be applied in the Costa Rican legal system, nor can it be the legal ground for determining the competence of this court."[117] Explicitly addressing Judge Lake's decision, he wrote: "The fact that Judge Lake deems it more convenient for the plaintiffs to have the case heard in another jurisdiction, even against their express will, is not relevant to the issue immediately at hand before this Court."[118] Judge González Mora's framing of Judge Lake's dismissal as immaterial to Costa Rican decision-making functions rejects the U.S. judge's foisting of the case onto the Central American nation's courts.[119] For the Costa Rican jurist, Judge Lake's FNC dismissal threatened the Central American nation's self-determination by enacting an externalization, or "exportation," of U.S. sovereignty. If Costa Rican courts accepted jurisdiction over the *Delgado* cases, they would in effect be carrying out the orders of a U.S. judge rather than making their own decisions. Rejecting such a sacrifice of self-determination, Judge González Mora took the first step toward sending *Abarca* plaintiffs' claims back to the United States.[120]

Before Judge Lake would accept the DBCP-affected workers' return to his Texas courtroom, however, the case would have to be dismissed by the highest court of Costa Rica. To meet this high bar, the plaintiffs had to carry out of the bizarre act of appealing a decision with which they had agreed.[121] It was a complicated appeals process, but by the end of the year plaintiffs had been granted a hearing by the First Division of the Costa Rican Supreme Court. The Supreme Court denied the plaintiffs' appeal on procedural grounds, issuing a decision that the lower court's finding "that adjudication of the present case does not correspond to a Costa Rican judge, is affirmed."[122]

All told, the Costa Rican rulings expressed resistance to the power of both the United States as a nation and the multinational corporations who were defendants in the case. By expressly arguing that the U.S. judge's ruling had no impact on the Costa Rican court's decision, Judge González Mora asserted the Central American nation's independence from and equality with its counterpart, despite the North American nation's more powerful position geopolitically. Yet his ruling also countered the power of the transnational corporation defendants, precisely by refusing them the mobility between forums that they sought in order to minimize their legal exposure. Judge González Mora's opinion and the Supreme Court's affirmation of its logic staked out Costa Rica's ground in the ongoing struggle—the nation would neither cave to the wishes of the U.S. judiciary nor make its courts available to corporations seeking to minimize their DBCP damages.

While *Abarca* was moving through the Costa Rican courts, DBCP-affected workers and their lawyers were following a similar path in other Central American nations. Cases brought in Guatemala, Honduras, Nicaragua, and Panama would also test their nations' court systems to see if the *Delgado* decision would stand in the Central American context. In the case of three of those nations, judges' responses were similar to that of the Costa Rican judges. By the end of 1995, lower courts in Panama, Nicaragua, and Guatemala would reject Judge Lake's FNC dismissal.[123] By March 1996, the case filed in Panama had progressed to that nation's Supreme Court, where it too was dismissed for lack of jurisdiction.[124] In contrast, Honduran courts accepted jurisdiction over DBCP cases dismissed from the United States, despite adopting legislation in 1996 declaring that cases dismissed by FNC in other nations would not have jurisdiction in Honduras.[125] Despite the outcome in Honduras, Central American opinion seemed largely unified on the topic of FNC—it was foreign to the isthmus, and U.S. dismissal would be considered invalid in the various nations' courts.

The anti-FNC rulings and rules in Central America seemed to open the way for a resolution of the lawsuits in the United States. However, a trial in a U.S. court remained beyond the grasp of the *Delgado* plaintiffs. When *Abarca* plaintiffs refiled in Judge Lake's court, he put off accepting their case while other appeals were being decided, leaving the case in a legal limbo that promised to last for years.[126]

In the meantime, lawyers brokered settlement agreements in many of the other *Delgado* cases over the course of 1997 and 1998. All companies except Dole and Dole-associated companies paid a total $52–53 million, to be divided between plaintiffs from all twelve countries involved.[127] According to plaintiff attorney Fred Misko, the settlement left many plaintiffs with a payout equivalent to one or two years' salary, much lower than what many had hoped for or expected.[128] Misko's colleague Charles Siegel found the negotiated payment disappointing, saying that the case had gone "off the rails" after the defendants had brought Dead Sea Bromine into the case.[129] Unsurprisingly, aspects of the litigation that were not included in settlement agreements would remain in U.S. courts—once again caught up in procedural runarounds—for years to come.

CONCLUSION

By the mid-1990s, the innovative transnational DBCP claims that began in 1983 with the *Sibaja* case were mired in delay and controversy over questions of location, representation, and translation. While the plaintiffs had struggled to find a court in the United States where they could bring their claims, they were met with fierce resistance from defendants, who articulated an exclusive vision of justice that would redraw national boundaries to exclude foreign claims. As the cases were caught in procedural matters, judicial opinion on whether they could or should be heard in the United States was mixed, with opinions expressing opposing ideas about the boundaries of the United States as a nation, the nature of transnational corporations, and the rights of foreign plaintiffs. By the time Judge Lake dismissed *Delgado*, however, the exclusionary logic of FNC seemed to hold sway, as thousands of plaintiffs were told to try their cases somewhere more "convenient" than the United States. These struggles over location were marked by difficulties and discord over the questions of representation and translation. As plaintiffs' attorneys brought banana workers' concerns into the realm of the U.S. legal system, they not only stood in for workers in proceedings and decision making, but also took charge of transforming workers' experience into the byzantine language of the courtroom, and vice versa. These processes could leave workers unaware of or unhappy with the progress of their cases, even as lawyers negotiated settlements that could be seen as positive legal outcomes.

The uncertainties and delays of the U.S. litigation would not go unaddressed by DBCP-affected workers in their own nations, however. From the mid-1990s on, Central Americans whose health had been compromised by the chemical had begun to organize. Movements of *los afectados* in Costa Rica and Nicaragua would bring the issue of DBCP-caused sterilization to the forefront of public debate in both countries, critique the history of litigation in the United States, and engage in globally oriented strategies that would achieve some successes in both securing some compensation for workers and countering the exclusionary legal landscape in the United States.

FIVE

Making a Movement

IN 1994, COSTA RICAN *AFECTADOS* faced a confusing scenario. On the one hand, about 11,000 Costa Rican men and women had filed claims in the United States, but few of them knew the progress of the cases or how much they might win.[1] Meanwhile, Standard Fruit had offered direct payments to some workers, in some cases totaling as much as 2 million Costa Rican *colones* (about US$4,400 at 1994 exchange rates).[2] Should *afectados* continue to place their hopes in the far-off litigation, or try to secure money directly from banana companies still operating in Costa Rica?

San José attorney Susana Chávez Sell assured her clients that any continuing direct payments from Standard would be "much smaller than those you can win in the [legal] process," warning, "DO NOT BE FOOLED."[3] But attorney assurances notwithstanding, by September it was clear that some *afectados* who had chosen the litigation route would be dropped from the suits. And plaintiff lawyers facing a tight squeeze from the defendants' Dead Sea FNC strategy had lost much of their former optimism that the cases could be brought to trial. The best alternative seemed to be a settlement; however, plaintiffs' weakened position meant the potential amounts were far less than they had hoped for as trial awards.[4] In addition, only those plaintiffs who met strict criteria for sterility would remain part of the deal: those workers with fewer than 4 children and under 20 million sperm per milliliter of semen, the accepted standard to indicate oligospermia, a lower-than-normal sperm count.[5] To plaintiffs who did not meet these criteria, Chávez Sell offered only the "hope of the possibility" that they could be part of an eventual settlement based on "simple exposure to DBCP . . . or because it is scientifically determined that this product causes disease other than sterility."[6]

Until 1995, workers' efforts at achieving compensation had mostly been limited to bringing suits in the United States. By mid-decade, *afectados* were increasingly vocal in their critiques of a transnational legal process that they found opaque and unjust. They began organizing at home, seeking compensation that—unlike the litigation—matched their own understanding of DBCP harms. Instead of simply placing their faith in a "hope of possibility," *afectados* initiated a new phase of protest and action, demanding compensation from the state and increasingly engaging with the science of DBCP in hopes of aligning compensation with their own understanding of problems caused by the chemical. They continued to participate in lawsuits, but also made their own national state a key arena of contestation over DBCP science and compensation, converting some state actors into allies in their efforts and/or objects of their demands for both compensation and medical care.

Placing the question of popular, scientific, and official interpretations of DBCP's health harms at the center of their struggles, Costa Rican *afectados* drew on national traditions of union activism and state welfare policies. They created alliances with sympathetic researchers and used the language and authority of science to expand the narrow litigation view of problems caused by the chemical and who could be compensated for those problems, working to include women and to expand the definition of harm to men beyond the issue of sterility. At the same time, they leveraged the power of science with street protests that placed pressure on state decision-makers by drawing national attention to their cause. Their actions constituted both an alternative and a supplement to the transnational lawsuit strategy, relocating their struggle closer to home and winning some material gains for some workers.

BEGINNINGS OF A NATIONAL MOVEMENT

In June 1997, news of a settlement in *Delgado* hit the Costa Rican press.[7] *La Nación* reported that Dow would pay $22 million to a total of 25,000 workers, including 6,000 Costa Ricans.[8] Before DBCP-affected workers received their share, however, almost $17 million would be deducted—$8 million in expenses and $8.8 million in lawyers' fees.[9] Ex-banana workers who had filed lawsuits but whose conditions did not meet the official definition for azoospermia or oligospermia (20 million sperm/milliliter) were not included in the settlement; instead, plaintiffs' attorneys offered them payment of $100 in recognition of their involvement with the case.[10] About half of *Delgado*'s

Latin American plaintiffs had abnormal sperm counts, meaning that the other half may have received only symbolic payment.[11] Critics of the settlement took issue mostly with the fact that plaintiffs with children or medical conditions other than sterility had been excluded: *Afectados* understood their exposure to the chemical to be the cause of a number of other symptoms and diseases, including cancer, bone and muscular pain, vision loss, skin problems, and kidney pain.[12] Some workers also found it unfair that the fact that they had had four or more children—even before exposure to DBCP—would exclude them from the litigation. One man claiming DBCP-induced damage, Plutarco Hernández, asked, "How is it possible that they disqualify people just because they had children before they used the toxin!"[13] Miguel Villalobos agreed: "It's an injustice. I am sterile," he said, maintaining that the fact of his sterility should have qualified him for compensation, regardless of children born *before* exposure.[14] The disjuncture between worker and lawyer definitions of DBCP-caused harm provoked protest on the part of labor and human-rights groups, who, according to *La Nación,* saw the settlements as "inhuman, irrational, insufficient, and unjust."[15]

Afectados' outrage signaled that competing cultural, medical, and legal definitions of DBCP harm had become a central ground of contention not only between plaintiffs and defendants, but between former banana workers and their own lawyers. While the attorneys and their clients agreed that DBCP caused sterility in banana workers, even *that* damage had contested significance. For a successful outcome in a U.S. trial, lawyers would have to establish both general and specific causation—that is, that DBCP generally *could* cause the injuries claimed by the plaintiffs, and that in fact in each individual case it specifically *had.* The fact that oligospermia and azoospermia were relatively rare conditions with few other causes meant that sterility cases were well suited to courtroom standards of causation. But lawyers would still have to convince a jury that *harm* had been done by the injury. For Costa Ricans like Hernández and Villalobos, sterility constituted a harm even though they had already fathered multiple children. Juries in the United States, however, where the average number of children per family was lower than among rural Costa Ricans, may not have seen sterility as a true injury to men who were already fathers. Thus, conflicting cultural norms regarding family size and the significance of male sterility could weaken even oligospermia and azoospermia cases.

The range of illnesses that *afectados* linked to DBCP went well beyond the signal problem of sterility. Many of the conditions were also implicated by animal testing and other human evidence: DBCP was a recognized animal carcinogen and considered a probable human carcinogen by the International Agency for Research on Cancer (IARC); research also reported damage to the kidney and liver and showed evidence of reproductive problems in women and birth defects in offspring.[16] At some point, plaintiffs' lawyers had seen some of these illnesses as bases for successful lawsuits; a document called "The Legacy," compiled by *Delgado* plaintiff attorneys, included an explanation of DBCP's link to cancer and promised that "cancer cases, of course, will themselves be tried."[17] But cancer cases were not as strong as sterility cases in terms of their potential for a successful trial. Cancer had a multiplicity of causes, making it difficult to link farmworkers' cancers to DBCP exposure. Other problems Costa Rican *afectados* attributed to DBCP—such as nonspecific bone and muscular pain—were not reflected in the scientific literature at all, meaning that lawyers could not construct the proof they needed for a successful lawsuit.[18]

The attribution of health problems to DBCP reflected social as well as scientific processes. Former banana workers were likely to have experienced a number of occupational exposures—not only to pesticides but also to long work days and hard physical labor. Those exposures left their mark,[19] but most banana worker health problems did not receive much public attention. Therefore, as *afectado* activism intensified and the health harms of DBCP gained attention—and litigation presented the chance of financial recovery—former banana workers with ailments could easily identify with the movement. Attributing a wide range of problems to the nematicide made common sense—these illnesses emerged from the same historical context as DBCP damage did: a system of banana production that had little regard for worker safety, paid poverty wages, and "used up" workers through intense and sustained demands on their bodies. Workers knew from experience that work had hurt their bodies; the focus on DBCP suggested a specific cause of that damage. Attributing a wide range of diseases to DBCP also presented some hope of financial gain: litigation presented a rare opportunity for banana workers to win compensation from the fruit and chemical companies who had so powerfully shaped their lives. It was not surprising that banana workers blamed a range of problems on the chemical, while lawyers, wedded to a narrow set of scientific and legal strictures for causation, focused on the signal disease of sterility.

This disjuncture between *afectado* understandings and medico-legal strictures spurred discontent and action in Costa Rica. In the mid-1990s, a number of activist groups (for example, Asociación Pro Defensa de los Trabajadores y del Medio Ambiente [Association for the Defense of Workers and the Environment] and Comité Nacional de Bananeros Esterilizados [National Committee of Sterilized Banana Workers]) took up the *afectado* cause.[20] These groups—some short-lived and leaving little record, others more enduring—faced a national context for collective action shaped by the national history of unionism, as well as Costa Ricans' widely shared sense that the state should provide them with a set of "social guarantees."

The development of the social welfare policies for which Costa Rica is often cited as an exception among its Central American neighbors dates to at least the 1940s. Then, under President Rafael Ángel Calderón Guardia, a social security program and labor code were established, and amendments to the national constitution provided "social guarantees" including health services and housing, the right to strike, and a minimum wage.[21] After social democrats triumphed over the Calderón-Catholic-Communist coalition in the 1948 civil war, the social democrat National Liberation Party (*Partido Liberación Nacional* or PLN) gained dominance. Under PLN leadership, the state increased involvement in economic and social policy, even as the labor movement was suppressed.[22] By the 1970s, as anthropologist Marc Edelman writes, the Costa Rican welfare state had not achieved "any fundamental, progressive redistribution of resources, [but] it did pay a 'social wage' that by regional standards was extraordinarily high," winning the state "a remarkably high level of legitimacy."[23]

In the 1950s and 1960s, banana unions remained relatively stronger than those in other sectors, and by the 1970s strike activity was up and unions negotiated an unprecedented number of bargaining agreements.[24] Then came pressures including spiking oil prices, insufficient success in efforts to industrialize and diversify the economy, and a growing debt and trade deficit. During the 1970s, state institutions adopted measures including upward wage adjustment and land redistribution, even as local security forces (the *Guardia Rural*) often did landowners' bidding in ejecting peasants from disputed lands.[25] Right-wing and business opposition to PLN policies grew, bringing Rodrigo Carazo, of the *Partido Unidad Social Cristiana*, to the presidential office.[26] In 1981, Costa Rica became the first Latin American nation to default on its loans, portending what would become known as the "lost decade" in Latin America. Costa Rican policy makers, like those in many other Latin American nations, followed neoliberal prescriptions for economic growth. The PLN was restored to power for the remainder of the decade, and during the administrations of Luis Monge and Óscar Arias, leaders agreed to cut social spending, auction off state-owned companies, pass banking reforms, and reduce subsidies, while raising taxes, utility rates, and interest rates in order to secure assistance from the IMF (International Monetary Fund) and USAID (United States Agency for International Development).[27]

The outlook for banana unions already looked dire in 1984, when a 72-day strike by Communist-led United Fruit workers in the south Pacific coast met with repression, including the death of two workers.[28] In a significant defeat for workers, the strike ended when United Fruit entirely discontinued banana production in the region.[29] Although the company was already in the process of replacing banana cultivation with that of less labor-intensive oil palm, popular opinion blamed workers for destruction of the banana industry in the Pacific. Another blow to unionists came the same year, when a new law strengthened the role of so-called "Solidarity Associations." These were workers' mutual associations that operated with a high degree of employer involvement and sought "labor-management harmony," in contrast with the unions' oppositional stance. Solidarity Associations could only enter into unenforceable agreements—not legal collective bargaining contracts—with employers, but the other services they offered, including credit, social and cultural programs, and yearly dividends, appealed to many workers, ultimately limiting union organizing and leading to a dramatic decline in union membership.[30] This changed context for organizing coincided with national efforts to attract foreign investment, including encouraging expansion of the banana industry. Critics noted that the expansion came largely without labor and environmental protections, leaving workers and their communities at increased risk.[31]

In response, banana unionists developed new alliances and approaches, partnering with each other, with environmental and religious organizations within Costa Rica, and with banana worker unions from other banana-growing nations in Latin America. In 1990, three banana worker unions formed the Coalition of Costa Rican Banana Unions (*Coordinadora de Sindicatos Bananeros de Costa Rica* or COSIBA-CR) a joint organization that would fight for labor.[32] The Coalition would in turn become part of Foro Emaús, a Costa Rican environmental, religious, labor, and human-rights network formed in 1992 and focused on banana workers. COSIBA-CR also became part of a groundbreaking regional grouping of other national banana worker union coalitions, the Coalition of Latin American Banana

Unions (*Coordinadora Latinoamericana de Sindicatos Bananeros Agroindustriales* or COLSIBA). COLSIBA broke new ground not only by organizing across national borders, but also by engaging with the larger global politics of the banana trade.[33]

The group that would take the lead in responding to what *afectados* saw as the injustices of the legal settlement announced its formation in the national press in March 1997.[34] The National Banana Workers' Council, or CONATRAB (*Consejo Nacional de Trabajadores Bananeros*), started with just a handful of individuals (by some accounts, three; by others, five) concerned about the settlements, but quickly grew into a national organization with chapters in the Atlantic banana-growing region, the two peninsulas on the Pacific coast (the Nicoya in the north, and the Osa in the south) as well as in the capital in the central highland.

CONATRAB drew on the history and language of banana unionist struggle, even as its aims and organization departed significantly from those of traditional labor unions. Studies of social movements since the 1970s have tended to contrast "old" social movements—usually characterized as class-based or concerned with the redistribution of resources, e.g., labor movements—with "new" social movements, usually described as based on shared identity and emphasizing struggle in the social, cultural, and representational realm rather than fighting for material gains.[35] This is, of course, a simplified typology, and many have complicated this duality, noting that many contemporary social movements do not fit neatly in either category.[36] This was certainly the case with CONATRAB. As the group gained recognition, it was most frequently referred to—by the media as well as its own leaders—as a *sindicato bananero*, or banana union, and the name of the group referenced radical union antecedents.[37] In a 2009 interview with sociologist Allen Cordero Ulate, the Secretary General of CONATRAB, Orlando Barrantes Cartín explained:

> The name is CONATRAB because we're trying to reclaim some of the spirit of the strike of [19]34 of Carlos Luis Fallas ... the first major strike in the Latin American banana industry. Carlos Luis Fallas' organization was called COTRAB, Workers Council [*Consejo de Trabajadores*], so we used Banana Workers National Council by the suggestion of Carlos Cortés.... I remember the first newsletter we put out, we reproduced a photo, which we then turned into a drawing, of the strike of [19]34.[38]

For Barrantes, CONATRAB was taking up the fight of the "Communist parties" that had "disappeared" from Costa Rica, and the group would "reclaim the aspirations and needs of a sector of the Costa Rican working class. . . . It is a class movement."[39] While CONATRAB drew on the history, imagery, and constituency of banana labor unions, its form and strategies were not the traditional labor union repertoire. In contrast to COTRAB's strikes and the centralized organization of left political parties, CONATRAB used a what is generally considered a new social movement protest repertoire, organizing in communities rather than strictly in the workplace, and using street protests rather than strikes.[40] Membership was not based on current employment in banana plantations, and unlike the traditional banana unions still organizing in Costa Rica, the group did not seek to represent workers in bargaining with their employers. While CONATRAB became part of Foro Emaús, it did not join the national banana union coalition COSIBA-CR (nor, by extension, COLSIBA, the transnational coalition).[41] At the same time, CONATRAB had some important continuities and overlaps with the Costa Rican peasant movements that mobilized in response to the deeply felt effects of the Latin American debt crisis of the early 1980s and subsequent neoliberal restructuring.[42] Like CONATRAB, these movements were "not . . . easily pigeonholed in an arid taxonomy of 'identity-based' versus 'class-based' movements," and their strategies (particularly the blockade) and members (many *campesinos* had at some point worked on banana fields) would inform and strengthen CONATRAB.[43] Peasant organizations fell on a political spectrum from the radical left to the liberal democratic, while in the realm of formal politics CONATRAB would be successful in garnering at least symbolic support from politicians across the political spectrum.[44]

The most salient feature uniting CONATRAB members was their shared experience of what they defined as DBCP-related health problems—not only sterility, but the broad range of illnesses former banana workers understood to be caused by DBCP: "different grades of sterility (azoospermia, oligospermia), testicular atrophy and pain, cancer of the liver, kidneys and stomach, severe allergies, bone problems, vision deficits, menstrual and hormonal alterations, children with very serious congenital problems, moral and psychological harm."[45] In addition to illness and disease, however, *afectados* were united by what Barrantes called their "other painful experience"—that is, with lawsuits filed in the United States. Although the phrase "affected persons" had been used as early as 1983 in Costa Rica, by the late 1990s, *afectado* became a politicized identifier signifying more than the presence of a specific injury or health problem. Instead, the term came to indicate a common experience of physical suffering attributed to (usually former) banana employment and DBCP exposure, as well as participation in transnational and national processes for seeking justice and compensation.[46]

THE *DEFENSORÍA DE LOS HABITANTES* AND *AFECTADOS'* TURN TO THE STATE

Anthropologist Marc Edelman has argued that despite the changes of neoliberalism, "state agencies remain absolutely central points of reference, foci of demands and sites of struggle" for Costa Rican activists.[47] This was certainly true for the *afectados* amassed in CONATRAB. As they looked for ways to hold their attorneys accountable, they drew on national political traditions and—despite the state's diminished role—directed their complaints and demands for assistance to various organs of the Costa Rican state, operating at a juncture between the postwar social welfare state and the market orthodoxy of neoliberalism. Costa Rican banana workers concerned with pesticide exposure had sought state intervention as early as 1942, when copper sulfate sprayers wrote to then-President Calderón complaining of health problems.[48] *Afectados* continued that tradition, looking almost immediately to state institutions for justice, even as their transnational lawsuits remained in play.

At the same time, *afectados'* early efforts also reflected how the state had been altered by globalization. In October 1997, CONATRAB filed a formal complaint with the *Defensoría de los Habitantes*, an independent ombudsperson office within the legislative branch of Costa Rica's tripartite political system which was responsible for promoting human rights in Costa Rica and monitoring the "legality, morality, and justice of the actions or omissions of the administrative activity of the public sector."[49] The *Defensoría* did not exercise direct governing powers, but rather made recommendations in the form of reports. Notably, the agency had its roots less in the decades-long Costa Rican history of social welfare than in the state's internalization of a global human rights regime. Established in Costa Rica through a process of institutional changes in the 1980s and early 1990s, the *Defensoría* can be understood as part of a globally emergent commitment to the norms of human rights that began in the post–World War II period with the adoption of the Universal Declaration of Human Rights in 1948.[50] Formation of the *Defensoría* signaled a commitment to internalizing global human rights norms even as the guarantees of the social welfare state were being eroded by

150 · CHAPTER FIVE

the emerging neoliberal, privatizing agenda that included the rise of Solidarity Associations and the renewed expansion of the banana industry.[51] *Afectados'* turn to the *Defensoría*, then, brought the national and the transnational into complicated relationship with each other, as they sought an alternative to, and accountability from, a transnational litigation project through a "watchdog" arm of the Costa Rican state with its roots in global norms rather than national traditions.

Even before the formation of CONATRAB, concerned *afectados* had voiced their concern to the office of the *Defensoría*. By 1995, the *Defensoría* had expanded state engagement with *afectado* demands, involving the Ministries of Labor, Agriculture, Exterior, and Health in tasks including seeking information on the status of the lawsuits, providing health services to *afectados* and controlling other pesticide use. CONATRAB's 1997 complaint initiated a period of intensified *Defensoría* action on the *afectado* question.[52] CONATRAB framed the problem primarily as a failure of the Costa Rican welfare state. Workers had been harmed by exposure to the chemical through a failure of the national regulatory apparatus; their health problems had not been adequately diagnosed, treated, or covered by social security; and the Costa Rican government had ignored their plight. Moreover, workers added that they had been poorly represented by their lawyers in legal demands in the United States and called for state support in that ongoing struggle. In sum, they maintained that the Costa Rican government had a responsibility to medically evaluate and treat them and provide for their basic necessities, as well as to support them in the ongoing transnational legal battle.[53] They articulated the state's responsibility in both national and transnational terms, and in modes that corresponded to the traditional role of the welfare state and also expanded that role into support for a legal struggle that went beyond national boundaries.

The *Defensoría's* response to *afectados* combined investigation and advocacy. In addition to a series of interviews with 1,700 banana workers and 400 of their spouses and *compañeras* (female partners), for which the *Defensoría* traveled to communities in Guápiles, Puntarenas, Nicoya, and Golfito, the *Defensoría* conducted a review of the scientific literature on DBCP, based largely on a report commissioned from researchers at the *Universidad Nacional de Costa Rica,* in Heredia. In addition, the office carried out a series of requests, meetings, and interviews with various government agencies (including the National Insurance Institute [INS] and the Ministries of Foreign Relations and Labor), the Costa Rican Bar, international organizations such as the International Labor Organization, and various medical and other experts. Finally, the *Defensoría* met with authorities to determine the status of direct settlements and cases brought in the Costa Rican Civil and Labor courts.[54]

The *Defensoría*'s "Final Report and Recommendations," released on October 20, 1998, retold the DBCP story, recounting the early toxicological testing, regulation in the United States, production, and use on banana plantations, and enumerating the Costa Rican laws (beginning with a 1954 executive decree) that should have protected banana workers from exposure to DBCP.[55] Largely affirming *afectados'* critiques of the legal process, it determined that the settlements achieved in U.S. litigation "could not be considered a success."[56] The challenges faced by plaintiffs were echoed in officials' own transnational failures: the *Defensoría* reported that even an honorary consul (named by the Minister of Foreign relations at the *Defensoría*'s request) had been unable to obtain information from Houston courts on the settlements. The report cited *afectado* statements that attorneys had mishandled the litigation and settlement process: clients were not provided with full explanations of cases in progress or results of testing, settlement paperwork had gone unexplained, and workers lacked clear information about the total amounts paid by defendants and retained by their own lawyers.[57]

Central to the *Defensoría*'s findings was the fact that half of the plaintiffs—those with nonsterility claims or men who had fathered children— had been dropped from the case. The *Defensoría* found that *afectados* had repeatedly told lawyers that "their suffering was not limited to sterility, but the effects were much broader and included [damages to] their compañeras and children." Not only were workers with nonsterility effects excluded from the settlement agreement, but that agreement also stipulated that those receiving compensation for sterility would have to agree not to pursue claims for any other conditions, including:

> Loss of or decrease in reproductive functions and capacity, sexual problems, greater risk of contracting illness including cancer, physical pain and mental anguish, greater risk of genetic damage and the loss of ability to enjoy a full and normal life, as well as other damages as a result of working with chemicals or other substances, or of being near or in contact with the same.[58]

These stipulations provide a good sense of the range of nonsterility complaints raised by *afectados*, acknowledged by the settlement only in their explicit exclusion.[59]

The *Defensoría*'s own consideration of the health effects of DBCP began from a broad conception of harm, finding that "the number of directly affected individuals likely exceed the number of workers who were exposed by mixing, applying or assisting in applying [the chemical], to whom should be added the compañeras and children born with physical and genetic problems that can be related to said product, as well as the children of workers" who brought lunch to their fathers in the field.[60] While this formulation included women in the ranks of *afectados*, it did not fundamentally challenge reigning ideas about DBCP's impact along gender lines; the *Defensoría* considered women to be affected by DBCP only insofar as they were spouses or partners of workers, who were assumed to be male. This assumption reflected the hiring practices of banana companies, which used male workers to handle and apply the chemical, and the potential for so-called "take home" exposures (such as through contaminated clothing). However, it did not account for women workers who may have been exposed to DBCP through contact with liquid or vaporized DBCP in the course of their own work on the plantations or by virtue of residence on or next to plantations where DBCP was used.

To understand the conditions this broad range of *afectados* and *afectadas* might have, the *Defensoría* reviewed the scientific literature and also spoke with *afectados* about the symptoms they experienced. Drawing in large part on a report commissioned from scientists at the National University (*Universidad Nacional*, or UNA) as well as other sources, the *Defensoría* summarized acute toxicity as well as chronic effects, including a higher-than-expected proportion of female children born to DBCP-exposed people.[61] The report noted that laboratory tests had found that DBCP caused cancers in laboratory animals, and IARC rated DBCP as a possible human carcinogen, its "2B" classification. Research in Costa Rica, carried out by UNA scientists during the years DBCP was in use, had found an unusually high incidence of a number of cancers in male and female banana workers (penile cancer in men, cervical cancer and leukemia in women, and melanoma and lung cancer overall). Because workers had been exposed to a number of pesticides, these effects could not be attributed to any one chemical; but by the same token, they could not be used to establish safety of any of the pesticides used by the cohort. As part of its investigation, the *Defensoría* had received 2,500 completed questionnaires, in addition to the direct interviews carried out with former banana workers and their spouses. Male workers "consistently" reported experiencing headaches; testicular pain; difficulty ejaculating; pain in the bones, joints, prostate, and penis; occasional stomach aches and gastrointestinal problems; arterial, cardiac, and pulmonary hypertension; skin problems including allergies; fungal infections; and dermal "stains."[62] Some workers also reported vision and hearing problems, severe neuropathy, lumbago, waist pain, finger- and toenail disfigurement, dizziness or nausea, and neck problems. Female spouses of affected male workers attributed a range of ailments to DBCP exposure, including skin, joint, leg, and kidney problems; head and bone pain; gastric and respiratory issues; diabetes; and endocrine disorders.[63] Parents also reported illnesses in children, ranging from "nerves" and skin allergies to severe physical deformities.

The *Defensoría* also reported psychological effects among former banana workers and their *compañeras*. Under the rubric of "reactive depression," the report discussed the emotional effects of sterility, childlessness, and health problems among children. First among the depression-caused effects was sexual impotence, understood as a "direct consequence" of sterility that damaged *afectados'* sex lives and relations with their partners, lowering self-esteem and causing tension, anxiety, and depression. Like early journalistic coverage, the report figured DBCP damage as threatening the family unit, as the "individual and the couple opted for solutions like changing partners multiple times or having children extramaritally." According to the *Defensoría*, health problems caused by DBCP were at the root of an increase in "human sorrow" and "alterations" in affect, self-perception, relations with others, and in the "meaning and significance of life."[64]

The *Defensoría* tread a careful line between scientific and *afectado* understanding of health problems caused by DBCP, noting: "The Defensoría does not intend to conclude in this report that all pathologies are associated with banana workers' exposure to DBCP, but rather to signal that there is agreement between the research studied and the conditions reported by the afectados." Many of the conditions reported by *afectados* were not at all similar to the findings reported in the scientific literature; however, the *Defensoría* avoided discounting workers' own understandings of DBCP health harms while at the same time affirming the value of scientific study, insisting that the banana workers' report signaled the need for "an epidemiologic study of this population and provision of adequate treatment" of those suffering from health problems.[65]

Whatever doubts about *afectado* claims may have emerged between the lines of the report, the *Defensoría* made a robust argument for the responsibility of the state toward DBCP-affected workers.[66] Referencing both human rights norms and Costa Rican law, the *Defensoría* found that *afectados* had suffered harm to "fundamental rights," including the "right to life, to physical integrity, dignity." The state's failure to uphold the basic rights of banana workers went contrary to laws and institutions central to the Costa Rican social welfare state, including the social security system and the "Social Guarantees" of the constitution (which had been amended in 1994 to hold that "every person has a right to a healthy environment"), as well as health, environmental, and labor regulations.[67]

The *Defensoría's* consideration of the state's duties toward banana workers affirmed their rights to medical attention and a pension, but raised more questions about the grounds for compensation for occupational injuries caused by DBCP. The *Defensoría* considered questions around statute of limitations and current employment status, finding that DBCP damage could be "discovered" by workers well after exposure, and that neither this fact nor lack of current status as a banana worker should prevent compensation. The question of causality was central to the granting of workers' compensation, however, and here the *Defensoría* injected a measure of doubt, noting that "it can be inferred that many of [the workers'] conditions and illnesses have not yet been diagnosed by specialists, or related to their exposure to DBCP."[68] While this statement left open the possibility that such conditions might be linked to DBCP in the future, it stopped far short of endorsing *afectados'* own understandings of causality. Despite the *Defensoría's* obvious sympathy to *afectado* claims, then, the controversy centering on competing explanations of DBCP health effects had not been resolved.

The *Defensoría's* final recommendations called for various state agencies to attend to the basic necessities of *afectados*, providing health care and pensions in accordance with established government programs. The report also called on INS to resolve the compensation claims already filed with that agency's Workers' Compensation department—all of them sterility claims—and hire a staff person to liaise with *afectados*. In addition, the report asked the executive cabinet, called the Governing Council (*Consejo de Gobierno*), to attempt to broker agreements between the fruit companies and affected workers, and directed the Ministry of Foreign Relations and Culture to seek information regarding lawsuits brought in Texas.

To answer the questions of who should be compensated, and on what grounds, the *Defensoría* invoked the authoritative power of scientific research, instructing the Ministry of Health to "promote clinical investigation" of DBCP ailments.[69] These instructions ensured that the science and politics of causality would remain at the center of the national discussion over what was to be done for *afectados*.

The first impact of the *Defensoría*'s report would be felt just a few days after its release. President Miguel Ángel Rodríguez, from the conservative *Partido Unidad Social Cristiana,* or PUSC, had taken office the previous May, and on October 20 his Governing Council issued an executive decree addressing *afectados'* plight. The decree officially recognized that DBCP had harmed the health of ex-banana workers and that the state bore responsibility toward them. "Given the magnitude of the problem and the potential harm done to Costa Rican citizens," it read, "the state must give them total support, offering all types of necessary assistance when it is needed, including taking measures to prevent the recurrence of such cases, all to protect the interests of workers and their families."[70] The Governing Council followed the *Defensoría*'s advice, creating an "Interinstitutional Commission" to study the situation of the injured workers.[71] The commission would include representatives from the Ministries of Labor, Exterior, Agriculture, Health, and Culture, as well as representatives from the national social security office (*Caja Costarricense de Seguro Social*) and the *Defensoría* itself. While the *Defensoría* was understood to participate as a representative of worker interests, CONATRAB would also have direct representation on the Commission.[72] CONATRAB's Orlando Barrantes acknowledged the import of the executive decree: "It constitutes a great advance because before we had no backing, and for the first time, the Government has committed to create a commission."[73]

STRENGTH IN THE STREETS

Actually achieving compensation from the state would require political as well as scientific action. As the Interinstitutional Commission began its work, CONATRAB used public protests to keep the DBCP issue in the public eye and apply pressure to state actors as necessary. Régulo Sánchez Barrantes, a longtime CONATRAB leader, explained the value of demonstrations like those the group would use in 1998–2000. "I would say that what has been the strength of CONATRAB," he recalled, is "the street."[74] On November 17, 1998, less than a month after the *Defensoría*'s report was released, CONATRAB organized a march down San José's Second Avenue.[75] Hundreds of banana workers marched through the capital, ending at the Legislative Assembly. Although *afectados* had carried signs protesting their "abandonment" by the state, as CONATRAB members filled the meeting space, representatives from across the political spectrum publicly affirmed their support of the *afectados*.[76] Their backing—expressed in general and sometimes nostalgic terms with references to the classic Costa Rican protest novel, Carlos Luis Fallas's *Mamita Yunai,* and the days (supposedly of the past) when "a man was worth less than a bunch" of bananas—seemed to indicate that opposing DBCP claims was politically a losing proposition.[77]

In practice, the political response was soon bogged down in bureaucratic procedure. At the end of November 1999, after meeting for over a year, the Interinstitutional Commission recommended that *afectados* receive specialized medical attention and pensions, and that a special fund be created to compensate them. Emphasizing the responsibility of the state, the Commission did not let transnational corporations off the hook; it recommended that the fund be created with resources obtained from the chemical and fruit companies that had produced and used DBCP.[78] Delays blunted the impact of this symbolic victory—it was March 2000 before another official group, called the Executive Unit, was formed to implement the Commission's recommendations.[79] By the end of July, compensation remained unpaid.

On July 29 2000, CONATRAB released a statement asserting that "the thousands of Nemagon-affected banana workers and their families have been deceived in the negotiations with the government."[80] In what the group called their "first warning," CONATRAB mounted a nationally coordinated protest that used protest tactics associated with *campesino* groups in the 1980s as well as with massive protests against the privatization of the national electric utility, known by the acronym ICE, earlier in 2000: the roadblock. Protestors, gathering in seven transit routes in the Caribbean and Pacific coast regions and on the street in front of the metropolitan cathedral in downtown San José, prevented the passage of traffic along key routes, including banana trucks on their way to the port of Limón.[81] The gathered protestors—their numbers reaching up to 300 in Siquirres—publicly demanded the promised compensation.

Reactions from the Ministry of Labor countered CONATRAB's calls for a "rapid response" by placing the blame for delays on the need to verify *afectado* claims.[82] According to *La Nación,* Bernardo Benavides, the vice minister of labor and a member of the Executive Unit, seemed "surprised" at the news of the protests, calling them "precipitate and unjustified."[83]

Benavides told the press that the Executive Unit had only recently begun to receive the first reports on *afectados*, and that the group still needed to confirm that all of them—in the words of the reporter, "in reality suffered some disability because of Nemagon."

CONATRAB continued to push for compensation. By December, the group had developed a plan, explicitly allying with organized *campesino* groups as well as with transportation and railroad workers who had received promises from the government but had not yet seen the fruits of their negotiations. On December 12, the allied protestors again blocked key points on the highway leading from San José, through the Caribbean lowland banana-growing region, to the port of Limón.[84] For CONATRAB, the heart of the protests was in Guápiles, where as many as 700 *afectados* blocked the highway.[85] There, fighting broke out between riot police and protestors, leaving more than 50 people injured. *La Nación* reported that four police officers were held "hostage" by protestors, who hoped to use them "as a key to open the door to dialogue with the Government and to demand the release of some 22 detained [protestors]."[86]

The conflictual protest tactics were successful in changing the terms of CONATRAB's and other protestors' negotiations with the state—at least in the short term. As Régulo Sanchez remembered the encounter,

> The police mistreated 18 compañeros, with [tear gas] bombs, and sent them to the hospital. But the Nemagon-affected workers hit 26 police with sticks and stones, and almost broke one policeman. It was an intense confrontation. And it was what led the government to say that CONATRAB is very strong, we have to do something to help them.[87]

By December 15, protestors had met with representatives of various government institutions, extracting promises to meet at least some of their demands.[88] In CONATRAB's case, the agreement established yet another commission to work on the DBCP problem.[89] The formation of another bureaucratic body could be read as a politics of appeasement and delay; after all, *afectados* had not achieved state commitment to compensation or any resolution of contested ideas about DBCP damage.[90] *Afectados* continued to strive for a response from the state, however, using whatever advantage had been gained from the December protests to initiate a return to key questions of causality and compensation. The next step would—at least temporarily—move the fight out of the streets and into the realm of law and science.

By April 2001, the postprotest negotiation between the government and CONATRAB had resulted in an agreement between INS and the *afectado* group. A "Memorandum of Understanding" signed by the two groups provided for compensation of 683,000 *colones* (US$1,500 at the 2001 exchange rate) for male CONATRAB members according to the familiar standard of fewer than 20 million sperm/milliliter.[91] At the same time, however, it launched a process that held the potential to open up the scientific and legal understanding of what constituted compensable DBCP harm. The memorandum set in motion two processes—one legal and one scientific—that would reshape the terrain of DBCP compensation in Costa Rica.

First, the memorandum called for the drafting and passage of a compensation law, with significant involvement from CONATRAB members. By May 2001 the bill, entitled "Determination of Social and Economic Benefits for the DBCP-Affected Population," had made it onto the agenda of the Legislative Assembly.[92] The legislation challenged and expanded established notions of who could be compensated and according to what criteria. In identifying the "population affected by DBCP," the law went beyond men with low sperm counts, effectively encoding a much broader definition of *"afectado."* First, the bill explicitly included women harmed by DBCP on the job. While the text usually referred to the phrase "banana workers" in the Spanish language's masculine generic, at least one reference to *trabajadoras* indicated that the *"afectados"* named in the law were both male and female. Also entitled to compensation under the law were children of *afectados*, as well as spouses and common-law spouses (and here the law specified *compañera de un trabajador o compañero de una trabajadora*, that is, "female partner of a male worker or male partner of a female worker," expanding the gender of the *afectado* while affirming only heterosexual pairings).[93] This language, explicitly egalitarian in terms of gender, not only acknowledged women's role as workers but also served to de-emphasize the equation between DBCP damage, sterility, and loss of masculinity. Including women as *afectadas* in their own right, not merely through heteronormative family ties, rhetorically weakened the narrow DBCP-equals-sterility equation. This not only allowed for the compensation of women, but also opened the way for the compensation of male *afectados* who didn't meet sterility criteria, and worked against the stigmatization of male *afectados* as emasculated men.

In addition to including women workers, the bill used broad language to define DBCP damage, affirming some *afectado* conceptions of harm, while also creating the terrain for future struggle. The bill left the definition of both moral and physical harm up to the INS's medical professionals, stating that "if the studies realized by [INS] demonstrate the existence of a physical damage or objective moral damage linked to or associated with the use of DBCP," INS would approve the payment.[94] Maximum compensation was set at 683,000 *colones*—the same amount as in the Memorandum of Understanding—with physical damage entitling claimants to up to 60% of that total, and moral harm to 40%. Compensation would be funded from INS's own coffers, in particular from the Workers' Compensation reserves. The amount, supposed by INS to reflect a banana worker's annual salary, was indexed to inflation.[95] By September 6, the bill had been approved by both the PUSC-majority Legislative Assembly and by President Rodríguez as Law 8130, giving most *afectados* three months to claim benefits under its auspices.[96] The lawmakers' approval officially formalized and expanded *afectados'* right to payment from INS, and it seemed that former banana workers—both men and women—and their heterosexual partners could now count on the state to compensate them for a broad range of health problems.

While the legal process was moving forward, the scientific process instigated by the Memorandum also progressed. The Memorandum had called for the formation of a Technical Commission of four medical expert members, two appointed by INS and two by CONATRAB, that would "determine the medical and legal circumstances to take into account to determine the indemnity paid for harms suffered." In other words, the commission would fill in the medical definitions left unspecified by the law. For *afectados*, a broad definition would fulfill their own sense of justice and secure compensation for a maximum number of people. For the state, an authoritative scientific definition, arrived at with the legitimacy conferred by the participation of CONATRAB-appointed scientists, would allow the state to contain or minimize conflict over future DBCP compensation.

Under the terms of the Memorandum, CONATRAB had the right to appoint two members to the four-member Technical Commission. The *afectados* reached out to scientists at the *Universidad Nacional* (UNA), in Heredia, the same group that had prepared the scientific report for the *Defensoría* in 1998. UNA researchers had established a Program on Pesticides in 1986, and by 1998 had established the Regional Institute for the Study of Toxic Substances (*Instituto Regional de Estudios en Sustancias Tóxicas*, or

IRET).[97] IRET personnel understood the purpose of their research activities to be to "understand and improve the environmental situation and the quality of life of workers and Central American society." By 2001, they had compiled a roster of research on the health effects of pesticides in Central America, and had not shied away from naming the dangerous failures of regulatory and corporate pesticide policy.[98] In other words, UNA researchers were not oriented toward meeting the goals of lawyers or industry, but rather placed worker and environmental health at the center of their scientific endeavors—and the *afectados* group asked Drs. Catharina Wesseling and Patricia Monge to serve on their behalf. For its part, the INS ended up appointing only one representative, Dr. Sonia Román, an occupational medicine physician at the Institute.[99] Over the course of a year, Wesseling, Monge, and Román reviewed the existing scientific literature on DBCP and developed recommendations for the compensation of women and of men who could not give sperm samples. The review of the literature covered much of the same ground as the *Defensoría* report had a few years before: acute effects of DBCP exposure caused a whole range of lesions and infections in experimental animals and humans alike. Studies on chronic effects of the chemical in humans and animals documented DBCP's impact on the male reproductive system (including interrelated changes in Sertoli cells, seminal vesicles, testicular volume, and sperm density) as well as "changes in the kidney and liver, depression of the central nervous system, an inversion in the masculinity index, increased incidence of miscarriage, low birth weight, genotoxicity, mutagenicity, ovarian and endometrial cysts, hormonal changes, ... nasal toxicity, atrophy of the spleen, and corneal opacities."[100] DBCP's carcinogenicity in animals had also, of course, earned it a classification as a "probable human carcinogen." Describing the evidence for this range of health problems, the Technical Commission report noted that the evidence was not conclusive and that not all studies agreed; the most definitive findings affirmed male reproductive problems and a change in the sex ratio of children born to affected adults (with a higher-than-expected number of females born), which also suggested the possibility that the chemical caused other genetic mutations.

Evidence of other harms was less definitive; in the words of the Commission, "in the industrialized countries with the capacity to carry out these studies, the exposed populations were very small ... [and] the few existing studies had very short follow-up periods,"[101] meaning that health effects that were relatively rare or took a long time to develop would be missed.

Studies on human females were especially scarce, despite cancer in mammary and reproductive tissues in laboratory animals, as well as suggestions that DBCP could cause a range of problems in fetuses and infants. And some chronic effects—such as kidney, liver, and skin problems—had not been studied in humans at all.

Given the suggestive but incomplete evidence on DBCP's health effects, the Technical Commission offered two alternatives for the compensation of women. In Alternative A, DBCP-exposed women should undergo diagnostic tests including a complete clinical history emphasizing work and reproductive history: a "detailed physical exam with special emphasis in vision, nose, reproductive, nervous, digestive and renal systems"; a full set of laboratory tests and medical imaging; psychiatric and neurological testing; and multidisciplinary clinical evaluations.[102] Such a set of medical care and tests, however, would obviously be extraordinarily expensive, as well as intrusive. As an alternative to requiring such tests to verify DBCP-exposed women's qualification for compensation from INS, compensation could be based on the amount of time they had been exposed to DBCP. Women exposed for less than one year would be entitled to 25% of 683,000 *colones*, with rates increasing with time of exposure; women with 5 years or more of exposure would be entitled to 100% of the compensation offered their sterile male counterparts. Whether and for how long a woman qualified as exposed would be determined by the Ministry of Labor.

As for men who could not provide a semen sample, the Technical Commission recommended a set of clinical and legal tests to substitute for the semen analysis. The commission considered "specialized medical tests" other than a semen analysis (presumably including the painful testicular biopsy) to be both costly and upsetting to workers.[103] Instead, the panel recommended that men unable to give a sample be compensated if exposure could be established and they had had fewer than one child since 1980. While this standard could be used to evaluate reproductive effects in individual workers, the commission expanded its recommendations to include a number of tests and studies that would not only reveal effects on individuals' health, but help to fill some of the gaps in the DBCP research. These included testing hormone levels, following groups of exposed workers to study the incidence of cancer over time, conducting psychological analysis for each worker, conducting liver- and kidney-function tests, and monitoring pregnancies in partners of DBCP-exposed men to measure the miscarriage rate.

The Technical Commission's decision did not affirm all aspects of *afectados'* understanding of DBCP damage. However, it did make scientific determinations from a position anchored in concern for workers' well-being and in line with broader Costa Rican affirmations of the responsibility of the state to provide social welfare. The Commission's recommendation for men who were unable to provide semen samples affirmed alternative proofs of harm rather than shutting out workers who were unable to provide evidence in the form privileged by courtroom science. In the case of women *afectadas*, the Technical Commission interpreted an unfinished scientific record in a manner that broadened, rather than foreclosed, the grounds for compensation. For both men and women, rather than try to radically minimize the number of people who qualified for compensation, as the competitive environment of the courtroom or the profit-oriented logic of a corporate settlement plan tended to do, the Technical Commission took as its goal the assurance of compensation for those that had been injured. If an individual had both exposure and symptoms, DBCP could be assumed to be the causal agent. As Dr. Monge explained, the decades of laboratory and epidemiological research on DBCP "sufficiently justified" the conclusion that verified exposure to DBCP was very likely to cause disease.[104]

NEW DELAYS, NEW OPPORTUNITIES?

Despite the achievements of the *afectados*, delays in compensation and continued controversy during 2002 and 2003 made it seem that promises of state compensation might be illusory, and transnational litigation once again emerged as a potentially viable strategy. In June 2002, almost a year after the passage of Law 8130 and two months after a closely contested presidential election had brought the conservative PUSC's Abel Pacheco a narrow victory, union leader Carlos Arguedas, Secretary of Occupational Health and Environment at the Union of Agricultural Plantation Workers (*Sindicato de Trabajadores de Plantaciones Agrícolas*, or SITRAP) and himself a former banana worker affected by DBCP, told President Pacheco that he was tired of waiting. Following the passage of Law 8130, SITRAP had been trying to help affected banana workers obtain compensation. "They have opened an Office in the Ministry of Labor," he recounted to Pacheco,

and since last year, thousands of workers have lined up at this office to once again present our claims, and now we are waiting for at least a call or a note telling us what is happening with our case. Meanwhile, we see that every year there are fewer of us in line ... not because [anyone] is giving up the case, but because each year many die from cancer or other illnesses caused by ... DBCP.[105]

Arguedas was not the only one who felt left out in the cold. Another group of workers—67 *afectados* affiliated with ASOTRAMA (*Asociación Pro Defensa de los Trabajadores Agrícolas y del Medio Ambiente,* an organization advocating for agricultural laborers and for the environment and one of the first organizations to reach out to the *Defensoría* in the mid-1990s) filed an appeal with the Costa Rican Supreme Court the following September.[106] Specifically, they asked the court's *Sala Constitucional*—the chamber charged with deciding constitutional and human-rights questions—to consider whether they were being treated unfairly. Charging that the *Unidad Ejecutora* or Executive Unit, the INS, and the Ministry of Labor had not addressed their claims for DBCP-related compensation, the 67 workers held that the INS was giving priority to claims filed by CONATRAB members. They even reported that an INS functionary had said she had "precise and clear orders" to attend only to claims filed by *afectados* associated with CONATRAB.[107] More than a simple critique of a slow-moving bureaucracy, their complaint revealed the fault lines that existed between CONATRAB and other groups.

INS denied claims of differential treatment. The Court's decision noted that INS had reported that 976 claims had been made under Law 8130.[108] Although none had been paid, some—including 38 of the claims filed by ASOTRAMA members—had apparently been approved. Over 600 claims were awaiting the scheduling of exams at INS, while another 200 or so were missing some information and awaiting further processing. The court's decision noted that an unspecified number of compensation payments *had* been made under the "sperm count project." Although that "project" was likely the compensation scheme for sterility spelled out in the Memorandum of Understanding, the decision did not explicitly state that, or seem to view it as evidence that CONATRAB members were receiving preferential treatment from INS.[109]

In the meantime, developments in both Costa Rica and the United States cast doubt on whether state compensation was really the best road forward for *afectados*. In mid-November, the newly appointed INS director Germán Serrano announced he had found an "accounting error" that transformed the surplus of 20 billion *colones* reported by his predecessor into a *deficit* of 1 billion *colones*.[110] After the dust settled, budgetary adjustments meant that only 502 million *colones* remained in the Workers' Compensation reserves from which DBCP compensation came, enough for only about 735 more people to be compensated at the 683,000-*colones* level. This was fewer than the number who had applied under Law 8130, and did not come close to being able to cover compensation for an affected population whose numbers some estimated to reach 13,500. CONATRAB mounted coordinated protests in seven towns in the Caribbean lowlands and northern and southern Pacific coasts, insisting that INS pay their compensation before more members of an aging population of *afectados* died.[111]

Meanwhile, in the United States, a landmark achievement by plaintiff lawyers signaled a renewed hope for successful litigation there—good news for Costa Ricans who still had active suits. In July 2002, while some *afectados* were speaking out against the slow payment of claims under Law 8130 in Costa Rica, a federal judge in Louisiana took the novel step of denying a *forum non conveniens* (FNC) motion brought by the usual roster of defendants in the case of *Canales Martínez, et al. v. Dow Chemical Company, et al.*[112] The settlements so critiqued by Costa Rican workers had multiple roots, but primary among them was the success of the defendants in using FNC to weaken the plaintiffs' position. Removing the threat of that doctrine would not solve all *afectados'* problems with plaintiff lawyers' translation and representation. However, a stronger legal position *could* lead to a bigger settlement or even a trial award, addressing *afectados'* concerns that earlier settlements had left them underpaid. The *Canales Martínez* case had originally been filed in Louisiana before that state had, following Texas, enacted a statute enabling FNC dismissals in 1999.[113] Like other DBCP cases, *Canales Martínez* had bounced between state and federal courts as the FNC question was disputed. By 2002, the case had found its way to the U.S. District Court for the Eastern District of Louisiana, where defendants predictably raised the FNC defense. In response, Judge Carl Barbier conducted a careful analysis of the availability and adequacy of courts in the plaintiffs' home countries: Costa Rica, Honduras, and the Philippines.[114] In the case of Costa Rica, the judge found that both the Civil Code and the Costa Rican courts' response to DBCP litigation sent "back" to that country after FNC dismissal in the United States[115] clearly indicated that Costa Rica was not an "available" forum. (While the judge found that Costa Rica would have been an "adequate" forum had it been available, he declined to dismiss cases to Honduras and the Philippines on the basis that they were neither available nor adequate.) Judge Barbier's decision reflected the success of Costa Rican jurists whose approach to the DBCP cases had emphasized their inadmissibility, as well as this particular Louisiana judge's respect for and careful analysis of the law of other nations.[116]

The import of the decision in Louisiana was complicated by another development; the Dead Sea Bromine strategy used by defendants in the *Delgado* group of cases had proven useful in other litigation as well, and defendants in Hawaii, Texas, and Louisiana had brought the Israeli company, a supposed "foreign sovereign," into the litigation, creating federal jurisdiction and removing cases from state courts. Plaintiff lawyers in *Patrickson v. Dole*, a case filed in Hawaii in 1997, contested removal, and by 2003 they had succeeded in taking their argument all the way to the U.S. Supreme Court.[117] There, justices made two important decisions. First, because Dead Sea Bromine was only indirectly owned by the state of Israel, it could not be considered an "instrumentality" or arm of that state, and as such its inclusion in a suit did not create grounds for removal to federal court. Second, the court ruled that questions about ownership and attendant sovereign privileges should be settled based on the facts of possession at the time a suit was brought, rather than when the acts being litigated occurred. (Israel's indirect ownership of Dead Sea Bromine had fallen below 50% by 1995).[118] This unappealable decision had the effect of remanding all cases back to the state courts where they had originally been filed. This included *Canales Martínez,* ironically exposing that case to dismissal under Louisiana's new FNC law. In most cases, however, the Supreme Court's *Patrickson* decision constituted a victory that would allow for the reactivation of this and other Costa Rican DBCP cases in the United States.[119]

MAKING THE PROMISE REAL

Meanwhile, *afectados* in Costa Rica continued to use protest and negotiation to keep DBCP in the public eye while confronting bureaucratic delays. On June 11, 2003, two hundred women occupied the INS office building in San José, demanding that they, like their male counterparts, be compensated for damage done by DBCP.[120] Protesters included grandmotherly María Gutiérrez, who had travelled 123 miles from Guanacaste to attend the protest. Fidelina Espinoza, who worked in the banana fields for almost 20 years, experienced allergies and problems breathing, while other *afectadas* described vision problems and pain they attributed to DBCP exposure. Now, they were demanding compensation.

The women's demands and the Technical Commission's report created a conundrum for INS. The costs of carrying out tests suggested in the Commission's report would far exceed the cost of compensation for each woman, an absurdity that seemed to make compensation of women both economically and politically impossible, especially in light of INS's budget crisis.[121] Yet Law 8130 specified that tests had to be carried out in order to compensate *afectados*. The impasse brought two institutions of the Costa Rican state into contact, as INS leadership sought input from the *Procuraduría*, or attorney general's office, which allowed it to both deflect some share of the responsibility and further delay compensation.[122] When the *Procuraduría* found that INS must conduct at least *some* testing,[123] INS director Germán Serrano hoped that office would support renunciation of the entire agreement with CONATRAB.[124] The agreement, he wrote in early September, was becoming a "source of conflict," and INS did not want to honor it.[125]

At least some of the "conflicts" Serrano referred to were the ongoing protests keeping *afectados'* efforts at compensation in the headlines. In July, CONATRAB had carried out another set of coordinated protests at various points in the nation, although police reportedly prevented a demonstration planned in Palmar Norte, Puntarenas. In September, just before Serrano made his bid to the *Procuraduría*, CONATRAB was taking part in a large roadblock protest in Limón, together with unionists and interest groups from other sectors.[126]

If the INS hoped to weaken CONATRAB, its strategy backfired. Attention to the issue led the *procurador* or attorney general to meet with Orlando Barrantes and, after considering the matter, he offered a nonbinding opinion that INS should honor its agreement with CONATRAB.[127] Faced with this opinion and ongoing protests, INS announced in early October that it would pay compensation—without medical tests—to the women protestors who had occupied its building in June.[128] The victory pertained to women covered under the Memorandum of Understanding only; the *Procuraduría's* first opinion meant that women who filed for compensation under Law 8130 could not be compensated without a test. But it was a significant achievement. Women had never before been compensated for DBCP-caused illness—in Costa Rica or perhaps anywhere.

Announcements were one thing, but cash in hand was another. In May of 2004, CONATRAB mounted protests at several points in the capital—in front of the Ministry of Finance, INS, the National Theater, and the offices of the *Defensoría*[129]—maintaining they were "tired of waiting" for payments under the Memorandum of Understanding. INS officials countered that funds would be paid gradually, and (with the exception for women) only after medical tests had established DBCP damage.[130] Two officials from the presidential Cabinet and Finance Ministry sounded a cynical note: "We don't want to seem unjust, but we know that many people pose as affected just to obtain compensation."[131] These questions about *afectado* authenticity gave new weight to the existing policy of testing each male *afectado* before compensation. Medical testing, of course, had always mediated the disjunctures between *afectados'* understanding of DBCP harm and the dominant medical understanding of that harm; the ministers' public accusations of fraud, however, threw into stark relief the gatekeeping function of those tests, reframing *afectado* protestors as potential cheats and liars. Such accusations worked to shift attention and blame from the fruit and chemical corporations' violations and the state's failures, instead placing the onus of corruption on the workers themselves.

On the second day of the May demonstrations, CONATRAB members mounted a hunger strike, attracting media attention and showing a level of commitment that powerfully countered accusations that *afectados* were only in it for the money. More than thirty hours into the strike, with four strikers already admitted to the hospital, the government agreed to pay 2,600 workers who had not yet been compensated. The next day, CONATRAB leaders met with the acting head of the *Defensoría de Habitantes*, José Manuel Echandí, who certified a list of *afectados* covered under the Memorandum of Understanding whose information would be sent along to INS to make testing appointments.[132] CONATRAB considered it a victory, with Orlando Barrantes calling it the "first time the government has recognized the list of *afectados* presented three years ago."[133] With the negotiations set to start, most of the protestors "collected their suitcases, cartons, and the plastic bags that had protected them from the cold, to return to their homes with the hope for a resolution."[134]

Though a victory for CONATRAB, this latest agreement with the government left out all of those *afectados* who were not included in the Memorandum of Understanding. As litigation proceeded in the United States and Law 8130 continued unenforced in Costa Rica, thousands of ex-banana workers awaited their money. COSIBA-CR, the confederation of banana worker unions in Costa Rica, sought to help all workers understand the status of both INS compensation and lawsuits. In June, the group called a meeting in Siquirres, which was attended by almost 1,500 DBCP-affected men and women. Those gathered unanimously authorized COSIBA-CR to ask for a meeting with lawyers and also to seek the support of international trade unions and solidarity organizations.[135]

Some believed, however, that the solution still lay in national law. When *afectado* Lino Carmona López asked in a letter to the editor why he still had not received compensation, a public relations person from INS maintained that the law did not allocate sufficient resources to fund the medical tests required.[136] The key to realizing payments, she wrote, lay in reforming Law 8130. In fact, such a reform was under way in the Social Issues Commission of the Legislative Assembly, where representatives including José Merino del Río—the single assembly member from the newly formed leftist party *Frente Amplio* (Broad Front)—worked with *afectados* themselves to craft a revision of the law.[137]

In 2005 and early 2006, CONATRAB and COSIBA-CR pushed for the passage of the bill.[138] In January 2006, unionists warned that representatives from the PUSC and the Libertarian Movement Party (*Partido Movimiento Libertario*, or PML) were trying to block the bill; but that threat was forestalled when the February 2006 elections narrowly returned former president Óscar Arias, of the PLN, to office, and removed the bulk of PUSC legislators in the wake of a corruption scandal. Whether due to banana worker pressure or electoral reconfiguration, by September 26 a legislative "debate" on the reform law saw only the enthusiastic support of legislators from not only the Frente Amplio, but also the PUSC, PLN, and other parties.[139]

The reforms to Law 8130 were a testament to the influence CONATRAB and the Technical Commission had exerted on thinking around testing and compensation. The amended Law 8130 included three categories of compensation—men who could give sperm samples, men who could not, and women.[140] For the first group, sperm counts would remain the *sine qua non* for compensation determination. For men who could not provide a sperm sample, the law stopped short of adopting the Technical Commission's criteria (documented history of exposure and no more than one child fathered after 1980), but provided a road to compensation and medical flexibility by allowing their compensation "in accordance with the results of [unspecified] physical, psychological and laboratory exams, and duration of exposure to the agrichemical." Women would be compensated according to the criteria established by the INS Technical Commission. Other changes in the law were also favorable to workers. They removed any deadline for workers to apply and extended the timeframe for family members to do so. For workers themselves, the compensation amount could potentially exceed past awards—683,000 *colones* would be converted from a maximum to a minimum amount, which would rise with inflation. Finally, the new law specified that the costs would be funded by INS's annual income.

Despite the fact that the reformed law represented a gain for *afectados*, its debate and passage received only cursory mention in the national press.[141] Perhaps not surprisingly, it did not result in rapid compensation for new groups of *afectados*. Unions and the left more generally—including CONATRAB and agricultural worker unions—were engrossed in the fight against Costa Rican ratification of the Central American Free Trade Agreement (CAFTA), a struggle that ended only in October 2007 when a national referendum approved the agreement by a razor-thin margin.[142]

While CAFTA opponents mobilized, *afectado* complaints of slow or non-existent payment continued. Carlos Arguedas, who was now serving as Coordinator of Occupational Health and Environment at the international banana union coordinating body COLSIBA in addition to his role at SITRAP, wrote in February 2007 that "the only thing that has changed is that all *afectados* can file a claim with [INS], but they are placing many obstacles in order to not pay the majority."[143] In February and March 2007—shortly after the reelection of former president and CAFTA supporter Óscar Arias—CONATRAB mounted demonstrations combining protest at the compensation delays with resistance to CAFTA ratification. More than two thousand people rallied in Guápiles, with another three hundred gathering in San Martín de Nicoya, Guanacaste.[144] *Afectados'* letters to the editor complained, "if one day we receive this money, it will be so devalued that it will not be enough to buy a newspaper,"[145] and asked, "Who will be on our side? Only death?"[146]

Over the following years, the compensation process lurched forward under the countervailing forces of *afectado* pressure and bureaucratic delay. In May 2008, the comptroller general approved a budget of 5 million *colones* to fund *afectado* compensation, according to INS enough to compensate 3,500 people.[147] The first 900 of those would reportedly be referred for testing and paid as early as June. INS told *La Nación* that 3,600 hundred cases had already been referred for testing and that the institute had already compensated 4,315 *afectados,* for a total of over 3 billion *colones* paid.[148] By the next year, however, compensations had once again stalled.

Afectados once again turned to the national court system for a remedy. In November 2009, assembly member José Merino filed in the *Sala*

Constitucional of the Supreme Court a complaint on behalf of 15 claimants against INS officials.[149] In addition to objecting to the stalled process, Merino charged that INS was failing to carry out adequate testing for "moral harm" and physical damage. For example, *afectado* Edgar Cabalceta Barrantes was denied compensation because his sperm count—at 24.5 million sperm/milliliter—was in the "normal" range, despite the fact that his sperm mobility and morphology were poor. INS's failure to take the latter aspects into account, argued Merino, violated the law that called for evaluation of sperm "volume, motility and morphology, among other aspects."

Finally, echoing *afectado* complaints over the previous decade, Merino's petition argued that it was unfair that *afectados* had been refused compensation by INS when they had fathered children after 1980, reasoning that DBCP testicular damage might show up after exposure had ceased—an interpretation not supported by research that had found sterility could actually reverse over time without exposure.[150] The complaint was given urgency by the specter of *afectado* mortality; Merino quoted *afectados* as saying, "in these months of waiting, four *compañeros* have already died without being able to take the tests mandated by law."

Notwithstanding Merino's plea that the *Sala Constitucional* resolve the matter quickly, it would be the following July before a vote was taken. The decision was mixed news for the *afectados*.[151] On the one hand, the bench took a strong stance that the INS must send many of the cases in question for testing "immediately" and complete the compensation process (determination and payment if appropriate) within two months. In other cases, the court required INS to provide information on its existing decisions and how to expedite stalled applications or appeal decisions. The position of the *Sala Constitutional* on psychological testing and the interpretation of semen samples, however, was more ambiguous. The court ordered the INS to immediately send 8 of the 15 cases named in the complaint for testing, but did not specify what kind. Psychological testing was mentioned only once, when the court directed INS to carry out the "physical or psychological examinations necessary to resolve" two of the cases at hand, "as set forth in Law 8130"—a directive that in fact left the institute a high degree of discretion. Despite these ambiguities, the court left no question that testing of some kind must be carried out, and quickly. The order to undertake testing immediately and complete the compensation process within two months was accompanied by a warning that failure to comply with the order could mean up to two years in prison for the INS official responsible for the program.[152]

The ruling had an almost immediate effect. When news of the court's decision was reported in September, INS reported that more than 10,000 *afectados* had been compensated—1,113 under the Memorandum of Understanding and another 8,903 under the Determination of Benefits Law.[153] Within just a few weeks, a total of 14,541 men and women had been compensated.[154] Thanks to the reform law that linked the compensation amount to inflation, men with sterility were receiving 1.6 million *colones*. The struggles of the *afectados* to include women and also men who could not give sperm samples were reflected in INS's report that those who based their claims on a history of exposure to the chemical—rather than medical tests— were receiving 25% of the amount awarded to sterile men. *La Nación* reported that the exposure-based benefits were paid "mostly" to women.[155]

The dual benefit levels both represented *afectados'* success in expanding the official definition of DBCP damage beyond sterility and marked the limits of their ability to compel the state to fully accept their own definitions of DBCP harm. Despite these limits, and despite persistent controversy regarding the pace of compensation, *afectados* had achieved a tangible victory for thousands of Costa Ricans.

CONCLUSION

CONATRAB's successes suggest that, faced with the intransigence of the fruit and chemical corporations in the U.S. legal sphere, and distanced from control of their own legal struggle by the mediation of lawyers, the *afectados* turned to making demands of the state in order to define their own movement, build political alternatives within Costa Rica, and achieve at least some level of material success. By engaging the Costa Rican state, *afectados* were able to define themselves and their struggle in their own terms— they contributed to legislation and chose scientific experts—a feat that had not been possible in the lawyer-dominated transnational legal contests. While compensation amounts were relatively low, the movement succeeded in securing at least some level of material gain for over 14,000 men and women.

Costa Rican *afectados* were, however, largely unsuccessful at projecting their struggle beyond national boundaries. Their initial turn to the state included a demand for assistance in the transnational legal process. Efforts at achieving compensation directly from the state, however, eclipsed this transnationally oriented demand, and holding corporations accountable dropped from the movement's practical agenda, if not from its rhetoric. The movement has not succeeded in penalizing the fruit and chemical companies, despite rumors that Costa Rica has contemplated a suit against the fruit and chemical companies to recoup its workers' compensation losses. The failure of the Costa Rican state to engage in sustained efforts to achieve justice from corporations like Dole, Dow, Chiquita, and Shell raises the question not only of how the successes of this workers' movement might be extended beyond the borders of the nation, but whether any state-oriented movement can project its claims into a transnational sphere. These are issues that the *afectado* movement in neighboring Nicaragua confronted head on.

SIX

National Law, Transnational Justice?

IN NOVEMBER 1999, THE MANAGUA daily *El Nuevo Diario* reported that hundreds of DBCP-affected agricultural workers were protesting in the capital: "Under the torrid sun of the capital, a hundred marchers carrying giant crosses, placards and banners, arrived with their faces sweating, their feet swollen, and crying in unison, 'We want Justice! End the exploitation of the transnational Standard Fruit Company!' "[1] Seven days before, the ex-banana workers and their family members had gathered in Chinandega, a town of about 120,000 in northwest Nicaragua, at the center of the hot and humid agricultural region where United Fruit and Standard Fruit had operated in the 1960s and 1970s, respectively.[2] They had come together to make the almost 90-mile journey to Managua—not by car or by bus, but on foot.

Like their Costa Rican counterparts, Nicaraguan *afectados* frustrated at the stalled progress and unsatisfactory outcomes of their legal cases had turned to protest aimed at their own government. But while the Costa Rican *afectado* movement placed the burden of recompense on the state, Nicaraguans firmly linked national politics to a transnational struggle. Mobilizing in a small, relatively powerless nation during a period of globalization in which state power was generally understood to be on the decline, Nicaraguans nevertheless hoped to use legislation and litigation at the national level to change the rules—and hopefully the outcome—of transnational litigation, and hold transnational corporations accountable for the disease they had caused.

The Nicaraguan strategy called the fruit and chemical corporations' bluff. The corporations had repeatedly argued that Nicaragua was an adequate forum for DBCP litigation—and now Nicaraguans welcomed them, but on local terms. The law at the center of the Nicaraguan *afectados'* movement set rules on causation that broadly defined DBCP disease, lowered the burden on plaintiffs, and provided for compensation similar to what U.S. plaintiffs' DBCP suits had brought in that country. Facilitating litigation in Nicaragua, the law simultaneously opened the door to successful litigation in the United States by emptying out the primary rationale for FNC dismissals—that is, defendants' desire to litigate in a forum with weak tort laws.

Nicaraguan Law 364, called the "Special Law for Banana Workers Victimized by the Use of DBCP-Based Pesticides," passed in 2000 and soon generated multimillion dollar verdicts in favor of *afectados* and opened the way for new lawsuits in the United States. In the wake of their successes, however, Nicaraguan *afectados* and their attorneys faced the full brunt of corporate responses. Using a timeworn strategy of U.S.-based corporations abroad, defendant corporations called upon the political power of the United States in an attempt to reverse the law. In U.S. courtrooms, they mobilized stereotypes of Latin American corruption. And, faced with mounting verdicts in Nicaragua, they turned to a newer strategy, invoking provisions of a trade agreement to thwart the Nicaraguan state's ability to carry out its own laws. *Afectados* mobilized nationally, using mass protests, anti-imperialist rhetoric, and visual representations of physical suffering in their struggle for transnational justice.

A NATIONAL MOVEMENT LOOKS OUTWARD

Rural organizing and resistance had a long history in Nicaragua, playing a key role in uprisings against the succession of strongmen that had controlled the nation for generations, including the movement led by Augusto César Sandino in the 1920s and 1930s.[3] Before the popular overthrow of the Somoza regime in 1978–79, these movements faced violent repression at the hands of governing elites. Rural workers—along with their urban counterparts—were also at the center of the Sandinista National Liberation Front (*Frente Sandinista de Liberación Nacional,* or FSLN), a military effort and revolutionary project of the late 1970s and 1980s. With the Sandinista revolution, Nicaraguans were able for the first time to participate in the political process in an open and democratic manner. Rural workers and producers took part in governing through their participation in the mass organizations that, especially in the first half of the 1980s, formed an integral part of the FSLN's strategy for participatory democracy. The mass organizations included the

Rural Workers Association (*Asociación de Trabajadores Campesinos*, or ATC) and the National Union of Farmers and Ranchers (*Unión Nacional de Agricultores y Ganaderos*, or UNAG).[4] The organizations were not perfect vehicles for the democratic expression of rural citizens' priorities, and enjoyed varying levels of autonomy from FSLN priorities; by the second half of the 1980s, their power and vibrancy had waned.[5] Despite this, the mass organizations gave many Nicaraguans an experience of active participation in democratic processes. Over the course of the 1980s, 500,000 Nicaraguans (or one-third of the adult population) were members of popular organizations; 125,000 of those belonged to UNAG, and 50,000 to ATC.[6] Even if the needs and desires of the rural workers were not fully understood or prioritized by the FSLN while it held power, the experience of participation in the governing process built skills and expectations that would carry over into the next decades. In the words of political scientist Richard Stahler-Sholk, "the Sandinista revolution made social subjects out of those who were once the mere objects of policy."[7]

In 1990, these social subjects exercised their power by voting the revolutionary Sandinista government out of power. The election of U.S.-backed presidential candidate Violeta Chamorro paradoxically signaled both the end of the revolution and one of its key achievements: the first-ever peaceful and democratic transfer of power in Nicaragua. However, the transition would not be an easy one for most Nicaraguans. The Sandinista government had already enacted austerity measures to control inflation in the face of a national crisis brought on by the Contra War and declining agricultural export fortunes. After taking office, Chamorro quickly enacted the neoliberal reforms prescribed by the United States, the IMF, and the World Bank. These policies increased the burden on the poorest Nicaraguans, sending under- and unemployment sky-high (to over 70%), severely limiting social services such as health and education, and creating uncertainty about the already-troubled land reform policies of the FSLN era.[8]

Added to these problems, many people in Nicaragua's northwest were noticing unusual health problems, including miscarriages, children born with birth defects, and failure to conceive. A potential reason for at least some of these disturbing health problems came from unionists and lawyers from Costa Rica, who shared news of how Nemagon had been found to cause sterility in men exposed to the chemical in the banana plantations of that nation.[9] The news resonated with many who had worked on or lived near banana plantations. Although bananas had never been as important to

Nicaragua's economy as the cotton or sugar crops, United Fruit and Standard Fruit had been present there during years when DBCP was in use. From 1960 to 1967, United worked with about 31 "independent producers," and Standard, invited to Nicaragua by Somoza in 1970, had employed 15,000 Nicaraguans by 1978.[10] Thus, in the 1990s, there were thousands of people with a history of working on farms or living in plantation housing during the years of DBCP use. Now many asked themselves if their exposure to DBCP during earlier decades was causing the health problems of the 1990s.

For those that did see a link between their own health and their historical exposure, litigation was not a medical cure, but did offer some kind of response to the disturbing news, as well as the possibility of compensation. In the first half of the 1990s, lawsuits in the United States still seemed quite promising, as the Texas Supreme Court decision in *Castro Alfaro* had (temporarily) defeated the threat of *forum non conveniens* (FNC) in Texas and therefore opened up the possibility of large verdicts in the United States. Nicaraguans were integrated into the existing strategy. By mid-decade, the rural workers' organization ATC, one of the most vibrant of the continuing mass organizations, was connecting *afectados* with lawyers in Nicaragua and the United States, and a number of Nicaraguans had been added to the *Delgado* litigation under way in Texas.[11]

But by the end of 1995, after Judge Lake's FNC dismissal of *Delgado*, the litigation in the United States seemed to have lost much of its potential. Although civil codes in most Latin American countries found FNC to be illegal, dismissals still left plaintiffs the burden of relitigating dismissed cases in their home nations expressly to achieve dismissal and reopen the U.S. forum. This strategy was risky, costly, and time consuming. So, lawyers and lawmakers dismayed with FNC dismissals in the United States came up with a new strategy: to enshrine in legislation the rules on forum already inherent in the civil code.[12] In the mid- and late 1990s, a number of Latin American legislatures considered or passed "blocking statutes," laws that clearly stated a nation's refusal to accept dismissed cases into its courts.[13] The intent of these laws was to defeat an FNC dismissal in the United States during the judge's "availability" analysis, by explicitly defining Latin American courts as "unavailable" to suits.

In Nicaragua, organized *afectados* and their allies staked out a novel approach to the transnational challenges of FNC dismissals. Although the news reports identified the November 1999 protestors only as "marchers," "banana workers," or "*campesinos*,"[14] in fact a new organization of *afectados* had taken shape. Led by Victorino Espinales—a former ATC leader and representative in the Sandinista government who himself had been sterilized by his work in the *bananeras*[15]—the new group was called ASOTRAEXDAN, an acronym that originally stood for *Asociación de Trabajadores y Ex Trabajadores Damnificados del Nemagón* (Association of Workers and Ex-Workers Hurt by Nemagon). Early on, however, a slight but important change in the name signaled the central purpose of this group—*damnificados* (victims) was changed to *demandantes* (plaintiffs). In the words of Espinales, the new group was an "association founded mainly . . . to push for the adoption of a specific law for those affected by the use of Nemagon, which will enable us now to sue multinationals."[16]

As *afectados* protested, allies including Daniel Ortega, president of Nicaragua from 1985 to 1990, were working to bring that law to a vote in the National Assembly, Nicaragua's unicameral legislative body.[17] They were working in a charged political context—Arnoldo Alemán, of the Constitutionalist Liberal Party (*Partido Liberal Constitucionalista,* or PLC), had been elected president in 1996 after a narrow victory over Ortega, continuing the neoliberal economic policies that gutted social services and reconcentrated wealth upward. The president's tenure was soon marked by conflict, incompetence, and what many saw as unprecedented corruption. Ortega was also under fire from critics, as he sought to build his own political power within an increasingly autocratic FSLN by negotiating a series of pacts with the president, boosting both politicians' power.[18] The DBCP bill was introduced not by Ortega, but by another FSLN *diputado,* or representative, along with two from the *Camino Cristiano* (Christian Path) Party, a center-right evangelical party with a tiny minority of three representatives in the National Assembly.[19] The legislation differed from FNC blocking statutes in that it expressly applied to DBCP cases only. It also went far beyond the simple refusal to accept FNC-dismissed cases already expressed in other national laws and championed by attorney advocates.

The bill set special conditions for DBCP-affected people to bring cases against the fruit and chemical transnationals in Nicaraguan courts. If passed, it would establish that if a plaintiff could prove both exposure and sterility, there was an "irrefutable presumption" of causation, sparing plaintiffs lengthy or invasive questioning about their personal or work lives. It provided for the inclusion of claims for any "physical, psychological, or pathological" issue caused by the chemical. Further, it designated minimum award amounts for oligospermia and azoospermia, specified an expedited procedure for trials, and guaranteed legal assistance to *afectados* unable to afford legal counsel.

The law also sought to impose obligations on defendant corporations that were strict enough to bring an end to any defense argument that Nicaragua was a more "convenient" forum than the United States. Making explicit its anti-FNC intent, the wording specified that it applied to "enterprises sued in the United States of America which have opted to have the lawsuits transferred to Nicaragua." Defendants would be obliged to post a bond of US$100,000 for each plaintiff in order to participate in the litigation, which would be used to cover the costs of the suits as well as to pay any awards, and post an additional bond if the case was not resolved within 90 days. And most definitively, it established that defendants who failed to pay the required bond "shall submit unconditionally to the jurisdiction of the Courts of the United States of North America … expressly waiving the exception of 'Forum Non Conveniens' raised before those Courts."

By making litigation easy for plaintiffs but difficult for defendants, the proposed "Emergency Law" suggested two paths—it could function as a blocking statute, defeating FNC dismissals of claims brought in the United States, or it could allow Nicaraguans affected by DBCP to effectively seek justice at home. Like compensation laws in Costa Rica, however, it would not be passed without pressure from *afectados* taking to the streets.

Also like its counterparts in Costa Rica, the emergent Nicaraguan *afectados* movement was centered around a shared sense of bodily illness and injury. In Espinales's words,

> Damages are many and enormous: there have already been 110 deaths from various causes, and many other comrades are only waiting for the end because the doctors have diagnosed that there is no cure. We're talking about tumors of the kidney, pancreas, spleen; early blindness in people 40 years of age who see almost nothing; bone fragility; disproportionate increase in body temperature; atrophy of the testicles; hematomas, rashes and deformations in the whole body; weight loss; loss of skin, hair and nails; nerve disorders, total and partial infertility, and damage to sperm that are resulting in the birth of deformed babies.[20]

However, the Nicaraguan movement did not primarily turn to science or scientists to bolster its claims of DBCP's effects on health. Instead, it created and disseminated vivid descriptions and images of suffering bodies that included injuries to men, women, and children, including a wide range of illnesses and the specter of a "slow death."[21]

Protests by *afectados* emphasizing bodily damage to workers built movement power by drawing the attention of the public and journalists. Sick and dying workers made for sensational news copy, and the press—especially the historically leftist daily *El Nuevo Diario*—reported extensively, and often supportively, on the evolving movement. For example, at a July 2000 protest in Chinandega, a group of female protestors defined themselves as victims of the "death spray."[22] Among their spokespeople was Melba Poveda, already recognized by *El Nuevo Diario* as "the most famous *afectada*," due to the fact that she had had more than 70 malignant tumors surgically removed. In September 2000, *afectados* pushing for the passage of the new legislation once again protested in Managua, reoccupying the grassy lot across from the National Assembly.[23] The press reported their claims that 82 people had died of DBCP-caused illness in recent months and that 22,000 workers and their family members suffered from some health problem caused by the pesticide. Hunger strikes at both June and September protests called attention to bodily vulnerability, rhetorically positioning lawmakers as complicit in protestors' illnesses if they remained unresponsive.

Despite protests, by early September the DBCP law still had not passed. Threatening a more unorthodox bodily protest, Espinales told the press that "if the government's soul still has not been touched and we have to wait longer, we are left no other option than to walk naked through the streets." ASOTRAEXDAN would march unclothed, he said, if the National Assembly did not pass the law within the week.[24] The threat of a nude protest gave political visibility to bodily damage that was physically *in*visible. Many of the problems attributed to DBCP—including male sterility, its best-known and least contested health effect—were not apparent to the naked eye. On the other hand, such exposure would also show *visible* bodily damage, such as discolored or scarred skin, or malformed limbs, that *afectados* also attributed to DBCP exposure. In Espinales's words, the naked march would expose "the damage that Nemagon caused to our bodies."[25] The presence of national and international media added some muscle to the threat—sensational images of nude *campesinos* would likely be broadcast worldwide.[26]

"Luckily," Espinales would later recall, "there was no need."[27] The threat of a naked march—along with the "takeover" of the National Assembly by a "complex multitude" of protestors—goaded the National Assembly into passing the DBCP legislation.[28] The National Assembly unanimously approved the bill on October 5, 2000.[29] Under pressure from business groups,

President Alemán delayed approving it for a few months, but added his signature in November.[30] Early the next year, the "Special Law to Process Lawsuits Filed by People Affected by the Use of Pesticides Manufactured with DBCP," more commonly known as "Special Law 364," was enacted. ASOTRAEXDAN had achieved a significant victory, using bodily display to build national political pressure and pass a law meant to solve a problem of transnational dimensions. Their first success would set the pattern for years to come.

THE PERILS AND PROMISES OF LITIGATION

The passage of Law 364 made Nicaragua a key site in the transnational struggle over DBCP justice and compensation. The stakes of the Nicaraguan litigation became crystal clear as the first cases were filed in Managua. In late February 2001, one hundred Nicaraguans filed claims for between US$300,000 and US$500,000 each for injuries caused by DBCP.[31] ASOTRAEXDAN, working with the Nicaraguan law firm known as OGESA (Ojeda, Gutiérrez, Espinoza y Asociados) and their U.S. associates from Engstrom, Lipscomb & Lack and Girardi & Keese planned to present one case of 100 plaintiffs per week for 37 weeks, until the almost 4,000 affected workers affiliated with the group had had their day in court.[32] By the end of May, they had filed 11 cases with demands totaling US$1.7 billion.[33]

These initial cases drew the attention of attorneys and former banana workers; soon, a proliferation of cases and law firms had raised not only the chance affected Nicaraguans would receive just compensation but also the potential for discord among *afectados*. The financial promise of the lawsuits contrasted sharply with conditions in Nicaragua, where widespread poverty combined with a shrinking state and corrupt administration.[34] In the face of brutal poverty, the potential of a payout undoubtedly sparked hope in individual plaintiffs and their families. The litigation also attracted new attorneys. The Texas-based firm of Provost Umphrey started researching cases and signing up clients in Nicaragua. Working with Managua-based attorney Jacinto Obregón, PU filed 1,200 cases by August 2001, and showed no signs of stopping.[35] Despite, or perhaps because of, the high stakes of litigation and the proliferation of legal options, by the end of 2001 the concerted popular action that had led to the passage of the bill gave way, in part at least, to factionalization. By October, OGESA lawyer Walter Gutiérrez publicly accused other (unnamed) attorneys of "unscrupulous" behavior, such as using false pretenses and small bribes to induce clients to sign contracts.[36] Soon it was apparent that the *afectados*, rather than remaining united under one umbrella, had "divided in two groups to sue the companies" they held responsible for their health problems.[37]

Defendants, on the other hand, were united in their opposition to the Nicaraguan litigation. When they had faced DBCP suits in U.S. courts, Dole, Dow, Shell, Chiquita (formerly United Fruit), Occidental Chemical, and Del Monte had argued strenuously that the cases should be heard in Nicaraguan courts. Now, faced with multimillion dollar verdicts, they changed their minds. Hoping to remove the threat of Law 364 altogether, they turned to a strategy infamously used by banana companies over the centuries: seeking the support and political muscle of the U.S. government. Dole, "spearhead[ing a] campaign to rescind" the legislation, soon won the support of U.S. embassy officials who took on the campaign as their own.[38] After Alemán's administration drew to an end, U.S. officials worked with the administration of President Enrique Bolaños, who ran on an anticorruption platform and had narrowly defeated Ortega, leaving the right-wing PLC in power. Law 364 was on the agenda when Assistant Secretary of State Otto Reich met with Bolaños's new Minster of Agriculture and Forestry in February 2002. Reich "raised the issue of the . . . law that targets U.S. companies in Nicaragua," suggesting that it was "damaging to foreign investment" in Nicaragua.[39] The message that Law 364 was harmful to the nation that had passed it was paternalistic and also embodied a subtle threat: while U.S. officials mentioned the potential of "worsening perceptions about the treatment of U.S. investors in Nicaragua,"[40] they themselves had much influence over these perceptions. For example, the U.S. Trade Representative's annual reports on trade barriers began reflecting company perspectives, noting that "U.S. firms have expressed concern that Nicaraguan legislation passed in 2001 may retroactively create additional legal liabilities for U.S. companies that used the chemical pesticide."[41]

Although Bolaños was, according to the U.S. Embassy, "initially prepared to facilitate an amendment to Law 364 with input from company lawyers," *afectados* had already built significant backing for the legislation among Nicaraguans, forcing embassy and corporate officials to look for less blatant ways to change the law. A March 2002 embassy cable to the secretary of state explained "[Nicaraguan] officials are extremely hesitant to take on the politically hazardous task of revoking the 2001 DBCP pesticide law." More simply put, to revoke the law would be "political suicide."[42] Although revocation of the law seemed impossible, U.S. diplomats believed it could be altered or replaced, with the cooperation of Nicaraguan officials and with a key role played by the very companies targeted by the law.[43] This approach resonated with Nicaraguan officials who shared—or at least saw the political benefits of affirming—the U.S. perspective on Law 364's effect on the investment climate. Both U.S. and Nicaraguan officials felt that the "the affected U.S. companies [should] develop substitute legislation to supersede the current law," and embassy staff agreed that "it is evident that the rewritten draft legislation will have to come from the U.S. companies in question."[44] Having decided on this approach, "the Department [of State and the] Embassy would continue pressing the [Nicaraguan government] on the issue." A March 18 cable from Washington suggested the pressure could come from the top: "The DBCP legal crisis could be among the topics of discussion during the Secretary [of State Colin Powell]'s April 5 meeting with [Nicaraguan] Foreign Minister Caldera."[45] A month later, another cable from Washington reiterated that "reminders of important bilateral issues, including the resolution of the legal quagmire of Nicaragua's pesticide law, will help keep [Nicaraguan] officials on task."[46]

As part of those efforts, Ambassador Oliver Garza lit upon a strategy that would set off a political firestorm around U.S. intervention and DBCP. In response to Garza's request, Bolaños had directed the acting attorney general, Francisco Fiallos, to examine the law."[47] By September, apparently at the urging of President Bolaños, Fiallos had sent his opinion to the head of the Nicaraguan Supreme Court, asking him to distribute it to all of Nicaragua's civil judges.[48] Although the attorney general's interpretation was not binding, he hoped it would serve as "further illustration for their respective sentences."[49] Fiallos's opinion must have pleased U.S. and Nicaraguan opponents of the law, as he found that "the law was unconstitutional in many respects."[50]

The fruit and chemical corporations seized on Fiallos's opinion to bolster their own cases. Litigation was moving forward on the first of the Nicaraguan cases, and defendants echoed the argument that Law 364 was unconstitutional, citing that as grounds for the cases against them to be dismissed.[51] Dole also cited Fiallos in the company's July 2002 quarterly report, apparently as the basis for the assertion that "the pending lawsuits are not expected to have a material adverse effect on Dole's financial condition or results of operations."[52] Behind the scenes, however, Dole and its codefendants were worried enough about the litigation to meet with Nicaraguan officials to discuss a possible settlement scheme that would circumvent the legislation.[53]

Although Fiallos's opinion went unnoticed by most Nicaraguans for months, in October 2002, *El Nuevo Diario* broke the story, tracing Fiallos's opinion back to high-level U.S. and Nicaraguan officials, and provoking anger at U.S. and corporate meddling in national affairs.[54] *Afectados* and other Nicaraguans reacted to the maneuvers of the corporations and the U.S. government by once again mobilizing at the national level. They understood and critiqued the affair in terms that drew on the long history of U.S. domination and interference in Nicaragua—from support for the Somoza regime to the funding of the violent Contra War and beyond. In postwar Nicaragua, despite the ongoing U.S. influence, for a politician to seem to capitulate to U.S. interests was politically fraught.[55] *Afectados* and their allies used anti-imperialist discourse to call on the state to represent the interest of its DBCP-affected citizens, rather than respond to the pressures exerted by the United States, which they saw as implicitly representing the interests of the transnational corporations.

The threat from the United States brought groups of *afectados* together in temporary alliance to put pressure on Bolaños and the National Assembly to support Law 364. On October 20, the factions joined together in "one single block" to press the Nicaraguan state to reject U.S. influence and Fiallos's opinion, as 4,000 people joined a protest and march in Chinandega.[56] On November 14, nearly 1,500 men, women, and children left Chinandega by foot, retracing the path to Managua.[57] Their targets included the U.S. embassy, where, in the words of Espinales, they would present "a document signed by all the plaintiffs, as a means of protest and a plea for justice."[58] But the real points of leverage for marchers were Nicaraguan decision-makers. After delivering another letter to the president's house, the mass protest would end at the National Assembly, where "we will stay as long as it takes to resolve this issue."[59] The park across the Assembly buildings, a flat corner lot green with trees, was soon transformed into a temporary encampment, with makeshift shelters made of wood and black plastic, shared cooking facilities, and hammocks strung to accommodate protestors who settled in for the long haul.

Mass mobilization and anti-imperialist rhetoric were a powerful combination. By the end of November 2002, President Bolaños declared null any initiative that called for the repeal or amendment of Law 364, and formed an Interinstitutional Commission on *afectados'* needs.[60] Sixty-three representatives signed a resolution supporting the DBCP-affected workers, and committing to protect Law 364. In addition, the Supreme Court issued an

FIGURE 3. *Afectados* along the roadside on a long-distance march from Chinandega to Managua, November 2002. Photo by Manuel Ángel Esquivel Urbina.

FIGURE 4. In November 2002, *afectados* protest in front of the U.S. Embassy in Managua, carrying a coffin labeled "Killed by Nemagon." Photo by Manuel Ángel Esquivel Urbina.

advisory opinion that Law 364 was not unconstitutional.[61] Despite differences, *afectados* had mobilized their combined power, forcing their government to stand up to the demands of the United States.

VICTORIES AND UNCERTAINTIES

In mid-December 2002, shortly after *afectados'* victory in defending Law 364, news of a gigantic verdict kept DBCP in the Nicaraguan headlines. In the first sentence declared under the new law, Judge Vida Benavente ordered Dow, Shell, and Dole and Standard Fruit entities to pay over US$489 million to 446 plaintiffs for physical and moral damage as well as punitive fines.[62] The verdict stated that exposure to the pesticide had contributed to "numberless diseases," including sterility in men and women, miscarriages, genetic malformations in children, visual problems, central nervous system problems, impotence and "sexual frigidity," prostate and uterine cancer, as well as liver, lung, kidney, skin, bone, and muscle problems. The largest awards, in the amount of US$4 million, were for women plaintiffs, and included moral and punitive damages.[63]

This case, like Law 364 and U.S. litigation before it, encompassed a debate over where cases should be heard. Law 364 said the defendants who had not posted a bond of US$100,000 were required to "submit unconditionally to the jurisdiction of the Courts of the United States of North America." However, plaintiffs in this case had chosen to pursue the litigation at home even when most defendants did not follow those rules.[64] Shell Oil, in contrast, had tried to transfer the proceedings to a U.S. court.[65] However, the judge was receptive to plaintiffs' arguments that the cases could be kept in Nicaragua. Keeping cases in Nicaragua seemed to resolve some of the problems of representation and translation that haunted litigation in the United States: the Nicaraguan law more closely reflected *afectado* conceptions of who was affected by DBCP and how. From the plaintiffs' perspective, that choice must have seemed a good one when the judgment was announced.[66] Plaintiff Francisco García responded, "It is moving to know that there are people who still see us as human."[67]

The glow of victory did not last long, however. The true significance of the Benavente verdict would hinge on whether it successfully led to compensation for the plaintiffs. Law 364 had, of course, been enacted in response to defendants' arguments that these cases were better tried in plaintiffs' home countries. But the corporations roundly refused the authority of the Nicaraguan court, declining to pay the gargantuan verdict. Although Shell gas stations were a common sight in the country, the defendants had no or few assets and business dealings in Nicaragua. The companies, therefore, had little to lose by simply flouting the authority of the court.

This unwillingness to respect the verdict prompted Nicaraguan plaintiffs' lawyers to look back to the United States. Although the U.S. State Department had taken a stance against the Nicaraguan lawsuits, the lawyers hoped that in the right jurisdiction, a judge in the United States would enforce the decision. In May 2003, lawyers from U.S. firms Engstrom, Lipscomb & Lack and Girardi & Keese filed a complaint in Los Angeles Superior Court asking for enforcement of the $489.4 million verdict.[68] Meanwhile, although corporate lawyers had previously argued that the Nicaraguan law and Nicaraguan courts were appropriate for the trial of the DBCP cases, they now began to argue that the cases could not be enforced in the United States because Nicaraguan law was inadequate. Specifically, they argued that Law 364 was deeply incompatible with U.S. legal culture. For example, Dow's spokesperson maintained that the law offended "virtually every notion Americans have of fair play and substantial justice"[69] because, among other reasons, it was retroactive, violated due process, and selectively targeted just a few companies. This move seemed especially cynical in light of the fact that the company had previously argued that U.S. legal norms *shouldn't* apply to Nicaraguan DBCP cases. Now, it sought to use those very norms to denigrate Nicaraguan laws.

On October 20, U.S. District Court Judge Nora Manella dismissed the enforcement case, based primarily on a series of errors made by lawyers.[70] These included misnaming corporate entities—for example, the Nicaraguan suit had named Shell *Oil* Company rather than Shell Chemical Company and "Dole Food *Corporation,* Inc." (a nonexistent entity) instead of Dole Food Company, Inc.[71]—and problems with notification and service of defendants in the United States. This outcome—bitterly disappointing for many—made the OGESA law firm an object of suspicion and criticism in Nicaragua.[72]

By the end of 2003, Dole had seized on the plaintiff attorneys' mistakes, amplifying the judge's findings by bringing corruption charges against the OGESA lawyers, a medical technologist whom they accused of falsifying sperm counts,[73] and almost 1,000 plaintiffs in suits filed under Law 364.[74]

Filing a case in Judge Manella's court under the U.S. Racketeer Influenced and Corrupt Organizations (RICO) Act, Dole's lawyers contended that the Nicaraguan defendants had falsified medical tests, falsified their translation of the Benavente verdict, and denied Dole the opportunity to defend itself in the Nicaraguan courts.[75] They hoped to classify the Nicaraguan lawsuits as fraudulent and to prevent the enforcement of Nicaraguan DBCP cases in United States courts. Although Judge Manella would eventually dismiss all the claims brought by Dole, as the counter-suit progressed, the idea that at least some of the DBCP plaintiffs were lying would shadow the *afectados'* ongoing struggle even as their movement reached its peak.[76]

Accusations of malfeasance percolated among *afectados* and their attorneys in Nicaragua as well. Shortly after Judge Benavente's December 2002 verdict, fighting between groups of lawyers and *afectados* had intensified. When DBCP plaintiffs represented by Provost Umphrey held a "general assembly" in Chinandega on February 9, 2003, detractors distributed "anonymous fliers inciting distrust" in that law firm.[77] The OGESA law firm publicly joined in the critique, accusing their counterparts of "confusing, disorienting, and deceiving" plaintiffs. Other charges of corruption were levied against Espinales, this time by *afectados* themselves, with a subset whose number would eventually grow to 500 alleging he had used their names without their permission.[78] Espinales, in turn, brought ASOTRAEXDAN to a dramatic rupture with its lawyers, including accusing OGESA of "excessive charges,"[79] and many of ASOTRAEXDAN's members turned to a newer set of lawyers on the scene, Los Angeles-based Juan José Domínguez and Chinandega's Antonio Hernández.[80]

The conflicts between law firms, between ASOTRAEXDAN and lawyers, and between ASOTRAEXDAN membership and leadership were evidence of competition over scarce resources in Nicaragua, a growing politicization of the DBCP cases there, and the corrosive impact of defendants' accusations of corruption. For *afectados* without direct access to lawyers or decision-making power, the mounting controversies indicated that the problems of representation that shaped the relationship between Central American clients and their lawyers in the U.S.-based litigation (as discussed in Chapter 4) were echoed in the relationship between *afectados* and their own leaders. As lawyers and leaders such as Espinales fought over what they hoped was a lucrative client base, most *afectados* remained with little power over choice of lawyers or strategy.

FIGURE 5. *Afectados* affiliated with ASOTRAEXDAN, including Julio Rivas (facing camera), gathered at the National Stadium in Managua in 2003 after aligning with the legal team
led by Juan Domínguez. Photo by Manuel Ángel Esquivel Urbina.

BUILDING POLITICAL PRESSURE

The years 2004 and 2005 were paradoxical, as *afectados* used marches and
bodily protests to fight off continued threats to Law 364 even as conflicts
between groups continued and negotiations with the government seemed to
spin off into a moribund cycle of protest and appeasement. In early 2004
about two thousand *afectados* affiliated with ASOTRAEXDAN began
another long walk from Chinandega, arriving in Managua on February 10.
Protesting Dole's ongoing RICO suit, they also sought government help with
other lawsuits and government-funded health-care coverage.[81] As before, they
visited the U.S. Embassy, delivering a letter to George Bush, before demanding that Bolaños and the National Assembly "condemn" the still-unresolved
RICO suit that, in Espinales's words, was "trampling national sovereignty."[82]
Repeating past strategy, they settled in to camp across from the National
Assembly until their demands were met.

Hovering in the background was the worry that Law 364 might be threatened once again. This time, concerns centered on the impact of the Central
American Free Trade Agreement. During negotiations, U.S. officials
from the Embassy and the State Department had "repeatedly expressed

to [government of Nicaragua] officials from President Bolaños down [U.S. government] concern [that Law 364] could complicate CAFTA talks."[83] Although Law 364 had survived initial Agreement negotiations, and by December 2003 CAFTA text had been mostly nailed down, workers and corporations alike still saw the agreement—if implemented—as a threat to Law 364. Like similar agreements, CAFTA was meant to open markets and facilitate international trade by reducing tariffs, duties, trade restrictions, and subsidies, and otherwise curtailing government power to regulate the economy. If the model of CAFTA's precursor, the North American Free Trade Agreement (NAFTA), was any indication, the emerging agreement also threatened to limit states' abilities to make labor, environmental, and health regulations, on the grounds that they constituted "non-tariff barriers to trade."[84] CAFTA also contained "investor-state dispute resolution" provisions that, like NAFTA's Chapter 11, allowed investors to sue governments for the enactment of laws or regulations deemed to "distort" free trade.[85] The U.S. Chamber of Commerce, a powerful corporate lobbying group with a board member from Dow, had already hoped CAFTA might be used to eliminate Law 364.[86] Espinales told the press, "We know perfectly well that . . . we were the subjects of negotiation" in talks over CAFTA.[87] Insult was added to injury when, soon after marchers arrived in Managua, Bolaños met with Florida governor Jeb Bush, visiting Nicaragua to promote the agreement. The foreign dignitary received a hearty welcome from pro-CAFTA Bolaños, who on the other hand refused to meet with afectados.[88]

Securing a meeting with the president would require building pressure, and once again protesters and the press together generated dramatic images of the suffering of afectados. Individual protesters would display parts of their bodies for reporters or photographers. Hands and fingers, legs, backs, or faces with discolorations, disfigurements, or scars were repeatedly photographed and reproduced by the press and supporters who hoped to show the bodily toll of DBCP on banana workers. Many of the symptoms shown would be difficult to scientifically link to DBCP exposure, but for workers and the public they represented the visible evidence of the largely invisible toll the chemical had taken. Sterility or cancer could not be demonstrated to the camera, but these other marks of occupational damage—many likely caused by other chemical exposures—were readily photographed. Afectados also presented more-symbolic demonstrations of their suffering. As the Easter season approached, protesters restaged the traditional Catholic procession representing Jesus carrying the cross on the way to crucifixion, with aging campesinos

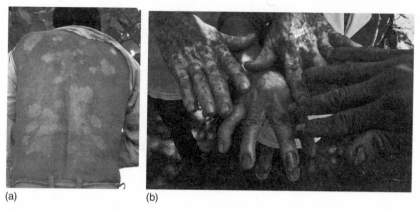

(a)　　　　　　　　　(b)

FIGURE 6. Nicaraguan *afectados* display discolored and stained back and hands for the photographer. (a) Photo by Giorgio Trucchi. (b) Photo by Manuel Ángel Esquivel Urbina.

carrying large wooden crosses over the course of the almost mile-long walk.[89] With displays such as these, sympathetic press coverage, and the ongoing protest presence, protestors won the support of the National Assembly, which unanimously agreed to sign a document protesting Dole's RICO suit; and of the Ministry of Health, which promised it would provide medical attention to DBCP-affected individuals, and latrines at the protest site.[90]

The 2004 protests, however, were marked by conflict between *afectado* groups. In February, a group of about 600 *afectados* arrived in the capital, calling themselves the "true plaintiffs," a clear attack on the authenticity of ASOTRAEXDAN.[91] Espinales lobbed his own counterattack, and when the new group also settled in to protest near the National Assembly buildings, the two contingents were soon separated by "a large contingent of riot police."[92] Other charges of *afectado* corruption came from observers and the press as doubts were cast on ASOTRAEXDAN's powerful bodily spectacle. The death of a protestor had prompted a public wake at the protest encampment, providing a moving sight that was reported by the press and generated sympathy among observers.[93] The wake, in the words of one observer, "impacted us all."[94] However, the empathy felt for protestors turned to doubt when *El Nuevo Diario* reported that the deceased had never worked on a banana farm and was merely accompanying an *afectado* friend. Such misrepresentation, some worried, "hurts a just struggle."[95]

In March, conflict and controversy were muted by legal and political achievements. On the March 4, the press reported the second *afectado* victory under Law 364, a US$82.9 million award to 81 women for injuries including cancer and other chronic illnesses (25 other plaintiffs had been excluded from the award for insufficient evidence of injuries).[96] It was another OGESA case, and the lawyers promised to bring the decision to the United States for enforcement, attesting that it did not contain the same errors as the first case. A few weeks later, President Bolaños finally met with the protestors. With an impending visit from Mexican and Central American heads of state adding to the pressures on the president, he invited both protesting *afectado* groups to the negotiating table.[97] Their meeting resulted in a series of measures that would come to be known as the El Raizón agreement.[98] The president promised legal and medical support, and pledged that no changes would be made to Law 364.[99] In exchange for the executive's commitments, the protestors agreed to go home, in buses provided by the government.

Perhaps unsurprisingly, the El Raizón agreement would prove disappointing for both the president and the *afectados*. Bolaños had hoped to forge unity between conflicting factions with an eye to opening the way for an extrajudicial settlement that would end conflict over DBCP.[100] Although the El Raizón agreement established an Interinstitutional Commission that became a forum where officials and *afectados* and their lawyers worked to "prepare the ground for . . . a possible agreement out of court,"[101] continued disagreements between groups made it impossible for "warring plaintiffs to agree on a common strategy."[102] For their part, *afectados* did not see much change in their circumstances despite government promise. Although the government had begun to provide some health-care services, its actions soon slowed.[103] By September, workers protested that they were going without medicine or health care, and continued to die from disease caused by exposure to DBCP.[104] In addition, the second, US$82.9 million verdict under Law 364 still had not been paid. In May 2004, Nicaragua, the United States, and four other nations signed CAFTA. While the treaty would not be ratified until April 2006, this step reminded *afectados* of the vulnerability of Law 364.

In response, *afectados* planned another protest—a long-distance march beginning in February 2005, followed by a protest encampment in Managua. Organizers—ASOTRAEXDAN's Espinales, in conjunction with colleagues from new *afectado* groups *Alianza Nacional* and *Asociación de Obreros Bananeros de Occidente Afectados por el Nemagón* (AOBON)—dubbed this "the march without end." This time, the march also included allied groups with overlapping populations and related aims, including sugarcane workers

FIGURE 7. In 2005, *afectados* mounted their largest camp yet in a lot near the National Assembly. Black tarps served as makeshift shelters for the more than eight months the protestors remained. Photo by AP Photo/Esteban Félix.

FIGURE 8. Protestors including Victorino Espinales (in hat with *R*) hold a banner protesting the ratification of the Central American Free Trade Agreement in 2005. Photo by Giorgio Trucchi.

who suffered from chronic kidney disease, which they traced to their exposure to pesticides used in cane cultivation. The inclusion of these groups swelled the number of long-distance marchers to five thousand. In addition, the broader tent meant a broader set of demands that incorporated the now-familiar claims for health care and government support of Law 364 but went much further, calling for measures including the strict enforcement of existing pesticide laws, environmental testing, reforestation, and fulfilling existing recommendations on integrated pest management as well as settling property disputes.[105] Protestors also allied themselves with the larger Nicaraguan anti-CAFTA movement, and banners militating against ratification appeared at the *afectados'* encampment.[106]

In 2005, public support of the *afectados* reached its peak. Acts of solidarity ranged from Nicaraguan poet Michele Najlis's support to a "Power Metal" concert organized by students hoping to support the Managua encampment.[107] In April, the Latin American Regional of the international Union of Food workers, known by its Spanish acronym Rel-UITA, launched an internet-based campaign in support of the Nicaraguan *afectados*.[108] The protests, including a dramatic demonstration in March when several *afectados* threatened to bury themselves alive, were covered in internet and print media based in multiple nations of the Americas and Europe.[109] Groups and individuals—including government officials—from Argentina, Uruguay, Brazil, Germany, Spain, Italy, Guatemala, the United States, and elsewhere wrote to the Nicaraguan government asking them to fulfill the protestors' demands.[110]

But though publicly the *afectado* movement was at its height, behind the scenes conflict was growing. For one thing, there was continued division between *afectado* groups, despite government efforts to bring the various factions together. As the combination of national and political pressure moved the government to negotiation, the OGESA group was excluded, and only ASOTRAEXDAN, AOBON, and *Alianza Nacional* members were represented in official talks. There was trouble within these groups as well. Espinales and attorneys Hernández and Domínguez fought over the payments lawyers made to Espinales to communicate with and identify new potential clients. When the issue could not be resolved, ASOTRAEXDAN and *Alianza Nacional* groups were left lawyerless.[111] The timing was awkward, coming as negotiations in Managua were bearing some fruit—including an August agreement with the National Assembly to leave Law 364 untouched and fulfill other promises reminiscent of earlier negotiations.

FIGURE 9. *Afectado* protestors threaten to bury themselves alive if their demands are not met. Photo by AP Photo/Esteban Félix.

More importantly, it intensified questions about leadership accountability to members.

By October, the impact and legacy of the "march without end" seemed uncertain. Over the course of a few days, most of the protestors would return home, although a smaller group would occupy the encampment for several more years. Ministry of Health officials claimed the protest had ended because the government had met all protestor demands.[112] Elsewhere, however, *afectado* leaders would say that the demobilization was part of an effort to find a "new negotiation mechanism" for moving multiple remaining demands forward.[113] The "march without end" had built awareness of *afectados'* plight in Nicaragua and beyond, simultaneously keeping Espinales and his colleagues at the center of national DBCP politics, but it had not fundamentally transformed the situation of Nicaraguan *afectados*, or quelled criticism of Espinales's tactics.

LIMITS OF STATE POWER

While the state of *afectado* organization was ambiguous, so was the progress of the Nicaraguan lawsuits—big verdicts in favor of the plaintiffs had accumulated, but so far the defendants had refused to pay. In April 2005, Dole cataloged for its investors the Nicaraguan decisions against DBCP defendants:

$489.4 million (nine cases with 468 claimants) on December 11, 2002; $82.9 million (one case with 58 claimants) on February 25, 2004; $15.7 million (one case with 20 claimants) on May 25, 2004; $4 million (one case with four claimants) on May 25, 2004; $56.5 million (one case with 72 claimants) on June 14, 2004; $64.8 million (one case with 86 claimants) on June 15, 2004 and $27.7 million (one case with approximately 36 claimants) on March 8, 2005. Active cases are currently pending in civil trial courts in Managua (4), Chinendega (7) and Puerto Cabezas (2).[114]

In August, another decision ordered Dow, Shell, Occidental, and the Standard/Dole entities to pay US$97 million to 150 plaintiffs.[115] Court orders notwithstanding, Dole's belief that "none of the Nicaraguan civil trial courts' judgments will be enforceable"[116] was evidently shared by other defendants, as all flouted the orders of Nicaraguan judges.

The challenge of enforcement brought the issue of location back to the center of DBCP litigation. Attorneys in the OGESA cases filed for enforcement of some cases in Ecuador, Colombia, and Venezuela,[117] where they undoubtedly hoped courts might be friendlier than those in the United States, and also have leverage over defendants who owned assets in their nations.[118] While enforcement cases in Ecuador and Colombia were unsuccessful or delayed, the United States remained a possibility for enforcement, especially after a 2005 appeals court ruling denying Dow and Shell's request for a judicial declaration that no Law 364 cases could be enforced in the United States.[119] Efforts at enforcement outside of Nicaragua left the future of the cases uncertain, while also reinforcing a sense—familiar to many Nicaraguans—that important national affairs would be decided elsewhere.

On January 13, 2006, Judge Benavente announced what seemed like a way to force the companies to respect Nicaraguan jurisprudence without looking beyond national borders. She would enforce her own verdicts by embargoing Dole and Shell's trademarks.[120] An expert would be called in to determine the value of the trademarks, which would be auctioned off if the company did not pay the US$550 million in judgments against it. Jubilant banana worker activists and their attorneys responded by demonstrating at Shell gasoline stations in Managua. Supporters expressed satisfaction that in Nicaragua even a "transnational with global prestige" was made to "respect the laws of this country."[121]

Shell and Dole, however, looked outside Nicaragua for a resolution to Benavente's action. Turning again to the U.S. government, Dole requested that Nicaragua be placed on the Trade Representative's "Special 301 Priority Watch List"—the short list of countries that might be faced with sanctions due to "inadequate" intellectual property laws.[122] The embassy recommended against this course, noting that "the practical import for Dole in particular seems rather low."[123] Shell used a newer—and what would turn out to be a more successful—tactic: invoking a bilateral investment treaty to bring the Nicaraguan government into arbitration.

Like many other treaties and trade agreements, the Nicaragua-Netherlands agreement allowed investors, including corporations, to bring national governments into "dispute resolution" processes. On May 17, Shell Brands International and Shell Nicaragua alleged that by seizing the Shell trademark, Nicaragua had violated the treaty, and filed a request for arbitration at the International Centre for Settlement of Investment Disputes (ICSID).[124] The ICSID, an organization established by a 1966 World Bank convention, serves as a forum for the resolution of disputes between investors and nations that have signed on to the ICSID convention; a number of investment agreements, bilateral treaties, and investment laws include a clause requiring consent to submit to ICSID arbitration in case of disputes.[125]

The call for ICSID arbitration forcefully underscored the limits of the Nicaraguan state's power to hold the chemical company accountable for the harms done by DBCP. Having already consented to ICSID arbitration by signing the investment treaty, Nicaragua was bound to respect the process or potentially endanger the treaty and its relationship with the Netherlands. This left Nicaragua in the ironic position of being brought to an international tribunal by a company that would not respect the verdicts of its national judicial system. The nation's situation was made desperate by the projected costs of ICSID arbitration. Hiring an arbiter to represent Nicaragua would cost about US$3,000 per day, with a total projected cost of twenty times that much, a significant figure for a nation with a GDP of US$5.3 billion in 2006.[126] (By contrast, Royal Dutch Shell had revenues of US$318.8 billion in 2006.[127])

The arbitration suit had the effect the corporations had hoped for. On June 13, 2006, exactly 5 months after Judge Benavente had instituted the embargo, her temporary judicial replacement granted a Shell motion to reverse the embargo.[128] Despite the reversal of the embargo decision, the threat of an ICSID suit would linger on for months in Nicaragua, as the government prepared to present a defense of the trademark embargo.[129] In March 2007, the Shell companies formally discontinued their suit.[130] Nevertheless, the situation did not bode well for the next big Nemagon verdict, a US$804 million suit brought by Provost Umphrey (PU) in collaboration with lawyer Jacinto Obregón Sánchez.[131]

DIRECT NEGOTIATION AND *AFECTADO* INTERESTS

As Nicaraguan officials at the highest levels worried over the ICSID suit, a game-altering shift was taking place among activists. While other plaintiff lawyers had negotiated with the defendants in the context of litigation,[132] in February 2006, Espinales announced that his group would bargain directly with Dole. It was a course change for ASOTRAEXDAN that signaled an end to its commitment to litigation in general and to Law 364 in particular. Espinales argued that the turn was based on a failure of legal representation: "The lawyers have been telling us a series of lies," he asserted, adding, "in the end [the lawsuits] have not served any purpose."[133] Dole, of course, had long preferred direct negotiation to litigation; by the end of the year the company would reach an agreement with the Honduran government to compensate *afectados* with sterility in an extrajudicial program.[134] In March, Dole's vice president, Michael Carter, traveled to Nicaragua to meet with Espinales, and in a press conference on March 21, the two men announced that they were in an "exploratory phase" of negotiations, which would be "accompanied" by the Minister of Health and representatives of the attorney general's office, also present at the conference.[135]

Espinales had legitimacy as a leader who had built the movement; on the other hand, he was increasingly the focus of criticism for alleged corruption and demagoguery. Could he legitimately broker a deal for all Nicaraguan *afectados*? These questions took on extra weight as it became clear that Law 364 might serve as a bargaining chip in an extrajudicial deal. Carter announced that direct compensation would hinge on the revocation of Law 364 (including retroactive annulment of decided cases).[136] ASOTRAEXDAN, of course, had organized much of the protest and political pressure that had previously made the idea of revoking Law 364 "political suicide" for Nicaraguan lawmakers. A reversal of that position signaled the possibility that the political calculus around the law could drastically change.

Not surprisingly, ASOTRAEXDAN's dealings with Dole provoked anger and criticism on the part of *afectados* not aligned with Espinales's group. To bargain away Law 364 would mean Nicaraguan legal cases would face the same FNC bind as they had before its passage, undermining any chances of compensation for *afectados* not included in a potential ASOTRAEXDAN-Dole deal. Other worker leaders said they would "repudiate any agreement with the government and the transnational"[137] and called Espinales a "traitor" for instituting talks with Carter, charging that he was "mocking all those *afectados* who marched barefoot, in the sun and in the cold nights, poorly fed [and] sick."[138] Espinales responded to accusations that he was an *entreguista* (sellout) with the counterargument that negotiation was the best way to put money in *afectados'* hands. Other groups and attorneys, he argued, "want the law, [but] we want our compensation, so we negotiate."[139]

In April and May, clients of PU/Obregón and Domínguez/Hernández rallied in Chinandega. The protests were organized by the law firms, showing that lawyers were doing what they could to keep political pressure on. But they also clearly demonstrated *afectados'* continued engagement and commitment to the legal process: 3,000 and 4,000 attendees were reported at the rallies, respectively.[140] The attorneys' efforts to protect their cases against the threat of an extrajudicial deal extended to the United States, where Provost Umphrey asked a Texas judge to order Dole to stop the extrajudicial negotiations with *afectados* who had already signed up with their law firm.[141]

The debate over representation and extrajudicial settlement was suffused with conflict over the health effects of DBCP. Tensions peaked as Dole personnel articulated a dramatically narrow stance on the issue of health problems caused by the chemical. In the press conference with ASOTRAEXDAN in March, Carter proclaimed, "There is no scientific evidence of the harmful consequences of Nemagon, except sterility in men who were exposed in production plants, which is why Dole has never lost a Nemagon lawsuit in the United States."[142] Dole's claim was, of course, the polar opposite of *afectados'* broad understandings of DBCP hazards; it also denied the substantial medical and scientific research on DBCP's effects on farmworkers.[143] Dole's blanket denial of DBCP-caused problems in farmworkers built on earlier attacks on Nicaraguan credibility, asserting that any nonsterility claims from former banana workers were nothing more than pure fabrication.

In mid-June, in response to PU's petition, a Texas judge issued a restraining order instructing Dole to stop negotiating with PU's banana worker clients, in accordance with U.S. law, which makes it illegal for a defendant to directly contact plaintiffs who are represented by a lawyer.[144] In addition, the order required Dole to abstain from publicly "disparaging" the lawsuits or Law 364 (or to stand by Espinales when he made such comments). Dole agreed to suspend the negotiations toward an extrajudicial settlement. The U.S. legal ruling quieted the controversy, once again showing how proceedings in the northern country could influence national politics in Nicaragua—this time to the benefit of *afectados* who remained in litigation and, of course, to the benefit of their attorneys. But Espinales noted the restraining order was temporary: "We maintain the position of continuing to talk with Dole, and we have already made positive advances regarding the central points."

Time proved Espinales correct. The restraining order did not bring an end to ASOTRAEXDAN's new settlement strategy. In May 2007, the group marched once again from Chinandega to Managua, but its demands no longer included support for Law 364. In Espinales's words, "this case can no longer be brought by lawyers and we want it declared a question of the state, for it to be on the national agenda."[145] The national agenda would depend in large part on newly-elected president Daniel Ortega, whose pacting with Alemán had resulted in electoral rules favoring his reelection.[146] With his Sandinista credentials, Ortega was arguably more responsive to workers' interests than his predecessors. However, Ortega and the FSLN had become increasingly autocratic over the years, spurring the formation of a left opposition party, the *Movimiento Renovador Sandinista* (MRS); Espinales had aligned himself with the MRS, even running unsuccessfully for a legislative seat under that party's banner.[147] And Espinales was already angered at the new administration's actions on DBCP, including the fact that, according to Espinales, money promised to ASOTRAEXDAN was instead going to other worker groups with close ties to the FSLN.[148]

Espinales, however, was working toward what he undoubtedly hoped would be a payoff much larger than the government could provide. He and leaders from the allied AOBON and *Alianza Nacional* groups were negotiating with Dole's Michael Carter to come up with an out-of-court settlement deal. The *afectado* leaders were expecting a percentage of any deal, about 10–15%, as recompense for their efforts.[149] On June 28, 2007, the negotiators sent a letter to President Ortega outlining their progress toward an agreement.[150] While compensation conditions and amounts remained unsettled, the letter—signed by Michael Carter and the seven Nicaraguan leaders of ASOTRAEXDAN, *Alianza,* and AOBON—showed that Espinales and his allies had abandoned litigation in favor of direct negotiation.

FIGURE 10. In May 2007, *afectados* march from Chinandega to Managua for the final time. Photo by AP Photo/Esteban Félix.

Behind the scenes, Dole also lobbied the Nicaraguan government. In an August 28 meeting with President Ortega, Carter and Dole's new president and CEO, David DeLorenzo, offered a US$15 million "settlement package" for a government-mediated settlement program.[151] Dole's proposal implicitly recognized the strength of farmworkers' sterility claims: men who were sterile or had decreased sperm counts would receive about US$6,000 and US$2,000, respectively—far less than U.S. workers with DBCP-induced sterility had won, but comparable to some awards under previous settlements and more than three times greater than awards under Costa Rica's state compensation plan. Other claims, which DeLorenzo called "unfounded," would be handled by a clinic Dole would fund in Chinandega. Although the Dole representatives found the discussion to be "interesting, good, friendly, and focused," Ortega demurred at the company's key requests; despite ASOTRAEXDAN's defection from the litigation camp, widespread support of Law 364 would still make its replacement "politically difficult."[152]

Ironically, as Espinales's camp veered away from litigation, ASOTRAEXDAN's former strategy was bearing remarkable fruit: a handful of Nicaraguan DBCP cases were set to go to trial in the United States. But rather than serving as cause for celebration, the planned trials deepened the divisions between banana worker groups. On July 11, 2007, Espinales was on hand as workers he represented lined up to sign documents revoking the powers they had granted to their lawyers. Espinales's remark to an Associated Press reporter—"It's certain they will lose the case because similar previous cases in these courts failed"[153]—seemed designed to turn *afectados* away from lawsuits and toward an agreement brokered by ASOTRAEXDAN, AOBON, and the *Alianza Nacional.*

FNC DEFEATED?

One of the cases scheduled for trial in U.S. District Court in Los Angeles during the summer of 2007 was *Tellez v. Dole;* others known as *Mejia v. Dole* and *Rivera v. Dole* would follow. This particular trio of cases had been filed in 2004 and 2005 by Domínguez and Hernández, along with California attorney Duane Miller, of the law firm Miller Axline, who had represented the Occidental production workers injured by DBCP in the 1980s; defendants were Amvac, Dole, Occidental, and Dow entities. *Tellez,* with 12 plaintiffs, was the first foreign DBCP case to ever go to trial in the United States, evidence of Law 364's power as a strategy to defeat FNC dismissal. In other words, even though Law 364 verdicts in Nicaraguan courts seemed unenforceable, the law ensured that defendants would not seek FNC dismissal for cases filed in the United States. Both plaintiffs and defendants realized that if *Tellez* was successful, others would follow.[154] The stakes were high: if a jury awarded Nicaraguans amounts similar to those the U.S. sterility plaintiffs had received in the past, the companies could eventually be looking at compensatory and even punitive damages in the hundreds of millions or more.

In July 2007, the *Tellez* trial began against Dow and Dole. (The case against Occidental had been dismissed, and Amvac had settled for a paltry $300,000).[155] The cases had landed on the docket of Judge Victoria Chaney, who had been appointed by Republican Governor George Deukmejian in 1990 and elevated to her position on the L.A. Superior Court by his successor and fellow Republican Pete Wilson.[156] Originally predicted to last two and a half months, the trial eventually extended over more than four. After closing arguments in October, the jury took nearly four weeks to reach a verdict. When they did, it was a measured victory for plaintiffs: they found that half the plaintiffs should receive compensation of $3.2 million between them, while an equal number received nothing (the jury found that these had not adequately proven that they had health problems caused by DBCP).[157] In addition, the jury ordered Dow and Dole to pay another $2.5 million in punitive damages for intentionally injuring the Nicaraguan plaintiffs.[158]

The significance of the outcome was disputed. Dole attorney Rick McKnight called it a "huge defeat" for the banana workers, because the nearly $6 million award wouldn't "even pay their costs, much less their bills."[159] But Dole's Michael Carter elsewhere admitted that the trial—the first he had ever attended—was disappointing, saying, "there was no way we should have lost this case."[160] On the other hand, attorney Duane Miller called it "a tremendous victory for the banana workers who were sterilized by DBCP."[161] And in Chinandega, hundreds of *afectados* "took to the streets to celebrate the news . . . with music, dancing, and gunpowder."[162] The victory in *Tellez* did not ease tensions among *afectado* groups in Nicaragua. In fact, these grew worse in the months following the verdict. By January 2008, Espinales had left the ongoing protest encampment amid charges of sexual assault and countercharges of violence.[163] Meanwhile, Dole and its codefendants, after more than twenty years of keeping foreign DBCP cases from going to trial, refused to capitulate.

THE "KILL STEP"

In early 2008, having already appealed the case, Dole and lawyers from defense firm Jones Day convinced Judge Chaney to overturn parts of the *Tellez* verdict. Among other actions, Chaney revoked punitive damages based on a logic that echoed the exclusionary intent of FNC by finding that such damages were appropriate only to the extent that they were in line with the "State's interest in protecting its own consumers and its own economy."[164] The judge held that California had no interest in punishing "a domestic corporation for injuries that occurred only in a foreign country." Like FNC decisions before it, the dismissal reinforced national boundaries to the benefit of transnational corporations by defining their activities abroad as outside the "interests" of U.S. law. It also let Dow and Dole off the hook for $2.5 million, a substantial part of the total verdict in *Tellez*.

Dole was not satisfied, however, and, dropping Jones Day, turned instead to the law firm of Gibson, Dunn & Crutcher (GDC), to pursue a strategy that used unusual—even extreme—tactics to bring an end to the litigation once and for all. They called it "the kill step."[165] This legal strategy was personal and brutal: discredit lawyers and make broad accusations of fraud to bring down cases.[166] Dole and GDC's "kill" targets included not only the litigation in process, but also the already-decided jury verdict in *Tellez*. For *Mejia* and *Rivera,* Dole and GDC's "kill step" built on Dole's history of charging plaintiffs with fraud, as well as the company's categorical denial that DBCP could cause disease in agricultural workers, by alleging that a large-scale fraud was taking place among DBCP plaintiffs, lawyers, and Nicaraguan judges. Dole had first made these allegations while trying to overturn *Tellez,* telling Chaney that a witness had come forward alleging at least two of the winning plaintiffs had never worked on a banana farm and had lied under oath.[167] Defense lawyers added that the witness had "received what was understood be a death threat" since working with them, and successfully petitioned Judge Chaney to grant an order keeping his identity secret.[168] Despite these protections, Witness X never testified about the fraud alleged in *Tellez*. Dole attorneys maintained he turned back from a scheduled appearance with the judge due to fears for his safety; three weeks later they revealed that he had refused to testify when Dole would not give him the $500,000 he demanded in compensation.[169] Without the witness's firsthand testimony, Chaney denied Dole's request for a new trial based on fraud.[170]

Over the course of 2008, defendants continued to make allegations of fraud in the *Mejia* and *Rivera* cases. Dole introduced more witnesses—their number eventually reached 27—whose testimony painted a dramatic picture of a conspiracy orchestrated by Hernández and Domínguez in which U.S. and Nicaraguan lawyers from opposing "camps" had worked together with Nicaraguan judges to present false evidence and obtain fraudulent verdicts under Law 364. Key testimony centered on a meeting that some witnesses reported had taken place in Monserrat, a wealthy Chinandega neighborhood:

those named as present included Domínguez and Hernández, a pair of PU lawyers, and Judge Socorro Toruño. The purported goal of the meeting was to "manufacture evidence of sterility and otherwise 'fix' those lawsuits in favor of plaintiffs."[171] Witnesses critiqued plaintiff lawyers' practice of hiring intermediaries called "captains" to locate and communicate with clients, alleging that captains recruited false plaintiffs and supplied them with artificial employment histories and other information needed to testify, including brochures and manuals explaining banana work, some of which were introduced into evidence. Some John Does also testified that plaintiff lawyers used violence to achieve their ends, and that they were afraid of those they accused as conspirators, including Juan Domínguez.[172]

Building on those allegations of violence, Dole's lawyers asked Judge Chaney to grant a protective order that not only gave the witnesses public anonymity but also placed strict limits on lawyers' actions in the case. The judge complied, and her protective order dramatically limited plaintiff lawyers' ability to cross-examine witnesses. Domínguez and Hernández, named as conspirators, were prevented from knowing the witnesses' identities or their testimony. Forced to proceed without the members of their team familiar with Nicaragua, Miller and his firm, Miller Axline, were also prohibited from conducting their own investigation specific to the charges raised by the witnesses.[173] Chaney's prohibitions included asking any plaintiffs "if they used any forged documents, [or] fak[ed] lab results."[174] She also prohibited attorneys from asking their own clients about specific individuals named by the John Doe witnesses.[175] The protective order left Miller Axline lawyers little room to seek help or gather evidence about the allegations made by Dole, making it nearly impossible for them to make a case in support of their clients.

Ironically, although Judge Chaney's order was meant to provide protection for witnesses in Nicaragua, it did little to dampen public discussion of the case there. Antonio Hernández still had some room for maneuvering in his home country. This time the boundaries between the two legal systems worked to Hernández's benefit, as he was not legally bound to obey a U.S. summons ordering him to submit to a deposition in the United States. *El Nuevo Diario* reported that Hernández had filed a slander suit in Nicaragua against an investigator, Francisco Valadez, hired by Dole through a company called IRI.[176] In March 2009, the *afectado* group ATBEN (*Asociación de Trabajadores Bananeros y Ex Bananeros de Nicaragua*) publicly accused Dole investigators of "bribe and blackmail,"[177] telling the press that the inspectors had offered bribes such as "50 million dollars, American visas for the *afectado* and their family, work and a home in the United States of America, and a protective order from an American judge" in exchange for their testimony that Hernández represented false plaintiffs.[178]

Domínguez was absent from hearings in Chaney's courtroom in April, as Chaney had stripped him of his rights and privileges as counsel due to her suspicions of his participation in fraud.[179] Miller Axline had tried to withdraw from the cases, but Chaney had refused.[180] Miller Axline lawyers barely spoke at the hearings on April 21 and 23, presenting no case in defense of their clients.[181] In an oral ruling made on April 23, 2009, Judge Chaney terminated *Mejia* and *Rivera* before they could go to trial on the grounds that plaintiffs had perpetrated a fraud on the court.[182]

The judge's ruling was remarkable for the disparaging terms in which it described Nicaragua and the people involved in DBCP-related activism and litigation. What she saw "down there" in Nicaragua was an "odd social ecosystem" with one mythical creature serving as a metaphor for the DBCP litigation: a "chimera that is really truly heinous and repulsive." U.S. and Nicaraguan attorneys were the "brains" of this beast, and "captains" who recruited clients were the eyes and the arms that "reached out and grabbed . . . men to make spurious claims."[183] Acknowledging her own lack of familiarity with Nicaraguan politics ("The Sandinista Revolution changed the system of government there. I'm not quite sure what it's been replaced with. I know there is a government there. I have no idea how well it's really functioning."), Chaney nevertheless painted a racist and xenophobic picture of Nicaragua as a lawless place whose childlike inhabitants were incapable of self-government and prone to corruption.[184] "There is a lack of respect for law, apparently, down there," she wrote, adding that she questioned "the authenticity and reliability of any documentary evidence presented by plaintiffs that comes out of Nicaragua."[185]

For Chaney, the problems she saw in Nicaragua rendered the DBCP cases irretrievably "tainted."[186] Discussing the possibility of allowing Miller Axline (which she considered unimplicated in any fraud) to continue the cases, she asked, "What are they going to do? Go back? Who are they going to talk to? They're going to talk to the same people that put up and claim [*sic*] that Joe worked with John on the banana plantation. But Miller Axline doesn't have the ability, any more than Dole or Dow or Amvac, to check the authenticity of that."[187] To back up her findings in this case, Chaney cited the U.S. State Department and Trade Representative's negative findings on Nicaragua's judicial system, which of course had been at least in part prompted by the DBCP defendants' own petitions to those agencies. In Chaney's view, Nicaragua presented such an impenetrable fog of fraud that the truth was simply not available there. The cases also posed a threat to U.S. jurisprudence, this time in the form of a metaphorical plant with "roots . . . in Nicaraguan courts but whose fraudulent shoots" had extended north. "There has been a strong attempt to bring the seeds of the Nicaraguan corruption here to this country," she warned.[188]

Chaney's dismissal of *Mejia* and *Rivera* was clearly a great victory for Dole. Not only had the company rid itself of these particular cases, but the judge's sweeping fraud ruling made it likely that no Nicaraguan DBCP case could ever be brought in the United States again. Miller Axline were finally allowed to withdraw from the litigation, leaving the *Mejia* plaintiffs without even an attorney to appeal the case. After years of popular struggle, Law 364 had opened the way for Nicaraguans to bring suit in the United States without facing FNC dismissal. Now, Dole and GDC's "kill step" had succeeded in doing away with Nicaraguan suits by disparaging Nicaragua and defining all cases originating there as suspect and unworthy.

Soon, another judicial decision would signal an equally bleak future for the other main achievement of Law 364: the multimillion dollar Nicaraguan decisions that plaintiffs hoped would eventually be enforced. Provost Umphrey lawyers had filed for U.S. enforcement of Judge Toruño's 2005 decision awarding US$97 million to 150 men for fertility-related claims, but in October 2009, U.S. District Court Judge Paul Huck issued an order denying enforcement based on jurisdictional, due process, and judicial impartiality grounds.[189] His decision enumerated concerns about Law 364, especially its irrefutable presumption of causation, required deposits and award, and narrow applicability.[190] Aspects of Huck's decision clearly reflected the success of the fruit and chemical corporations at setting the terms for a critique of Law 364; it discussed at length the controversial Fiallos memo instigated by the fruit and chemical companies with the support of the U.S. government, as well as official U.S. evaluations of the Nicaraguan judiciary that had also been responsive to corporate requests. Notably, Huck also repeated Dole's long-held but unfounded contention that DBCP had never been proven to cause even azoospermia in agricultural workers.[191] At the same time, the judge articulated other sources of authority for a finding that the Nicaraguan judiciary was not impartial and that Law 364 did not mesh with what he called an "international concept of due process."[192] Huck's order, finding that "defendants have established multiple, independent grounds" for nonrecognition of the claim, left plaintiffs little room for hope.[193] If the fate of *Mejia, Tellez,* and *Rivera* showed Law 364 strategy was unsuccessful in securing a trial in the United States, Huck signaled that the Nicaraguan verdicts under that law might also be rendered unenforceable and therefore meaningless.

VICTORY LOST?

After their victory in *Mejia* and *Rivera,* Dole's lawyers turned their attention back to the groundbreaking *Tellez* case. Not satisfied with Chaney's reversal of the punitive damages, they hoped to reverse the entire verdict via a seldom-used legal instrument called a "writ of error *coram vobis,*" which allowed for review of a decided case based on evidence that was not available during the original trial but may have changed its outcome. Dole's petition was based in large part on the John Doe testimony in *Mejia,* meaning it relied on anonymous and unchallenged allegations.[194] In July 2009, the court granted the *coram vobis* request, and issued an "Order to Show Cause" that the case should not be dismissed.[195] The order placed the burden on the Nicaraguan plaintiffs. As those plaintiffs were without a U.S. lawyer, it seemed more than likely that Chaney would vacate *Tellez* as Dole lawyers had requested.[196] However, the corporation's aggressive stance against its opponents had attracted attention to the case. Dole had filed a defamation suit against the director of a documentary on the history of DBCP and the *Tellez* case called *Bananas: The Movie.*[197] The suit attracted the attention of appellate lawyer Steve Condie, who stepped in to defend the *Tellez* plaintiffs.[198]

Over the course of almost a year, Dole continued to hammer home its fraud argument, while Condie, the alleged conspirators, *afectados* in Nicaragua, and others countered with their own evidence about what was happening in Nicaragua. As redacted testimony became available after the completion of the *Mejia* trial, several people named as taking part in the Monserrat meeting asserted the contrary. In Nicaragua, Judge Toruño denied ever being part of such a gathering; she was joined by other accused participants, including a PLN legislator, a laboratory owner, worker leaders, plaintiff lawyers, and the supposed host of the meeting.[199] When the two U.S. lawyers whom witnesses had named as participants in the Monserrat meeting were finally made aware of the accusations against them, they produced their passports showing they were never in Nicaragua at the same time during much of 2002 and most of 2003, when the conspiracy was supposedly being planned and the Monserrat meeting supposedly took place.[200] Others pointed out that evidence submitted to Nicaraguan courts did not follow the dictates supposedly agreed to at Monserrat.[201]

In both the United States and Nicaragua, some of the John Does themselves also countered Dole's story. Contrary to the defense's characterization of them as terrified witnesses testifying for altruistic reasons despite fear of plaintiff lawyers, they publicly admitted they had testified and been given money in exchange. On August 7, the *Los Angeles Times* reported that Nicaraguan Irving Castro Agüero had told them he was one of Dole's John Does[202] and that the company had paid him US$200 to testify. As reported in a correction to the story, Dole denied that it had paid him, but not that he was one of the John Does.[203] By May 2010, a legal action brought by José Francisco Palacios Ramos, one of those accused of conspiring at the Monserrat meeting,[204] had resulted in testimony from at least seven Nicaraguans—including Castro Agüero—that they had given payments of 4,000–6,500 *córdobas* (US$220–US$360 at 2007 exchange rates) in exchange for their testimony or written declarations (in some cases with unknown content) for Dole investigators.[205] The equivalent of one month's earnings, these amounts were much more modest—and therefore more credible—than the generous bribes some Nicaraguans had reportedly been offered by Dole. Most of the witnesses testified that the original offers for testimony had been lofty, but actual payments were much lower. One of the witnesses, Juan José Herrera Jarquín, said he had been "intimidated" not by plaintiffs' lawyers but by Dole's investigator Luis Madrigal and "some gringos."[206] In July 2010, *El Nuevo Diario* reported that Witness X was a now-deceased *Alianza Nacional* leader named Sergio García and that he had confessed to a journalist and *afectado* activist in November 2009: "I was Witness X. They tricked me, they brought me to the court, they didn't give me anything, and they sent me home without paying me anything. They used me."[207]

More information came from Jason Glaser, a U.S. filmmaker who, along with his business partner, worked as an undercover investigator for Provost Umphrey as he was making his own documentary on DBCP. Glaser told Condie that he had visited two of the John Does, whom he knew to be Jaime González and his cousin, in a Costa Rican hotel, where they had been moved at Dole's expense.[208] These two, known as Does 17 and 18, had been two of three witnesses who described the Monserrat meeting.[209] González, along with Sergio García, was a leader of *Alianza Nacional*, which had been negotiating with Dole since at least 2006. Along with Michael Carter as well as Espinales and other *Alianza* and AOBON members, they were signatories of the June 2007 letter to President Ortega.[210] Visited by Glaser and a colleague at his Costa Rican hotel, González reported that he had made a "commitment [with Dole] to an out of court settlement for compensation" and that Dole had set him up in Costa Rica because he had a "key piece" in their favor that could help them "g[e]t rid of" plaintiffs' lawyers.[211] Furthermore, González reported that he was in Costa Rica not because he feared violence from dangerous plaintiffs, but because there was a warrant for his arrest in Nicaragua.[212]

Although the statements by González and the other "outed" John Does eroded Dole's story of a thoroughgoing conspiracy maintained by violence, they did not explain away all the improprieties cataloged by Dole and its lawyers. How to understand, for example, the "training manuals" detailing processes and places that should have been known to legitimate workers? Condie did not now deny the existence of false *afectado* claims, but he did not link them not to a conspiracy of judges, lawyers, and plaintiffs. Instead he presented a picture of the captains gone wrong; although these intermediaries were paid a monthly salary by the law offices, Condie argued that "some of [them] had developed a sideline business—creating 'study guides' to sell to men they signed up who had never actually worked on banana plantations, and holding mandatory meetings at which admission was charged. . . . The greedy captains also sold tickets for bus rides to assemblies and rallies, and in various other ways fleeced their 'clients.'"[213] Condie acknowledged deep flaws in the system, but ascribed them to only a subset of Nicaraguans who were working against the interests of the lawyers who employed them.

The two competing explanations of what was happening in Nicaragua would be presented to the judge at the *coram vobis* hearings in May and July 2010. According to Dole, there was a violent conspiracy among plaintiffs. The plaintiffs had a different story, in which there was no wide-ranging conspiracy or organized violence and intimidation, but in which some Nicaraguans were indeed working the system, while Dole representatives were using a range of bribes to generate false testimony that was perfectly suited to Dole's final "kill step."

Judge Chaney's ruling on the *coram vobis* petition showed that her interpretation of the evidence had been strongly filtered through her existing opinion that a terrible and dangerous fraud permeated the Nicaraguan DBCP litigation. The evidence presented by Condie had clearly influenced her: she exonerated lawyers Mark Sparks and Benton Musselwhite based on evidence that they had not been at the Monserrat meeting, but despite the evidence disputing the defendants' allegations, she continued to regard the John Does as credible witnesses whose testimony provided legitimate evidence of fraud. However, the judge simultaneously disregarded all testimony given under Nicaraguan laws. Remaining convinced that Domínguez and Hernández had orchestrated a fraud, she found that "two of the prevailing plaintiffs were manufactured claimants, not having worked on Dole-contracted farms," and "all of the Tellez plaintiffs directly benefitted from this fraud."[214] Accordingly, on July 15, 2010, she vacated the jury verdict and dismissed the case with prejudice.

In Nicaragua, the news added to existing feelings of disappointment and disillusionment regarding the struggle of the *afectados*. In July 2010, in the run-up to Judge Chaney's order vacating *Tellez,* José Adán Silva, a journalist who had covered the *afectados'* story with sympathy for years, published a series of articles that seemed a sort of journalistic postmortem of the *afectado* struggle.[215] Poet Michele Najlis, formerly a very public supporter of the protestors, lamented what she saw as the movement's corruption by Espinales's conversion from "leadership to *caudillismo*," and the protestors' "deterioration" from dignified resistance to partisan support for the FSLN, including a stint of praying on street corners under promises of pay from the party.[216] Despite the fact that *afectados* still camped out in front of the National Assembly, it seemed that the Nicaraguan movement had lost much of its former authority and energy. In the context of this dampened enthusiasm, the news of Chaney's decision was met with anger and frustration directed at false claimants, John Does, Dole, Judge Chaney, *afectado* leaders, and above all, attorneys—but also with resignation that "as always the most affected are humble people with few resources . . . that are left with no hope, dying little by little," while "Dole achieved what it has sought for so many years, to remain unpunished."[217]

CONCLUSION

Following the grand defeat in *Tellez* there was a series of limited victories, partial solutions, and tenuous hopes. In July 2010, the Ortega administration announced that, as part of its efforts to provide housing for Nicaragua's most impoverished, it would build a swath of *casitas* for the protestors who had been living across from the National Assembly for so long.[218] The next year, 72 small houses—reported to be underwritten by Venezuelan funds—were transferred to protestors as Ortega was seeking reelection despite constitutional term limits.[219] The housing grant won praise from its beneficiaries, who cited the "care and solidarity of *comandante* Ortega."[220] The limited number of houses could be given to only a very few of the *afectados*, rendering the project something more complicated than a clear victory after years of protest.

The same year the *casitas* were built, a bigger but perhaps just as ambivalent payout came in the form of a settlement negotiated by Provost Umphrey. The agreement ended 38 cases against Dole, most of them filed in Nicaraguan courts, in exchange for payments to about 3,000 Nicaraguans, 1,000 Hondurans, and 700 Costa Ricans.[221] While exact amounts remain confidential, it is easy to see the settlement as a victory for Dole—the settled Nicaraguan suits alone had claimed US$9 billion in damages, but the company reported that the "settlement will not have a material effect on Dole's financial condition, results of operations or cash flows."[222] However, it did allow cases to continue against the other defendants, so it both put cash in *afectados'* hands and allowed them to pursue lawsuits and hope for more.

The hope that Law 364 will deliver on its promise seems ever slimmer. In March 2014 a California appeals court upheld Chaney's dismissal of *Tellez*, holding that Condie's arguments did not merit a reversal.[223] In May, his petition for review by the state Supreme Court also failed, leaving the U.S. Supreme Court as the only option for further appeal.[224] While a few Law 364 cases continue to wend their way through Nicaraguan courts, verdicts have slowed and filing of new cases has ground to a halt.[225] Provost Umphrey, waiting for its cases to clear appeals in Nicaragua, plans to try to enforce them—against Dow, Shell, and Occidental only—in a nation where the defendants have assets and the court might be less hostile to Nicaraguan jurisprudence than the United States proved to be.[226] Regardless of the future of the litigation, what seems clear is that the *afectado* movement has lost its former power and appeal to Nicaraguans.

Despite their defeat, the Nicaraguan *afectados'* history provides some important lessons on how people's movements can challenge corporate power. Working with lawyers and lawmakers, *afectados* passed a law that accomplished what the litigation strategy so far had failed to do—bring fruit and chemical corporations to trial for the harms caused by DBCP. In Nicaragua, their law defined DBCP damage according to *afectados'* own illness experiences and sense of justice; this in itself was an accomplishment of direct democracy achieved through sustained and visually powerful protests and engagement. The law had repercussions in the United States as well, forcing open courts previously kept out of reach by the doctrine of FNC. Before 2007, no DBCP case had ever come to trial in the United States, much less resulted in substantial verdicts in favor of Nicaraguans. Law 364, in other words, temporarily leveraged democratic power at the level of the Nicaraguan state to change the opportunities for litigation elsewhere. The early history of DBCP shows how U.S. laws and regulations profoundly shaped conditions for use of the chemical in Latin America; Nicaraguans showed that the impact of national laws could be felt in the opposite direction as well.

Why did the *afectado* movement fail to win compensation and hold corporations accountable? It is easy to conjecture that strategic errors led to the failure of Law 364's promise: if the law had been more measured in its definition of DBCP harm or presumptions of causality or requirements of defendants; if lawyers had not made careless errors; if conflict and misrepresentation had not emerged among *afectados* themselves, perhaps the Nicaraguan cases would have been upheld in U.S. courts, or the *Tellez* and *Mejia* cases would have survived. But there is a deeper answer. At the heart of the *afectados'* movement were efforts to redress some of the historic economic and geopolitical inequalities that characterize the banana industry and, indeed, the relationship between Nicaragua on one hand, and the United States and transnational corporations on the other. Law 364 was an assertion of Nicaraguan agency in a transnational sphere where the nation had too often been the object of corporations' and other nations' actions. The litigation itself would, plaintiffs hoped, return some of the wealth banana corporations had extracted from Nicaragua—indeed from workers' own labor. Faced with *afectado* efforts to right historical inequalities, Dole and the other corporations countered those efforts with all the weapons in their well-stocked arsenal. Corporations' vast monetary resources, the influence of the U.S. government, a transnational supragovernmental regime heavily weighed against national policies that might threaten capital, and ruthless legal tactics were used to discipline Nicaraguans seeking justice. Another advantage corporate defendants had over plaintiffs was the practical immortality of the corporate form. Corporations—although legally considered people—do not have the same bodily vulnerability as human people. As a revolving cast of human personnel serve the interests of the corporate entity, *afectados* and their families will continue to sicken, age, and die, ultimately removing any threat to the corporations.

While litigation holds the potential to help reverse such inequalities by redistributing resources and discouraging exploitive behavior, it also remains an unequal playing field marked by the problems of representation, translation, and location. In Nicaragua, *afectados'* actions were engaging and visceral for protestors and observers alike. In contrast, *afectados* had minimal and highly mediated influence on what happened in far-off courtrooms or supranational arbitration venues. At times, like *afectados* from other nations, they felt little control or understanding of even their local lawyers' actions. After Law 364 was passed and protected, and national lawsuits proved unenforceable, the measure of the Nicaraguan *afectado* movement came to hinge on distant all-or-nothing outcomes (would they win in the United States? would corporations pay?), leaving little hope when lawsuits and enforcement failed. The successes of the Nicaraguan movement were strongest where people could use protest and direct action to compel state action. However, the Nicaraguan government ultimately did not have the power to enforce its own courts' orders or assure that Nicaraguans would receive a fair hearing in the United States. Without the benefit of deep pockets, geopolitical clout, or near-immortality, Nicaraguans faced an uphill struggle against their corporate opponents. The same deep and broad inequalities that shaped the contours of DBCP exposure also limited *afectados'* success in holding corporations accountable.

Despite *afectados'* significant losses, the intensity of corporate efforts against the Nicaraguan *afectado* movement is itself a testament to its power. *Afectados'* successes and failures alike suggest that national movements, domestic policy, international relations, and transnational economic governance are important and intersecting sites of struggle for democracy and health justice. In a world with too few challenges to corporate power, even the fleeting achievements of the Nicaraguan *afectado* movement are remarkable and instructive.

Conclusion

THE STORY OF DBCP SHOWS how corporate and state actions produced inequalities in chemical exposures, and how banana workers sought accountability for their health problems, garnering some victories but also facing significant barriers. Central to the creation of and response to DBCP harms were fights over science and law as workers, corporations, state actors, and attorneys sought to define or contest acceptable risks, DBCP damage, and the terms of justice for harms done. This story is important to understanding the spatial organization of systems of production and trade, as well as their intersection with the political boundaries and regulatory scope of national states. And perhaps more important, it is central to understanding how workers and others might contest inequalities in a transnational political economy that is stacked against them. This history holds important lessons for scholars and activists interested in understanding the uneven distribution of occupational and environmental risk in the context of globalization, as well as the potential for national and transnational strategies to hold corporations accountable and prevent future harms.

The DBCP problem first took root in Hawaii at midcentury, when scientists developed the chemical to control nematodes, the soil-dwelling pests already known to bite into pineapple roots—and profits. Chemical companies hoped to transform the new compound into marketable products, but to do so they would need to both create a market and secure regulatory approval. Nematodes were near-microscopic roundworms unfamiliar to many farmers, and the corporations used the language of science to teach potential customers about both the pest and its chemical antidote. They pitched Nemagon and Fumazone (their branded DBCP products) using "experimental" test plots, scientific workshops, and republication of research findings as advertising.

Marketing materials also promised financial abundance, painting a rosy picture of profits through chemistry.

Scientists conducting toxicological trials were faced with a very different picture, however, as they discovered that DBCP caused testicular damage and other problems in experimental animals. Corporate personnel knew that these problems signaled similar effects in humans, endangering the companies' chances of securing government approval for DBCP under regulations still in development. Contrary to the basic tenets of toxicology and the industry's own avowals of safety, corporate scientists and marketers argued that DBCP was safe for humans. In 1964, regulators at the United States Food and Drug Administration (FDA) and the United States Department of Agriculture (USDA) accepted the corporations' interpretations, officially registering the chemical and clearing the way for DBCP application on scores of crops. Dow and Shell's scientific sleight-of-hand, combined with regulators' quiescence, endangered any worker with DBCP exposure—within or outside of the United States, in farm or factory.

While DBCP was used for nematode control in the United States even before 1964, it would not be widely used in Central American banana plantations until the 1970s, when people, plants, and processes at the local, national, and transnational levels combined to create conditions for its adoption. Nematodes themselves were international travelers, moving between banana-growing nations on plant materials sent between various and sometimes far-flung plantations. Arriving in a hospitable locale, the nematode would feed on the underground portions of the banana plant, leading to rot, structural weakening, and toppling of the plants. The problem became worse after midcentury, when banana corporations changed land use and production practices in response to political, environmental, and labor conditions in Central America, including slipping influence over local governments, increased worker militancy, and intractable plant diseases. Banana industry retooling included increased use of "independent producers" who grew bananas according to company specification but assumed all the risks of plant disease or labor unrest. Another change was the industry-wide adoption of a new banana cultivar, well suited for transport and U.S. consumers' tastes, but susceptible to nematode damage. In addition, growers' techniques for maximizing yields of marketable fruit made banana plants more top-heavy, and consequently more vulnerable when tiny pests ate into the anchoring root system.

Corporate and regulatory actions in the United States also helped open the way for DBCP use on Central American banana plantations. The major piece of U.S. pesticide control legislation, the Federal Insecticide, Fungicide, and Rodenticide Act (FIFRA), explicitly exempted exported pesticides from its requirements, so U.S. regulators' sphere of authority nominally ended at national borders. Nevertheless, their decisions played a large role in allowing DBCP use on bananas. Although the crop was not grown commercially in the United States, it was consumed in large quantities there, and regulators had authority over how much chemical residue could remain on the fruit. The residue limit, meant to protect U.S. consumers, also served as an implicit authorization for use outside of U.S. borders by allowing banana companies to use DBCP without fear that residue-bearing bananas would be stopped at the border.

Once regulators green-lighted DBCP residues on bananas, chemical corporations marketed to the fruit growers, even developing special formulations and application methods suited to the various places bananas were grown. The banana companies experimented with a number of nematode control measures, but by 1972 Standard Fruit was using DBCP in plantations in Nicaragua, Costa Rica, and Honduras; newcomer Del Monte was buying it by 1973; and United Fruit, despite reservations, expanded DBCP use between 1972 and 1977. In plantations with irrigation infrastructure, DBCP formulations were periodically applied with the water shot out by overhead towers; where regular rainfall made irrigation unnecessary, it was injected into the soil by crews using special handheld applicators.

Workers charged with applying DBCP received little or no scientific information on its toxicity, and little protection from exposure. They did, however, develop bodily knowledge of the chemical, gaining familiarity with its smell, look, and feel by working directly with it. Although U.S. regulations had made DBCP use practical and profitable for banana growers, the safety of workers applying it to that crop remained outside of U.S. purview. And although Central American banana-growing nations instituted national regulations on pesticides and occupational health at different points, those regulations often went unenforced. Workers' recollection of their contact with DBCP reflected their experiential understandings of DBCP's acute toxicity—it stank and stung and made them dizzy, could be used to kill fish, and occasionally was chosen as a means of suicide. These signs of the chemical's dangers remained in tension with workers' limited control over working conditions—opportunities for resisting chemical use were few and inadequate. For their part, managers at both Standard and United saw DBCP application—particularly the injector-based variety—as an important site for labor discipline and cost control.

The calculus of DBCP use changed after October 1977, when a group of California production workers very publicly linked their sterility to DBCP exposure. Their experience, covered widely by the media, sparked a complicated and sometimes cumbersome U.S. regulatory and corporate response that controlled some exposures to DBCP but not others, creating or exacerbating asymmetries in risk. In the United States, regulation of factory workers' exposures by the Occupational Safety and Health Administration (OSHA) was relatively quick and protective, while the Environmental Protection Agency (EPA), charged with protecting farmworkers and the environment, lagged behind with a partial and drawn-out response. This discrepancy manifested global inequalities within the nation—the abysmal occupational health protections afforded farmworkers suggest disparities in exposure were not strictly delineated by national borders, but were shaped by racism, xenophobia, and transnational political economies, as the largely immigrant farmworker population faced continued hazards. Corporate influence over controls and a legacy of U.S. imperialism was evident in the special exception granted for use on pineapples in Hawaii, where transition from "territory" to state was recent and where Dole entities had wielded great economic and political power for nearly a century.

The first cases of DBCP-related sterility in Central American banana workers emerged around the same time as the U.S. production worker cases, but in contrast to the U.S. cases, would not become public knowledge for nearly a decade. Although concerned Costa Rican physicians linked an unusual series of sterility cases to DBCP quite rapidly, officials dealt with the problem quietly, in direct negotiation with Standard Fruit personnel. As was the case in the parallel process taking place in the United States, this negotiation resulted in inequalities of exposure across national boundaries: most notably, Standard exported Costa Rican DBCP stores to Honduras. Costa Rican officials' diplomatic approach to regulation undoubtedly prevented media attention and controversy, and the dangers of DBCP would not make the news in Costa Rica or the rest of Central America until the 1980s.

U.S. regulations had a profound effect on DBCP exposures beyond U.S. national borders after 1977. Heightened protections for factory workers but incomplete controls on DBCP use (and therefore a continued market for the chemical) led one corporation, Amvac, to begin production in Mexico, where the company would, for a time, evade strict—and expensive—protections such as those mandated by OSHA. As Dow continued DBCP sales to Standard, the chemical company ironically invoked U.S. regulations to justify its actions, and the banana company insisted that there was no evidence DBCP could harm farmworkers the same way it had production workers. Although Dow and Standard wrote U.S. regulations into a contract for ongoing sales, those unenforceable rules were ignored by the fruit company, as managers found them to be "not operationally feasible."[1] Implicit U.S. regulatory approval for DBCP use in Central America came when, despite the formally national scope of regulatory decision-making, the EPA rubber-stamped Amvac's efforts to continue selling the nematicide to banana producers. By 1986, however, it became clear that the transnational impact of U.S. regulation could cut both ways: the U.S. revocation of a residue tolerance level was the "death rattle" for DBCP on bananas.[2] This action, taken to protect consumers in the United States, showed how national regulation can have a salutary effect on worker health transnationally.

The end of DBCP use did not bring an end to conflicts over science or geography. Beginning in the early 1980s, plaintiff lawyers from the United States teamed with their Central American counterparts and banana workers to file lawsuits in U.S. courts. The suits claimed that DBCP had caused sterility in male banana workers, just as it had in production workers in the United States. These cases were part of a larger trend of transnational litigation that seemed to promise that lawyers in the United States could use their access and expertise, combined with the authority of U.S. courts, to win compensation for injured plaintiffs from other nations. DBCP plaintiffs and lawyers took seriously globalization's promise of open borders: if the pesticide could flow from the United States to Central America, and U.S.-based banana companies could operate there, then Central Americans could reverse the route, bringing cases in the home of the corporations they held responsible for their injuries.

Globalization had not, however, flattened out the historical power differences between Central America and the United States. In the 1980s and 1990s, courts in the United States overwhelmingly accepted corporations' arguments to use a legal doctrine called *forum non conveniens* (FNC) to dismiss the banana worker cases on the grounds that they would more "conveniently" be tried elsewhere. Their success in this regard could largely be considered a defense victory. In most of the banana workers' home nations there were various procedural, economic, and technical barriers to such cases that all but ensured the litigation would not be pursued there—or, if pursued, could only result in monetary verdicts inconsequential to the corporations being sued. Lawyers achieved two major settlements in the 1990s, negotiated under very different conditions in 1992 and 1997–98. In both settlement deals, however, amounts given to plaintiffs were much smaller than awards to U.S. production workers sterilized by DBCP, and by 1997 corporate defendants had managed to regain the upper hand in ongoing lawsuits.

The corporations' success in winning dismissals on *forum non conveniens* grounds reveals some of the contradictions of contemporary globalization. Where U.S. policy at the EPA and FDA had historically affirmed the movement of DBCP across national boundaries—that is, from the United States to Central American nations—most U.S. courts where DBCP cases were brought agreed with corporate defendants that Central American cases should not (as a first resort at least) be allowed to cross the border into the United States. The negotiation and ratification of the Central American [and U.S.] Free Trade Agreement (CAFTA) in the first decade of the twenty-first century emphasized the double standard in regulating flows between the United States and Central America, as it sought to eliminate so-called "barriers to trade" without eliminating barriers to court access. Judges' decisions to exclude DBCP cases served the interests of capital—corporate defendants were spared the costs of legal verdicts against them—and enacted an exclusionary national identity by, in the minds of FNC supporters, defending U.S. courts from an onslaught of supposedly "foreign" cases.

In achieving these dismissals, the corporations ironically both depended on the nation-state's exclusionary powers *and* capitalized on their own transnational nature—that is, their presence in multiple nations and therefore their ability to claim the "convenience" of multiple court systems—to effectively choose their preferred legal forum. This move has parallels in corporations' search for sites of production where labor costs are lowest and costly regulations are minimal or unenforced, as states' efforts to attract investment devolve into a competition to achieve the cheapest—and therefore usually least protective—regulatory standards. Effectively, the fruit and chemical companies were able to choose the legal forum where they were surest of escaping any liability for their actions.

Litigation and transnational solidarities alike present challenges in negotiating differences of language, priorities, and knowledge. As the litigation met the FNC roadblock, tensions between *afectados*—as DBCP-affected people in Central American had come to be known—and their lawyers grew. *Afectados'* dissatisfaction with the failures as well as the more basic terms of the litigation crystallized not only around the relatively low amounts of compensation (U.S. production workers had won up to US$2 million for sterility claims in similar litigation while awards to Central American plaintiffs had rarely reached US$25,000 and were usually much lower), but around the percentage deducted for plaintiff lawyers' fees and costs, and perhaps more fundamentally, in reaction to what many Central Americans felt was an overly narrow definition of DBCP harms. While the U.S. cases had overwhelmingly focused on male sterility, many former banana workers believed that the chemical was responsible for a wide range of problems, including cancer, kidney problems, skin conditions, and more.

Costa Rican and Nicaraguan *afectados* responded to the failures and disjunctures of the transnational movement by developing national movements that turned to the state to supplement or intervene in the transnational litigation strategy. In both nations, *afectado* movements pushed for direct decision-making power for *afectados*, showing the limits to liberal democracy and the need for participatory forms of democratic engagement. Their actions also showed the ongoing relevance of the state as an object and tool of labor and social justice movements, even in the context of a neoliberalism that prescribed a diminished role for the state. Both movements prioritized workers' own understandings of DBCP harms and drew on national democratic traditions, demonstrating an alternative to transnational efforts that deprioritize local interests to fit with foreign or "global" values.

In Costa Rica, *afectados* mounted a national movement that combined direct action and strategic scientific alliances. While the transnational legal process seemed to have little space to accommodate *afectados'* own definitions of DBCP harms, Costa Ricans built a parallel movement on a national level that worked to create a democratic and scientifically authoritative definition of DBCP damage and to demand compensation from the Costa Rican government, which they held at least partially responsible for the injuries caused by the chemical. Over the course of 15 years, they built political pressure through street protests and strategic partnerships, securing places for *afectado* leaders and allied scientists on government commissions. Through concerted nationwide action, they were able to reach agreements and pass legislation that mandated compensation to *afectados* with a wide range of medical problems, perhaps most notably including women with historical workplace exposures to DBCP and medical complaints including cancer. By 2009, the Costa Rican *afectado* movement had been successful in winning compensation for a total of more than 14,000 people, with individual awards of about US$1,000 (mostly to women) and US$4,000 (mostly to men).

The Costa Rican process was remarkable because it incorporated both popular understanding of DBCP-caused illness and scientific expertise and decision-making. It also achieved what is probably the largest and most equitable payout in DBCP history: although amounts were small, they reached a large number people, and *afectados* themselves had a hand in setting the terms of compensation. These achievements are particularly remarkable in the context of changing political culture in Costa Rica, as the social welfare policies and institutions established in the mid-twentieth century have been increasingly replaced by neoliberal modes of governance. What the Costa Rican *afectado* movement did not do, however, was exact any price from the chemical and fruit corporations that produced and used DBCP; instead the state absorbed the health costs of the banana-growing business.

In Nicaragua, the national *afectado* movement that formed in the late 1990s hoped to do precisely that—hold transnational corporations accountable through organizing and acting within their own country. This strategy was important and groundbreaking on a global scale because it would use democratic action from "below"—a people's movement in a relatively powerless developing nation—to force action on the part of transnational corporations and change the policy of courts in the powerful United States. Nicaraguan *afectados* did not engage with scientists as their Costa Rican counterparts had. Instead, they built visibility and solidarity by dramatic bodily displays and protests meant to graphically illustrate the health toll of DBCP. At the same time, they passed national legislation that effectively countered corporate mobility as expressed in the success of the defendants' FNC strategy. Passing legislation that facilitated plaintiffs' DBCP cases in Nicaragua, they created the potential for judicial decisions there that reflected their own definitions of DBCP harm and promised acceptably high compensation amounts. With the changed litigation situation in Nicaragua, defendants were unlikely to seek dismissal from U.S. venues, effectively forcing the companies to face litigation in one nation or the other. The power of the Nicaraguan strategy was evidenced by the opposition to it on the part of corporations and the U.S. State Department, which lobbied for the repeal of the *afectados'* law. Mobilizing street protests and drawing on a long anti-imperialist tradition, Nicaraguan *afectados* forestalled these challenges.

The Nicaraguan strategy seemed victorious as judges there returned multimillion dollar decisions beginning in 2002. Another milestone came in 2007, when *Tellez v. Dole* became the first DBCP case with international plaintiffs to go to trial in the United States. However, corporate defendants losing lawsuits in Nicaragua ignored judgments against them because the state had no leverage against them. Without significant investments in the country, they feared no loss from flouting Nicaraguan courts, despite the fact that they had previously argued DBCP cases should be tried there instead of in the United States. When a Nicaraguan judge tried to enforce her verdict, defendants used the "investor dispute settlement" provisions of a trade agreement to bring Nicaragua into expensive arbitration in a supranational forum allied with the World Trade Organization, effectively forcing judicial reversal. For new litigation and enforcement suits in the United States, defendants disparaged the Nicaraguan *afectado* movement and Nicaraguan political culture more generally, successfully convincing U.S. judges that the lawsuits were not fit for enforcement or trial in the United States.

Corporate defendants have little to lose in contesting *afectado* claims at every opportunity possible. Corporations, though legally considered "people," rarely die, whereas DBCP-affected people are aging and dying—from a myriad of causes including diseases they attribute to their DBCP exposure. One painful loss was the 2010 death of Carlos Arguedas, a Costa Rican union and environmental activist and *afectado* who told me in 2004 that he wanted to "dedicate what is left of my life" to fighting for the environment and workers, and that he wanted to do it "for love," not money. While Arguedas's energy and commitment earned him a special place in the banana union movement, his death is sadly one of many. As activists and plaintiffs die, the litigation is likely to follow suit, proving the corporations' legal maneuvers to be an effective stalling tactic. The situation points to the urgency of policy change that would ensure that all plaintiffs who can achieve jurisdiction in the United States are able to pursue their cases against transnational corporations there if they so choose.

The key promise of Nicaragua's Law 364 was that democratic action at the national scale—even in a relatively powerless nation like Nicaragua—could be leveraged into meaningful change transnationally. At this point at least, it seems that this promise is unlikely to be realized. Even allowing for failures of the Nicaraguan movement and legal strategy, it seems clear that the outcome of the litigation was deeply shaped by inequalities of the "dispute resolution" process as well as by blatant disregard for Nicaraguan democracy and governance on the part of both corporations and the U.S. judicial system and State Department. Attempting to project the effects of the democratically achieved and protected Law 364 into the transnational sphere, Nicaraguans were caught between a rock and a hard place: subjugated to U.S. hegemony in international relations, and subjugated to corporate power in the governance of so-called trade agreements.[3] While transnational corporations disregarded Nicaraguan law and U.S. courts judged it inadequate, Nicaraguans—and Central Americans in general—had no parallel opportunity to judge the propriety of U.S. pesticide regulations and corporate actions that produced disproportionate pesticide risks in their own country. While *afectados* in Nicaragua and Costa Rica alike were able to transform popular protest into legislative change, the litigation process was more resistant to democratic participation due to persistent challenges of location, representation, and translation. At the same time, the investor-state dispute resolution process that supposedly regards states and corporations as equals is clearly stacked against nations that are too poor to robustly defend their interests.

The outlook for plaintiffs in the fight over DBCP accountability in the United States is not rosy. Attorney Steve Condie's petition to the California Supreme Court to review *Tellez* (now renamed *Laguna*)—a case in which a Nicaraguan victory was reversed subsequent to corruption charges—met with defeat in May 2014; an appeal to the United States Supreme Court is the only remaining long-shot option for review.[4] While other litigation is ongoing in the United States, mostly with plaintiffs from nations other than Nicaragua, it continues to be marked by defense challenges to plaintiffs' choice of forum, as well as other procedural obstacles.[5] In light of these problems, some plaintiff lawyers are also exploring litigation in new places, including some Central American nations. In addition, lawyers are seeking forums to enforce the still-pending decisions of the Nicaraguan courts.[6] Settlements—both inside and outside the formal judicial process—remain part of the legal landscape. In 2013, the law firm Provost Umphrey negotiated a settlement of all their cases against Dole, the terms of which have been kept strictly secret. The potential of "direct settlement" programs—that is, corporate payout outside of litigation and without lawyer representation, have long shadowed the litigation, with the largest such program negotiated between 6,000 members of the Honduran union SUTRASFCO (*Sindicato Unificado de Trabajadores de la Standard Fruit Company*) and Dole in 2006.[7] As of 2013, apparently only a handful of Honduran *afectados* had been compensated under this program, perhaps because medical determinations are made by physicians hired by Dole.[8] The Honduran experience has understandably dampened much enthusiasm for extrajudicial settlements by many workers and their allies. While the story of DBCP litigation is not over, the outlook seems much less hopeful for banana worker plaintiffs than it did in 2007, just after the historic yet temporary *Tellez* victory.

The dismissals of U.S. cases with Nicaraguan plaintiffs could have a profound impact on transnational litigation in the United States more generally. The judges' findings on Nicaraguan judicial culture were so sweeping that it seems likely that Nicaraguans pursuing litigation of any kind in the United States could face accusations of corruption. And it appears that the so-called "kill step" adopted by Dole and its attorneys at Gibson, Dunn & Crutcher may become defendants' usual tool for dealing with litigation from Latin America and elsewhere. For example, ChevronTexaco has worked with that law firm, using similar tactics—including successful corruption charges against a U.S. plaintiffs' attorney—in a suit over oil-related illness in the Ecuadorean Amazon, and their recent victory over plaintiffs has been called a "model for other companies."[9]

Meanwhile, DBCP damage has faded from the national agenda in Costa Rica, Nicaragua, and other Central American banana-growing nations as activism peters out, *afectados* age and die, and other issues take precedence. Today, concern with another health problem among farmworkers in Nicaragua, Guatemala, El Salvador, and Costa Rica has eclipsed the DBCP-related problems: an epidemic of Chronic Kidney Disease not related to traditional causes (CKDnT) whose toll has likely reached or exceeded 20,000.[10] There is much debate and uncertainty regarding the causes of the epidemic, but researchers hypothesize it may be related to occupational heat stress combined with exposure to a range of possible other agents in sugar cultivation.[11] Nicaraguans with CKDnT protested alongside *afectados* in Managua in 2005; as DBCP activism has died down, sugar worker protests have continued as outrage has mounted along with the CKDnT death toll.[12]

Readers discouraged by inequalities and injustices of this story—as well as those encouraged by the resistance of Central Americans affected by DBCP—may be asking in unison, "What is to be done?" Written from (and mostly for) anglophone North America, this book shows how U.S. actors and institutions played a central role in the "over-risking" of Central America, giving those of us who live in the United States a role to play in addressing these wrongs. What is that role? Many efforts to address the dangers of work in the global economy have called on people in the United States, Europe, and elsewhere to change their shopping habits. Consumer-based responses include fair trade or certification campaigns, targeted boycotts, and more general choices to "eat local" or "buy organic." These approaches seek to directly affect corporate behavior without the intervention of states that may be too weak, under-resourced, or unwilling to regulate employers. However, consumption-based approaches to transnational solidarity may be partial, transitory, and not necessarily effective, and they risk deprioritizing local needs in favor of choosing issues and frames with broad global appeal.[13] Organic production spares workers contact with pesticides, but especially as the field is increasingly dominated by large corporations, does little to undermine the larger power imbalance between owners and workers, who may still be subject to other occupational hazards, low wages, and other forms of exploitation that characterize industrial agriculture.[14] Small-scale agriculture at its best invigorates local economies and creates alternatives to the dominant model of food production. However, even extricating oneself completely from "big ag" (a nearly impossible task for nonsubsistence existence in the twenty-first century) does little to help farmworkers in the United States and globally.

The achievements in protection and accountability in DBCP's history point to some other potential strategies for improving worker health globally. These achievements include U.S. regulators' (albeit halting and uneven) phasing out of DBCP use in the United States and the related revocation of the residue tolerance, as well as Costa Ricans' attainment of the compensation law defining DBCP harm in a democratic and scientifically authoritative manner. Although their successes were ultimately reversed, I also count Nicaraguans' passage of Law 364 and their subsequent achievement of verdicts in both U.S. and Nicaraguan courts. Taken together, these successes show the centrality of citizen mobilization and state action in struggles over health and accountability, at both the national and transnational scales. In other words, the history of DBCP shows that although corporations have used state actors to achieve their own ends—lenient chemical regulation and exclusionary jurisprudence, for example—the state remains a site of struggle where people can leverage their power as citizens to protect themselves and push for policy changes that promote transnational health justice.

Speaking of the role national states have played in furthering the neoliberal economic program, Saskia Sassen has asked, "Does it have to be this way? Could national states instead pursue a broader international economic agenda, one that addresses questions of equity and mechanisms for accountability among the major global economic actors?"[15] The same question could be asked regarding states' engagement with an international occupational- and environmental health agenda. The DBCP history suggests the answer could be yes. Such a shift would require democratic engagement and limiting corporate power to recalibrate regulatory practice and pursue a broader international *health* agenda that "addresses questions of equity and mechanisms for accountability." Given the centrality of U.S. policy to DBCP history, I end this book by looking at points where citizen action in the United States could push for the U.S. government to adopt such an agenda.

First, any such agenda would have to prioritize a precautionary approach to the regulation of chemicals. The precautionary principle, advocated by many public health proponents, holds that "if there is a potential for harm from an activity and if there is uncertainty about the magnitude of impacts or causality, then anticipatory action should be taken to avoid harm."[16] Now as in the time of DBCP's first registration, the EPA uses a risk-benefit approach to decision-making. This approach is meant to "balance" risks—usually to health and the environment—against benefits, both biological (e.g., pest control or public health impact) and economic (e.g., decreased cost of production). In contrast, the precautionary approach would place human health and the environment, not profits, at the center of policy making and regulation. El Salvador offers a model in this regard, having recently passed legislation banning 53 pesticides suspected of possibly playing a role in the CDKnT epidemic there.[17]

Key to an international health agenda would be a revision of national regulatory policy to account for health effects beyond borders. Current U.S. policies for the export of unregistered pesticides require labeling in the importing country's language(s), including a notice that the pesticide is not registered in the United States, as well as consent of the foreign purchaser and notification of the appropriate official in the receiving country.[18] This system can be useful in sharing information across borders. However, there is no guarantee that information on toxicity will reach the final user of the pesticide, and analyses of the policies, including by the U.S. Government Accountability Office, have determined that the policy is inadequate in both concept and practice.[19] Even a well-functioning information exchange and consent system would not necessarily prevent the export or use of banned or restricted chemicals, or change any of the contextual factors that may lead to dangerous pesticide exposures. Information on the pesticide's toxicity may or may not be made available throughout the chain of distribution and use within an importing country, likely leaving farmworkers with the least information on risks. Banning the export of highly restricted, banned, or unregistered pesticides is essential to extending to other people the protections afforded to those within the United States.[20] Amending FIFRA to end the exemption of exported pesticides from most requirements of the law should be a high priority for activists and lawmakers.[21]

A vigorous international health agenda would also include dependable court access for plaintiffs alleging harms by transnational corporations, in order to provide justice to injured people and to raise the floor for health and safety conduct globally. Litigation serves an important role as a *post hoc* accountability mechanism for injured people, but the presence of a robust litigation environment can also affect corporate decision-making before harm is done. As legal scholar Cassandra Robertson has noted, corporations do keep an eye on potential litigation risks, including "what is the most expensive, least convenient forum in which they may be called to defend?"[22] She notes that "even if only some courts within a particular forum will accept the case while most would dismiss it, companies may well structure their conduct to avoid even the possibility of being subject to suit in an undesired forum."[23] Unfortunately, the actual trend in the United States is making it harder to bring transnational cases here. The Supreme Court has limited the use of the Alien Tort Statute in cases regarding conduct taking place outside of the United States,[24] and interpreted FNC in such a way as to pave a "fast track to dismissal" in U.S. federal courts.[25]

Ensuring court access for foreign plaintiffs will require abolishing FNC altogether, or changing its application or review with a commitment to ensuring equal access to justice for all plaintiffs.[26] Alternately, legislation at the state or federal level could articulate an affirmative right to court access for certain broad classes of plaintiffs, such as those seeking to hold transnational corporations accountable for occupational or environmental injuries. U.S. courts could also take an active role in enforcing health and environmental judgments from other nations. Court access is not enough, however; rules on the treatment of scientific evidence in the courtroom must be modified to better account for the uncertainty around causation in many environmental and occupational cases, opening space for the consideration of issues where lay understandings of health harms done conflict with narrow legalistic notions of scientific proof and "expertise."[27]

Of course, improving court access would not solve problems of representation and translation such as those that arose between DBCP lawyers and clients. These require thought on how attorneys can work with clients to come to a truly shared understanding of the goals, processes, and outcomes of litigation. Likewise, activists should carefully consider the place of litigation within their larger struggle. Transnational litigation may well require the development of more-robust forms of legal representation and careful approaches to translation in order to mediate different ideas of justice and science across not only political boundaries but cultural differences.[28] There are other ways for the United States to "raise the floor" on health and environmental standards worldwide. One is through the ratification and implementation of international treaties and conventions. Some of these, such as the Rotterdam Convention, speak specifically to trade in hazardous chemicals. The Rotterdam Convention surpasses U.S. export notification schemes with a legally binding informed consent process for trade in a number of hazardous substances, including pesticides. Another, the Stockholm Convention, aims to reduce or eliminate the use of certain persistent organic pollutants. Each of these conventions has its limits, but institutes stricter standards than currently in place in the United States. Although the United States signed these treaties in 1998 (Rotterdam) and 2001 (Stockholm), it has ratified neither. U.S. failure to pass implementing legislation has weakened the treaties' global effects and kept less stringent regulations in place.[29] The same can be said of a number of other environmental and health treaties, including the Basel Convention on transboundary waste transfer and the Kyoto Protocol on emission reductions.

Trade policy is perhaps the most important place where state actions can promote global health and environment.[30] To date, these issues are marginalized in trade policy: for example, North American Free Trade Agreement (NAFTA) negotiations largely relegated labor and environmental issues to "side agreements" that have largely failed to ensure protection of workers and environments.[31] Trade agreements to date have eroded or threatened national governments' ability to legislate health and safety, and integrated "investor dispute resolution" systems that favor corporations. Trade negotiations have a profound impact on a range of health issues including health and environmental protections, tobacco and alcohol sales, prescription medicine access, and food production, but "fly below the radar" of many people who are passionately concerned about health and the environment. The U.S. Trade Representative is supposed to ensure that "U.S. trade policy and trade negotiating objectives adequately reflect U.S. public and private sector interests," but the advisory committee system tasked with providing that input (and given access to much information deemed confidential to negotiations) is heavily slanted toward the corporate sector.[32] This is true even on committees with significant public health impacts: those dealing with food, alcohol, tobacco, pharmaceuticals, and health insurance. At last tally, these committees had 42 industry representatives but only 3 representing the interests of public health.[33] Mass political action in the United States and trading-partner nations is needed to prevent passage of harmful agreements and push for new directions in trade policy that would promote rather than endanger public health. In recent negotiations over the Transpacific Partnership (TPP), a "free trade" agreement under negotiation by twelve Pacific Rim nations including the United States, Malaysian negotiators proposed that national tobacco-related health and labeling laws should not be subject to challenges under TPP rules. This "carve out" won support in the United States from both proponents and critics of "free trade" deals, showing that health concerns are an important point of debate and opposition to the implementation of such deals: the specter of corporate challenges to national health laws does not sit well even with many of those who like the idea of "free trade."[34]

Taking each of these steps—instituting the precautionary principle in pesticide regulations, regulating pesticide exports, ensuring court access for non-U.S. plaintiffs, and leveraging treaties and trade policy to protect public health worldwide—will require controlling corporate influence over policy making at multiple levels. Pesticide regulation in the United States is still full of loopholes, overly dependent on corporate science, influenced by industry lobbying, and underfunded when it comes to enforcement.[35] Corporations have argued vociferously to shrink access to courts for foreign plaintiffs, both through pursuing precedent-setting cases such as *Kiobel* and through attempts to influence policy through aggressive lobbying.[36] The U.S. Trade Representative's public input overwhelmingly comes from the private sector, with those who serve on advisory committees gaining access to information kept secret from the public.[37] More broadly, corporations exercise an immense amount of influence over the full range of policy areas in the United States, through lobbying, campaign contributions, positions on advisory committees, the "revolving door" between industry and government posts, and more. This influence often runs counter to the interests of public health, as corporate influence is geared toward meeting the central aim of the corporate entity—maximization of profit.

Converting the U.S. state into a proponent of a global health agenda that would control corporate predations on public health cannot happen without a struggle. Indeed, it would require a sustained citizen engagement well beyond the formal democracy of the ballot box or the "vote with your dollars" orientation of consumer-oriented politics. Such engagement runs counter to the perception and realities of the state's diminished role in the context of neoliberal globalization. However, nothing about the trajectory of these political changes is inevitable. The political trends in Latin America over recent years have shown some ways in which citizens unhappy with the neoliberal status quo can both elect leaders and hold them accountable. Unlike corporations—or even nongovernmental organizations—democratic states maintain at least a formal commitment to structures of accountability and representation. In the United States, as in Costa Rica and Nicaragua, activist pressure on various state institutes and practices could result in meaningful changes to protect people's health. This is true in making change on a national scale, but also important to demanding domestic actions that could provoke global change. National states have been key actors in instituting policies that have reconfigured the global economy over the last forty-odd years, and still have an important role to play in shaping global economies. Successful grassroots demands of the national state could, especially in developed and globally powerful nations like the United States, leverage state power to secure positive change far beyond national boundaries.

Such an agenda may seem both wildly overambitious and woefully inadequate. Overambitious because the current political climate in the United States and globally can seem to offer little space to effectively control corporate power, open courts, change the terms of regulation, and transform the basis of trade and foreign policy. Inadequate because even these changes are only a first step toward the fundamental restructuring of local and global economies that will be necessary to do away with the multiple inequities and injustices so central to creating unequal exposures to DBCP and other chemicals. Avoiding the repetition of such histories will ultimately require deep and lasting transformations in economic and political life, not only in the United States and Central America, but globally. The story of DBCP underscores the importance of continuing efforts to reshape global capitalism's toxic exposures into a new terrain of health and justice.

NOTES

INTRODUCTION

1. Victoria Chaney, "Special Verdict Form," *Jose Adolfo Tellez et al. vs. Dole Food Company, Inc. et al.,* No. BC312852, (LASC, November 5, 2007).

2. Estimates on the number of DBCP-affected people in Central America vary, but the number claiming health problems caused by the chemical exceeds 30,000, even when considering only Costa Rica and Nicaragua. In 1999, U.S. attorneys for a large subset of plaintiffs tallied over 16,000 Central Americans; epidemiologists found a 50% incidence of reduced or zero sperm count among all Latin American plaintiffs in this group (including Central Americans and Ecuadoreans). M. Slutsky, J. L. Levin, and B. S. Levy, "Azoospermia and Oligospermia among a Large Cohort of DBCP Applicators in Twelve Countries," *International Journal of Occupational and Environmental Health* 5 (1999): 116–22. In Nicaragua in 2005, activists claimed over 22,000 people there had been harmed by DBCP, and over 2,000 had received verdicts in their favor in Nicaraguan courts. See Valeria Imhof, "La Dole 'Dobla el Brazo' de algunos ex bananeros," *El Nuevo Diario* 2006, http://www.elnuevodiario. com.ni/2006/02/10/nacionales/12367; *Sánchez Osorio y Otros vs. Standard Fruti [sic] Company y Otros;* Sentencia #0271–2005; Demanda #0214–0425–02cv, 8 de Agosto de 2005 (Nicaragua), JOSF; Dole Food Company, Inc., *Form 10K for the Fiscal Year Ended January 1, 2005* (Washington, DC: Securities and Exchange Commission, 2005); Róger Olivas and Valeria Imhof, "Otra victoria legal para afectados de Nemagón," *El Nuevo Diario,* December 5, 2006, http://impreso.elnuevodiario. com.ni/2006/12/05/nacionales/35648. By 2010, the Costa Rican National Insurance Institute had found DBCP-caused illness in over 14,000 men and women. Ximena Alfaro M, "Afectados por Nemagón consiguen pago del INS," *La Nación,* October 9, 2010, http://www.nacion.com/2010-10-09/ElPais/NotasSecundarias /ElPais2549464.aspx.

3. Fiona Tam, "Foxconn factories are labour camps," *South China Morning Post,* October 11, 2010, http://www.scmp.com/article/727143/foxconn-factories-are-labour-camps-report; Jason Burke, "Bangladeshi workers still missing eight months after

Rana Plaza collapse," *The Guardian,* December 25, 2013; Joe Berlinger, *Crude: The Real Price of Oil* (2009; Entendre Films), DVD.

4. For a useful review of these, and a critical evaluation of various theoretical and analytic approaches, see Robert J. Foster, "Tracking Globalization: Commodities and Value in Motion," in *Sage Handbook of Material Culture,* eds. Chris Tilley et al. (London: Sage Publications, 2008).

5. For ongoing campaigns against ChevronTexaco and Dow, see "ChevronToxico: The Campaign for Justice in Ecuador," http://chevrontoxico.com/; "International Campaign for Justice in Bhopal," http://bhopal.net/.

6. A book-length journalistic treatment that covers both use and accountability is Vicent Boix Bornay, *El parque de las hamacas: El químico que golpeó a los pobres* (Barcelona: Icaria Editorial, 2008).

7. David Harvey, *Spaces of Hope* (Berkeley: University of California Press, 2000), 57.

8. Ibid., 57–58.

9. For example, application of some pesticides may endanger workers without leaving any measurable traces on food. See Angus Lindsay Wright, "Rethinking the Circle of Poison: The Politics of Pesticide Poisoning among Mexican Farm Workers," *Latin American Perspectives* 13 (1990): 26–59; Angus Lindsay Wright, *The Death of Ramón González: The Modern Agricultural Dilemma,* 1st ed. (Austin: University of Texas Press, 1990).

10. David Weir and Mark Schapiro, *Circle of Poison: Pesticides and People in a Hungry World* (San Francisco: Institute for Food and Development Policy, 1981), 4, 79.

11. Lori Ann Thrupp, "Sterilization of Workers from Pesticide Exposure: The Causes and Consequences of DBCP-Induced Damage in Costa Rica and Beyond," *International Journal of Health Services* 21 no. 4 (1991): 731.

12. See Karl Polanyi, *The Great Transformation: The Political and Economic Origins of Our Time,* 2nd Beacon Paperback ed. (Boston, MA: Beacon Press, 2001). Polanyi's work is useful in considering the role of the state as integral to a "double movement" in market societies toward, on the one hand, "free" or laissez-faire markets, and on the other, protections from the worst ravages of those markets.

13. Leslie Sklair, "Social Movements and Global Capitalism," *Sociology* 29 no. 3 (1995): 502.

14. The campaign against Dow seeks accountability for the Bhopal chemical leak disaster perpetrated by Union Carbide, which Dow purchased in 1999. Campaigns against Nike and other apparel brands have focused on occupational health and other workplace abuses in textile sweatshops, and the campaign against Nestlé challenged that company's marketing of infant formula as a substitute for breast milk.

15. Other strategies also have aimed to create alternatives to those abuses, by for example creating alternative systems of production and distribution, including fair trade and local agriculture movements. See Laura T. Raynolds, Douglas L. Murray, and John Wilkinson, *Fair Trade: The Challenges of Transforming Globalization* (London; New York: Routledge, 2007).

16. Adriana Petryna, *Life Exposed: Biological Citizens after Chernobyl* (Princeton, N.J.: Princeton University Press, 2002).

17. Michelle Murphy, "Chemical Regimes of Living," *Environmental History* 13, no. 4 (2008): 699.

18. See, for example, John A. Guidry, "The Useful State? Social Movements and the Citizenship of Children in Brazil," in *Globalizations and Social Movements: Culture, Power and the Transnational Public Sphere*, eds. John A. Guidry, Michael D. Kennedy, and Mayer N. Zald (Ann Arbor: University of Michigan Press, 2000); G. Seidman, *Beyond the Boycott: Labor Rights, Human Rights, and Transnational Activism* (New York: Russell Sage Foundation, 2007); David N. Pellow, *Resisting Global Toxics: Transnational Movements for Environmental Justice* (Cambridge, Mass.: MIT Press, 2007).

19. David Harvey, *A Brief History of Neoliberalism* (New York: Oxford University Press, 2005), 2.

20. Wright, *The Death of Ramón González*.

21. For example, Julie Sze, *Noxious New York: The Racial Politics of Urban Health and Environmental Justice* (Cambridge, MA: MIT Press, 2007); Martin Melosi, *Effluent America: Cities, Industry, Energy, and the Environment* (Pittsburgh: University of Pittsburgh Press, 2001); Claudia Clark, *Radium Girls: Women and Industrial Health Reform, 1910–1935* (Chapel Hill, NC: The University of North Carolina Press, 1997); Gerald E. Markowitz and David Rosner, *Deceit and Denial: The Deadly Politics of Industrial Pollution* (Berkeley: University of California Press, 2002); David Rosner and Gerald Markowitz, eds., *Dying for Work: Workers' Safety and Health in Twentieth-Century America* (Bloomington and Indianapolis: Indiana University Press, 1989).

22. Phil Brown, *Contested Illnesses: Citizens, Science, and Health Social Movements* (Berkeley: University of California Press, 2012). See also Petryna, *Life Exposed*; Michelle Murphy, *Sick Building Syndrome and the Problem of Uncertainty: Environmental Politics, Technoscience, and Women Workers* (Durham, NC: Duke University Press, 2006); Nancy Langston, *Toxic Bodies: Hormone Disruptors and the Legacy of DES* (New Haven, CT: Yale University Press, 2010); J. Stephen Kroll-Smith, Phil Brown, and Valerie J. Gunter, *Illness and the Environment: A Reader in Contested Medicine* (New York: New York University Press, 2000).

23. Works that take a transnational approach include Carlos Eduardo Siqueira, *Dependent Convergence: The Struggle to Control Petrochemical Hazards in Brazil and the United States* (Amityville, NY: Baywood Publishing Company, 2003); Kim Fortun, *Advocacy after Bhopal: Environmentalism, Disaster, New Global Orders* (Chicago: University of Chicago Press, 2001).

24. See Walter LaFeber, *Inevitable Revolutions: The United States in Central America*, 2nd ed. (New York: W.W. Norton, 1993); Lars Schoultz, *Beneath the United States: A History of U.S. Policy toward Latin America* (Cambridge: Harvard University Press, 1998); Greg Grandin, *Empire's Workshop: Latin America, the United States, and the Rise of the New Imperialism*, 1st ed. (New York: Metropolitan Books, 2006); Thomas David Schoonover, *The United States in Central America,*

1860–1911: Episodes of Social Imperialism and Imperial Rivalry in the World System (Durham: Duke University Press, 1991); Juan Pablo Pérez Sáinz, *From the Finca to the Maquila: Labor and Capitalist Development in Central America* (Boulder, CO: Westview Press, 1999); Daniel Faber, *Environment under Fire: Imperialism and the Ecological Crisis in Central America* (New York: Monthly Review Press, 1993).

25. This brief account of the banana industry draws on the following works: essays by Mark Moberg, Steve Striffler, Marcelo Bucheli and Laura Raynolds in *Banana Wars: Power, Production and History in the Americas,* eds. Mark Moberg and Steve Striffler(Durham and London: Duke University Press, 2003); John Soluri, *Banana Cultures: Agriculture, Consumption, and Environmental Change in Honduras and the United States* (Austin: University of Texas Press, 2005); Thomas Lindas Karnes, *Tropical Enterprise: The Standard Fruit and Steamship Company in Latin America* (Baton Rouge: Louisiana State University Press, 1978); José Roberto López, *La economía del banano en Centroamérica* (San José, Costa Rica: Departamento Ecuménico de Investigaciones, 1988); Richard P. Tucker, *Insatiable Appetite: The United States and the Ecological Degradation of the Tropical World* (Berkeley, CA: University of California Press, 2000); Paul J. Dosal, *Doing Business with the Dictator: A Political History of United Fruit in Guatemala, 1899–1944* (Wilmington, DE: SR Books, 1993); Steve Marquardt, "'Green Havoc': Panama Disease, Environmental Change, and Labor Process in the Central American Banana Industry," *American Historical Review* 106 no. 1 (2001); Charles D. Kepner, *Social Aspects of the Banana Industry* (New York: AMS Press, 1967 [1936]); Charles D. Kepner and Jay H. Soothill, *The Banana Empire: A Case Study of Economic Imperialism;* May and Plaza.

26. Todd Gordon and Jeffery R. Webber, "Honduran Labyrinth," *Jacobin,* no. 10 (2013), http://jacobinmag.com/2013/04/honduran-labyrinth/.

27. See Mary Finley-Brook, "Geoeconomic Assumptions, Insecurity, and 'Free' Trade in Central America," *Geopolitics* 17, no. 3 (2012): 629–57, http://search.ebscohost.com/login.aspx?direct = true&db = aph&AN = 78192274&site = ehost-live.

28. Soluri, *Banana Cultures;* Moberg et al., *Banana Wars;* Dana Frank, *Bananeras: Women Transforming the Banana Unions of Latin America* (Cambridge, MA: South End Press, 2005); Henry J. Frundt, *Fair Bananas: Farmers, Workers, and Consumers Strive to Change an Industry* (Tucson, AZ: University of Arizona Press, 2009); Karla Slocum, *Free Trade and Freedom: Neoliberalism, Place, and Nation in the Caribbean* (Ann Arbor: University of Michigan Press, 2006).

29. Frundt, *Fair Bananas;* Frank, *Bananeras;* Laura Raynolds, "The Global Banana Trade," in *Banana Wars: Power, Production, and History in the Americas,* ed. Steve Striffler and Mark Moberg (Durham and London: Duke University Press, 2003).

30. Steve Marquardt, "Pesticides, Parakeets, and Unions in the Costa Rican Banana Industry, 1938–1962," *Latin American Research Review* 37, no. 2 (2001); Soluri, *Banana Cultures,* especially chapters 5 and 7; Kepner, *Social Aspects of the Banana Industry,* especially chapters 6 and 8; Foro Emaús Coordinating Committee, "Bananas for the World—and the Negative Consequences for Costa Rica?: The

Social and Environmental Impacts of the Banana Industry in Costa Rica" (San José, Costa Rica: Foro Emaús, 1998), http://members.tripod.com/foro_emaus/2ing .html.

31. The term "slow violence" is that of Rob Nixon, who uses it in an effort to prioritize questions of temporality and resistance in considering the practice and effects of what has often been referred to as "structural violence." See Rob Nixon, *Slow Violence and the Environmentalism of the Poor* (Cambridge, MA: Harvard University Press, 2011).

32. Carlos Luis Fallas, *Mamita Yunai* (San José: Editorial Costa Rica, 2010), 152. "Yunai" is a reference to United Fruit.

33. Soluri, *Banana Cultures*; Marquardt, "Pesticides, Parakeets, and Unions"; Marquardt, "Green Havoc."

34. Soluri, *Banana Cultures*, 195.

35. Ibid, chapter 7.

36. Ibid., 75.

37. See for example, V. J. Gunter and C. K. Harris, "Noisy Winter: The DDT Controversy in the Years before *Silent Spring*," *Rural Sociology* 63, no. 2 (1998): 179–98; Pete Daniel, *Toxic Drift: Pesticides and Health in the Post-World War II South* (Baton Rouge: Louisiana State University Press, in association with Smithsonian Institution, Washington, DC, 2005); Edmund P Russell III, "The Strange Career of DDT: Experts, Federal Capacity, and Environmentalism in World War II," *Technology and Culture* 40, no. 4 (1999): 770–96.

38. Samuel Hays, *Beauty, Health, and Permanence: Environmental Politics in the United States: 1955–1985* (Cambrigde: Cambrigde University Press, 1987); Nicholas A. Ashford, *Crisis in the Workplace: Occupational Disease and Injury* (Cambridge, MA: MIT Press, 1976); David S. Egilman and Susanna Rankin Bohme, "A Brief History of Warnings," in *Handbook of Warnings*, ed. Michael Wogalter (Mahwah, NJ: Lawrence Erlbaum Associates, 2005), 16–17.

39. Sandra Piszk, "Expediente No. 250–23–98: Informe Final con Recomendaciones" (San José: Defensoría de los Habitantes de la República de Costa Rica, 1998), CQLF, 17–21; Reglamento de seguridad sobre empleo de sustancias tóxicas en la agricultura [Decreto No. 6], *La Gaceta* [Costa Rica] 215 (21 December1968); Ley General de Salud [Ley No. 5395], *La Gaceta* [Costa Rica] 222 (24 November 1973); Lori Ann Thrupp, "Pesticides and Policies: Approaches to Pest-Control Dilemmas in Nicaragua and Costa Rica," *Latin American Perspectives* 15, no. 4 (2004): 46–48, 53–57; Reglamento sobre importación, distribución y uso de productos químicos y químico-biológicos para la industria agropecuaria, *La Gaceta* [Nicaragua] 104 (12 May 11960), http://legislacion.asamblea.gob.ni/Normaweb.nsf/164aa15ba012e5670 62568a2005b564b/0438aca0f5ba26e1062571bc005ff422?OpenDocument&Highli ght = 2,quimicos; Ley reguladora sobre importación, elaboración, almacenamiento, transporte, venta y uso de pesticidas [Decreto No. 43–74], Diario de Centroamérica [Guatemala] 93 (5 June 1974); Derivados del Artículo 80 del Decreto Ley No. 20 del 1 de septiembre de 1966 sobre protección fitosanitaria [Decreto No. 384], *Gaceta Oficial* [Panamá] 16.016 (21 December 21 1967).

40. DBCP litigation documents have enjoyed a relatively wide circulation, and many of the secondary sources I cite, including works by Cathy Trost, Lori Ann Thrupp, Vicent Boix, and Beth Rosenberg, also depend on some subset of this archive. Similar documents have constituted a rich archive for other historians of occupational and environmental health, notably Gerald Markowitz and David Rosner, whose work on these sources has resulted in both scholarly publications and expert testimony in various toxins-related cases, making them a target of defendants but clarifying the historical record and moving juries to protect against and compensate for occupational and environmental disease. See David Rosner and Gerald Markowitz, "The Trials and Tribulations of Two Historians: Adjudicating Responsibility for Pollution and Personal Harm," *Medical History* 53 (2009). Markowitz and Rosner note an expanding role for historians in toxic tort litigation, on both the defense and plaintiff side.

CHAPTER ONE

1. Steven Stoll, *The Fruits of Natural Advantage : Making the Industrial Countryside in California* (Berkeley: University of California Press, 1998); Richard P. Tucker, *Insatiable Appetite: The United States and the Ecological Degradation of the Tropical World* (Berkeley: University of California Press, 2000).

2. W.L. Nelson, "Pioneer of Soil Fumigation Dies," *Down to Earth,* Fall 1951.

3. Ibid.

4. E.P. Caswell and W.J. Apt, "Pineapple Nematode Research in Hawaii: Past, Present, and Future," Journal of Nematology 21 no. 2 (1989): 148, 151.

5. Edmund P. Russell, *War and Nature: Fighting Humans and Insects with Chemicals from World War I to "Silent Spring,"* Studies in Environment and History (Cambridge; New York: Cambridge University Press, 2001), 49.

6. Nelson, "Pioneer of Fumigation."

7. Nelson, "Pioneer of Fumigation"; Russell, "Speaking of Annihilation: Mobilizing for War against Human and Insect Enemies: 1914–1945," *Journal of American History* 82 no. 4 (1996): 1512; Ehud Yonay, "The Nematode Chronicles," *New West,* May 1981, 69.

8. Tucker, *Insatiable Appetite,* 95; Yonay, "The Nematode Chronicles," 69.

9. M.A. Manzelli, "What's New in Nematocides," *Farm Chemicals* 1956, 41–43.

10. Bert Lear and N.B. Akesson, "Applying Nematocides," *Farm Chemicals,* December 1959, 39–41.

11. Ibid., 39.

12. Carey McWilliams, *Factories in the Field: The Story of Migratory Farm Labor in California* (Boston: Little, Brown, 1939).

13. Fred Magdoff, John Bellamy Foster, and Frederick H. Buttel, "An Overview," in *Hungry for Profit: The Agribusiness Threat to Farmers, Food, and the Environment,* ed. Fred Magdoff, John Bellamy Foster, and Frederick H. Buttel (New York: Monthly Review Press, 2000); Paul K. Conklin, *A Revolution Down on the Farm:*

The Transformation of American Agriculture Since 1929 (Lexington: University Press of Kentucky, 2008), 99–107.

14. Kenneth A. Dahlberg, "Government Policies That Encourage Pesticide Use in the United States," in *The Pesticide Question: Environment, Economics, and Ethics*, ed. David Pimentel and Hugh Lehman (New York and London: Chapman and Hall, 1993), 283.

15. See Stoll, *The Fruits of Natural Advantage*, chapter 4.

16. Louis Ferleger, "Arming American Agriculture for the Twentieth Century: How the USDA's Top Managers Promoted Agricultural Development," *Agricultural History* 74, no. 2 (2000); Louis Ferleger, "Uplifting American Agriculture: Experiment Station Scientists and the Office of Experiment Stations in the Early Years after the Hatch Act," *Agricultural History* 64, no. 2 (1990). See also Christopher Henke, *Cultivating Science, Harvesting Power* (Cambridge, MA: MIT Press, 2008).

17. Russell, *War and Nature*.

18. Russell, "The Strange Career of DDT: Experts, Federal Capacity, and Environmentalism in World War II," *Technology and Culture* 40, no. 4 (1999): 782–83.

19. "DDT," *Time*, June 12 1944, cited in Russell, "The Strange Career of DDT," 783.

20. Frank Carey, "War Develops Powerful Attack against Insects," *Mobile Press Register* 1943. Cited in Russell, "The Strange Career of DDT," 782.

21. Russell, "The Strange Career of DDT," 789–90.

22. "DDT," *Time*, June 12 1944; Russell, "The Strange Career of DDT," 783.

23. "DDT is Good for Me-E-E," *Time*, June 30, 1947, http://www.mindfully.org/Pesticide/DDT-Household-Pests-USDA-Mar47.htm.

24. "Fight Insect Enemies," *Agricultural Chemicals*, June 1954, 51.

25. Ibid.

26. Roy Hansberry, "Role of Chemicals in the Future of Insect Control," *Agricultural Chemicals* (1954), 43.

27. Rachel Carson, *Silent Spring* (Cambridge, MA: The Riverside Press, 1962), 16–17.

28. *Open Door to Plenty* (Washington DC: National Agricultural Chemicals Associations, 1958), 8.

29. David E. Price, "The Relation of Pesticides to Health," *Agricultural Chemicals*, 1954, 34.

30. Arturo Escobar, *Encountering Development: The Making and Unmaking of the Third World,* Princeton Studies in Culture/Power/History. (Princeton, NJ: Princeton University Press, 1995), 3–4.

31. Hansberry, "Role of Chemicals," 43.

32. "DDT," *Time*, June 12 1944, cited in Russell, "The Strange Career of DDT," 783.

33. A. L. Taylor, "Soil Fumigation: Today. . . . And Tomorrow," *Agricultural Chemicals*, August 1954; C. E. Dieter, "Techniques for Collecting and Isolating Plant Parasitic Nematodes," *Down to Earth*, Fall 1954.

34. C.J. Nusbaum, "Nematode Assays—a Program to Aid Agricultural Workers," *Down to Earth* 19, no. 2 (1963): 2.

35. A.L. Taylor, "Nematocides and Nematicides—History," accessed April 27, 2007, http://flnem.ifas.ufl.edu/HISTORY/nematicide_his.htm; Khuong B. Nguyen, e-mail message to author, April 27, 2007.

36. For a discussion of how advertising teaches consumers to need and use products, see Susan Strasser, *Satisfaction Guaranteed: The Making of the American Mass Market,* 1st ed. (New York: Pantheon Books, 1989), 16–17.

37. Taylor, "Nematocides and Nematicides—History."

38. "Turning on the Worm," *Shell News,* March 1957, 22.

39. "Chemical Control for Nematodes Seen Growing," *Agricultural Chemicals,* February 1956, 61.

40. C.C. Compton and S.H. Benedict, "Nemagon (1,2-Dibromo-3-Chloropropane)," *Agricultural Chemicals,* March 1956; Martin T Hutchinson, "The Nematode and You," *New Jersey Farm and Garden,* September (1956), reprinted by Shell Chemical Corporation; C.W. McBeth and Glenn B. Bergeson, "1,2-Dibromo-3-Chloropropane—a New Nematocide," *Plant Disease Reporter* 39 no. 3 (1955): 223–25, reprinted by Shell Chemical Corporation, Agricultural Chemicals Devision, Denver, Colorado.

41. "Turning on the Worm," 22.

42. *Proceedings of the Shell Nematology Workshop: St. Louis, Missouri* (Shell Chemical Corporation, October 9–10, 1957); *Proceedings of the Shell Nematology Workshop: Portland, Oregon* (Shell Chemical Corporation, January 27–28, 1959); C.W. McBeth and Glenn B. Bergeson; "Shell Workshop Offers Nematode Training to 200," *Agricultural Chemicals* February, 1957, 44+.

43. McBeth and Bergeson, "Shell Workshop Offers Nematode Training," 61.

44. Ferleger, "Arming American Agriculture."

45. Taylor, "Nematocides and Nematicides—History."

46. C.J. Nusbaum, "Soil Fumigation for Nematode Control in Flue-Cured Tobacco," *Down to Earth,* Summer 1960, 15.

47. Compton and Benedict, "Nemagon," 46.

48. This was the case with Nusbaum, "Soil Fumigation for Nematode Control."

49. For example, Shell reprinted Compton and Benedict, "Nemagon" for promotional distribution outside the magazine format.

50. "Turning on the Worm," 22.

51. McBeth and Bergeson, "Shell Workshop Offers Nematode Training," 44.

52. Shell Chemical Corporation, "Portals of Progress in Pesticide Research: For the Benefit of Agriculture [Advertisement]," *Agricultural Chemicals,* May 1958, 6–7.

53. Cathy Trost, *Elements of Risk: The Chemical Industry and Its Threat to America* (New York: Times Books, 1984), 39.

54. *Soil Fumigation Handbook,* (Midland, MI: Dow Chemical Company, 1958), cited in Trost, *Elements of Risk,* 39.

55. See Russell, "The Strange Career of DDT."

56. Samuel Hays, *Beauty, Health, and Permanence: Environmental Politics in the United States: 1955–1985* (Cambrigde, MA: Cambrigde University Press, 1987), 174.

57. "Nemagon Soil Fumigant Commercially Available," *Farm Chemicals,* May 1955, 19.

58. For early concerns about pesticide dangers, see V. J. Gunter and C. K. Harris, "Noisy Winter: The DDT Controversy in the Years before 'Silent Spring,'" *Rural Sociology* 63, no. 2 (1998); Russell, "The Strange Career of DDT"; Pete Daniel, *Toxic Drift: Pesticides and Health in the Post-World War II South* (Baton Rouge: Louisiana State University Press in association with Smithsonian Institution, Washington, DC, 2005); Linda J. Lear, *Rachel Carson: Witness for Nature* (New York: Henry Holt, 1997); Joshua Blu Buhs, *The Fire Ant Wars: Nature, Science, and Public Policy in Twentieth-Century America* (Chicago: University of Chicago Press, 2004).

59. *Federal Insecticide, Rodenticide, and Fungicide Act,* 7 USC § 135–135k (1947).

60. F. C. Bishopp et al., *Insects: The Yearbook of Agriculture* (Washington, DC: United States Department of Agriculture, 1952), 302–309.

61. *Federal Insecticide, Rodenticide, and Fungicide Act.*

62. Paul Dunbar, "The Food and Drug Administration Looks at Insecticides," *Food-Drug-Cosmetic Law Quarterly* June 1949, 233.

63. *The 1938 Food, Drug, and Cosmetic Act,* Pub. L. No. 75–717 52 Stat. 1040 (1938).

64. Dunbar, "The FDA Looks at Pesticides," 235.

65. *An Act to Amend the Federal Food, Drug, and Cosmetic Act with Respect to Residues of Pesticide Chemicals in or on Raw Agricultural Commodities (Miller Amendment),* Pub. Law No. 518, 68 Stat. 559 (1954).

66. *An Act to Amend the Federal Food, Drug, and Cosmetic Act, as Amended, to Provide for Labelling of Economic Poisons with Registration Numbers, to Eliminate Registration under Protest, and for Other Purposes,* Pub. Law No. 88–305 (1964).

67. Gregory Hooks, "From an Autonomous to a Captured State Agency: The Decline of the New Deal in Agriculture," *American Sociological Review* 55, no. 1 (1990): 29–43; Grant McConnell, *The Decline of Agrarian Democracy* (Berkeley: University of California Press, 1953).

68. See Daniel, *Toxic Drift;* Linda J. Lear, "Bombshell in Beltsville: The USDA and the Challenge of 'Silent Spring,'" *Agricultural History* 66, no. 2 (1992): 151–70.

69. See Daniel, *Toxic Drift.*

70. See Daniel, *Toxic Drift;* Buhs, *The Fire Ant Wars.*

71. Edmund Russell offers an alternative explanation of the federal government's failure to regulate DDT in the 1945–1972 period. He argues that scientific and regulatory concern about DDT existed during the World War II years when the army and the Public Health Service tightly controlled the substance, just as the Environmental Protection Agency would after 1972. He says the important question is why the government permitted the use of DDT in the intervening years, and answers that a "more decentralized economic and political system" meant that scientists did not have control of the decision-making power and that the "clamor of public praise for DDT's wonders" meant there was no government mandate to tightly regulate

sales. Referring to the USDA's inability to prevent the sale of a pesticide (due to the registration-under-protest clause), Russell asserts that DDT sales during this period were "governed by the market." Russell notes that some DDT manufacturers were concerned about health risks, and some only entered the market after the lawsuits they expected did not materialize. However, he does not investigate the relationship between chemical companies and regulators during this period. "The Strange Career of DDT."

72. John D. Conner, "Legal Aspects in Marketing Agricultural Chemicals," *Agricultural Chemicals*, May 1954, 41.

73. John D. Conner and George A. Burroughs, *Manual of Chemical Products Liability; an Analysis of the Law Concerning Liability Arising from the Manufacture and Sale of Chemical Products* (Washington: Manufacturing Chemists' Association and National Agricultural Chemical Association, 1952).

74. Jerry Harrington, "The Midwest Agricultural Chemical Association: A Regional Study of an Industry on the Defensive," *Agricultural History* 70, no. 2 (1996): 415–38.

75. *Open Door to Plenty*, 3.

76. Ibid., 4–7.

77. Ibid., 7.

78. Ibid, 9.

79. Ibid, 29.

80. *An Act to Amend the Federal Food, Drug, and Cosmetic Act with Respect to Residues of Pesticide Chemicals in or on Raw Agricultural Commodities (Miller Amendment)*, 512.

81. Frederick Rowe Davis, "Pesticides and Toxicology: Episodes in the Evolution of Environmental Risk Assessment (1937–1997)" (PhD diss., Yale University, 2001), 174.

82. Christopher Sellers, *Hazards of the Job: From Industrial Disease to Environmental Health* (Chapel Hill: University of North Carolina Press, 1999).

83. Davis, "Pesticides and Toxicology," 196; Sellers, *Hazards of the Job*.

84. *Open Door to Plenty*, 32–33.

85. Ibid., 31.

86. David S. Egilman and Susanna Rankin Bohme, "A Brief History of Warnings," in *Handbook of Warnings*, ed. Michael Wogalter (Mahwah, NJ: Lawrence Erlbaum Associates, 2005), 11. The problem with the rational individual model is that it does not account for the social, economic, or cultural context in which an individual sees, interprets, and/or acts on the information contained in the warning.

87. Ibid.

88. *Warning Labels: A Guide for the Preparation of Warning Labels for Hazardous Chemicals, Manual L-1*, (Washington, DC: Manufacturing Chemists' Association of the United States, 1949).

89. *Federal Insecticide, Rodenticide, and Fungicide Act.*

90. Daniel, *Toxic Drift*, 136.

91. Price, "Relation of Pesticides to Health," 86.

92. Lear, "Bombshell in Beltsville," *Agricultural History* 66, no. 2 (1992): 166.

93. McBeth and Bergeson, "Shell Workshop Offers Nematode Training."

94. "Agricultural Bulletin: Nemagon, a New Soil Fumigant" (Agricultural Chemicals Sales Division, Shell Chemical Corporation, 1956), document no. 17036, SHLF, 6.

95. Ibid., 4–5.

96. C. H. Hine et al., *Confidential Report: An Evaluation of the Degree of Toxicity of 1,2-Dibromo-3-Chloropropane; Acute and Chronic Vapor Exposure of Rodents* (San Francisco: Department of Pharmacology and Experimental Therapeutics, University of California School of Medicine, January 12, 1955), document no. 17032, SHLF, 2. Another report found DBCP to be "slightly toxic": H. H. Anderson, C. H. Hine, and L. G. Rice, *Confidential Report: An Evaluation of the Degree of Toxicity of 1,2-Dibromo-3-Chloropropane; II—Percutaneous Toxicity and Irritation Studies* (San Francisco: Department of Pharmacology and Experimental Therapeutics, University of California School of Medicine, January 7, 1955), document no. 17031, SHLF, 8.

97. T. R. Torkelson, *Summary of a Chronic Vapor Toxicity Experiment: 1,2 Dibromo-3 Chloropropane (1956)* (n.p. : Dow Chemical, 1957), document no. 16448, SHLF.

98. N. G. White to V. K. Rowe, June 26, 1956, document no. 17037, SHLF; V. K. Rowe to Norman G. White, July 18, 1956, document no. 17038, SHLF.

99. V. K. Rowe to C. H. Hine, September 13, 1956, document no. 1114, SHLF.

100. C. H. Hine to V. K. Rowe, March 4, 1957, document no. 1115, SHLF.

101. F. W. Fletcher, "Proposed Release to Sales" (Agricultural Chemicals, The Dow Chemical Company, February 26, 1957), document no. 17196, SHLF.

102. "Fifty Years in Agricultural Chemicals," *Down to Earth,* Winter 1957, 9.

103. *Label Bulletin: Specimen Label, Nemagon EC-1* (New York: Shell Chemical Corporation, Agricultural Chemical Sales Division, 1957), document no. 17040, SHLF.

104. L. G. Smith, Memo to Technical Service Department Manager: Export Label—87.2% Nemagon Soil Fumigant Emulsible Concentrate, October 3, 1958, document no. 1680, SHLF.

105. *Agricultural Chemical Development Data Sheet* (Midland, MI: The Dow Chemical Company, 1956), document no. 3581, SHLF.

106. *Warning Labels: A Guide for the Preparation of Warning Labels for Hazardous Chemicals, Manual L-1.*

107. H. H. Anderson, C. H. Hine, and R. J. Guzman, "U.C. Report No. 278: Dibromochloropropane: 50 Vapor Exposures and Ancillary Blood Studies" (San Francisco: Department of Pharmacology and Experimental Therapeutics, University of California School of Medicine, April 21, 1958), document no. 1139, SHLF, 4–9.

108. Ibid., 6.

109. Ibid.

110. *Results of Repeated Exposure of Laboratory Animals to Various Concentrations of 1,2-Dibromo-3-Chloropropane* (Biochemical Research Laboratory, The Dow Chemical Company: July 23, 1958) document no. 1106, SHLF, 2

111. Ibid., 8.

112. Louis Lykken to Mitchell Zavon, June 4, 1958 (Plaintiff's Exhibit 2), CHLF.

113. *Results of Repeated Exposure of Laboratory Animals to Various Concentrations of 1,2-Dibromo-3-Chloropropane*, 2.

114. M. R. Zavon and C. A. Wilzbach, "Radiation Control Activities in a Local Health Department," *Public Health Reports* 74, no. 5 (1959): 439–40, http://www.ncbi.nlm.nih.gov/entrez/query.fcgi?cmd = Retrieve&db = PubMed&dopt = Citation&list_uids = 13658332; Daniel, *Toxic Drift*.

115. Mitchell R. Zavon to Lou Mitchell, June 15, 1958, document no. 5889, SHLF.

116. W. A. McGilvray, Memo to Operations Department Manager: Atmospheric Contamination Survey—Nemagon Operations—Building 471, December 31, 1958, document no. FL-08659, SHLF.

117. H. R. Hoyle, *Vapor Exposures Encountered by Workmen While Fumigating Soil at Wellston, Michigan* (Biochemical Research Laboratory, The Dow Chemical Company, June 4, 1958), document no. 4222, SHLF, 1.

118. Ibid.

119. Mark A. Wolf to F. W. Fletcher: *Toxicological Information on Fumazone Suitable for Use in a Bulletin* (Midland, MI: Biochemical Research Laboratory, Dow Chemical Company, July 10, 1959), document no. 1105, SHLF.

120. *Fumazone 70E* (Midland, MI: Agricultural Chemical Development, The Dow Chemical Company, 1959), document no. 165, SHLF.

121. *Nematocide, Plant Regulator, Defoliant, and Desiccant Amendment of 1959*, P.L. 86–139 (1959).

122. M. J. Sloan to George Lynn, September 27, 1960, document no. 3053, SHLF; Louis Lykken, Memo to Sales Development Manager, June 5, 1959, document no. 3073, SHLF.

123. J. J. Lawler to George P. Larrick, January 11, 1960, document no. 3064, SHLF; J. K. Kirk to J. J. Lawler and Calvin A. Campbell, March 4, 1960, document no. 4049, SHLF.

124. Deputy Commissioner of Food and Drugs, "Tolerances and Exemptions from Tolerances for Pesticide Chemicals in or on Raw Agricultural Commodities: Extension of Effective Date of Public Law 86–139 as it Affects Section 408 of the Federal Food, Drug, and Cosmetic Act," 26 *Federal Register* 3352 (April 12, 1961).

125. Ken Olson, *Results of Range Finding Toxicological Tests on M-1678 (Fumazone 70E)* (Biochemical Research Laboratory, The Dow Chemical Company, 1961), document no. 16451, SHLF, 2.

126. V. K. Rowe to Barrett L. Collier, September 6, 1960, document no. 17397, SHLF.

127. L. E. Mitchell to M. R. Sprinkle, October 14, 1960, document no. 36372 (Plaintiff's Exhibit 18), CHLF.

128. Ibid.

129. *Laboratory Analysis (Af-1859)* (Shell Chemical Company, Union Technical Service Laboratory, Agricultural Chemicals Department, November 17, 1960), document no. 5887, SHLF.

130. Mark A. Wolf to F. W. Fletcher: *Toxicological Information on Fumazone Suitable for Use in a Bulletin.*

131. Mitchell R. Zavon to Charles Poucher, October 27, 1960, document no. 93955, SHLF.

132. *Information Sheet: Application of Fumazone 70E in Irrigation Water* (Midland, MI: Agricultural Chemical Development, The Dow Chemical Company, 1960), document no. 4590, SHLF.

133. S. H. McAllister and H. H. Mcintyre to Chief, Plant Pest Control Division, February 6, 1961, document no. 3044, SHLF.

134. For example, see J. K. Kirk to J. J. Lawler and Calvin A. Campbell, March 4, 1960; G. E. Lynn to F. W. Fletcher, March 23, 1961, document no. 4063, SHLF; Justus C. Ward to S. N. McAllister, March 30, 1961, document no. 3033, SHLF.

135. W. W. Sunderland Econ-O-Gram to G. E. Lynn March 30, 1961, document no. 4034, SHLF.

136. Thomas H. Harris to Dr. C. C. Compton, April 4, 1961, document no. 4033, SHLF.

137. Justus C. Ward to S. N. McAllister, March 30, 1961.

138. Ibid.

139. C. C. Compton, Memo, July 21, 1961, document no. 1634, SHLF.

140. Ibid., 3.

141. Ibid.

142. Ibid.

143. Louis Lykken, Memo, August 21, 1961, document no. 1631, SHLF.

144. M. R. Sprinkle, Memo, November 9, 1961, document no. 1627, SHLF.

145. Lykken, Memo, August 21, 1961.

146. Sprinkle, Memo, November 9, 1961.

147. Of course, this ethical standard was sometimes egregiously violated, as with the now-notorious Tuskegee Syphilis Trials and with the lesser-known DDT "feeding" study that the Public Health Service's Wayland Hayes carried out in a population of prisoners, in violation of the Nuremberg Code. See Daniel *Toxic Drift*, 34. The idea that animal testing is the best proxy for human health effects has frequently been disputed, often by defendants in lawsuits arguing that their products do not have the same deleterious effects on humans as they do on animals. However, animal testing remains a cornerstone of toxicology and an important indicator of human toxicity. For a discussion of this controversy, see Ronald Brickman, Sheila Jasanoff, and Thomas Ilgen, *Controlling Chemicals: The Politics of Regulation in Europe and the United States* (Ithaca: Cornell University Press, 1985); J. Ziegler, "Health Risk Assessment Research: The OTA Report," *Environmental Health Perspectives* 101, no. 5 (1993); M. Weideman, "Toxicity Tests in Animals: Historical Perspectives and New Opportunities," *Environmental Health Perspectives* 101, no. 2 (1993).

148. Dunbar, "The FDA Looks at Insecticides," 236.

149. T. R. Torkelson et al., "Toxicological Investigations of 1,2-Dibromo-3-Chloropropane," *Toxicology and Applied Pharmacology* 3 (1961).

150. Ibid., 557.

151. Ibid., 558.

152. Louis Lykken to Charles Hine, March 5, 1962, document no. FSH-CR2000034, CHLF.

153. C. H. Hine, *The Toxicology and Pharmacology of 1,2-Dibromo-3-Chloropropane: Draft with Comments* (1962), document no. FSH-CR2000019–28, CHLF, 2.

154. Ibid, 6.

155. M. R. Zavon to Louis Lykken, September 28, 1961, FSH-CR2000030, SHLF.

156. *Summary of Use Experiences in Manufacture and Application of Nemagon Soil Fumigant* (Shell Chemical Company, Agricultural Chemicals Division, 1962), CHLF.

157. Sprinkle, Memo, November 9, 1961.

158. *Summary of Use Experiences in Manufacture and Application of Nemagon Soil Fumigant*, 1.

159. L. E. Mitchell to M. R. Sprinkle, October 14, 1960; *Laboratory Analysis (Af-1859)*.

160. *Summary of Use Experiences in Manufacture and Application of Nemagon Soil Fumigant*, 2.

161. Ibid.

162. Ibid.

163. Ibid.

164. Trost, *Elements of Risk*, 47.

165. *Report to Shell Chemical Company: Acute Dermal Toxicity Study on Nemagon Soil Fumigant* (Northbrook, Il: Industrial Bio-Test Laboratories, Inc., 1962), CHLF.

166. In 1977 and 1978, it was discovered that IBT had falsified data with the knowledge of at least some of its clients. Their DBCP test data was not implicated in the discovery. On the IBT scandal, see "The Darker Side of a Laboratory," *Chemical Week*, May 1986.

167. William Stokes to S. N. McAllister and Calvin A. Campbell, September 27, 1961, document no. 3014, SHLF.

168. Ibid.

169. D. W. Ullrich, memo, December 26, 1961, document no. 3006, SHLF; G. E. Lynn to Don. G. Koehlinger January 9, 1962, document no. 4098, SHLF; Robert S. Roe to G .E. Lynn and C. C. Compton, March 27, 1962, document no. 2994, SHLF.

170. Commissioner of Food and Drugs, "Tolerances and Exemptions from Tolerances for Pesticide Chemicals in or on Raw Agricultural Commodities," 28 Fed. Reg. 3021 (May 18, 1963).

171. *Notice of Registration under the Federal Insecticide, Fungicide, and Rodenticide Act: Registration Number 201–140, Nemagon C Soil Fumigant* (U.S. Department of Agriculture, Agricultural Research Service, Pesticides Regulation Division,

March 5, 1964), document no. 1823, SHLF; *Notice of Registration under the Federal Insecticide, Fungicide, and Rodenticide Act: Registration No. 464–322, Fumazone 121* (Washington, DC: U.S. Department of Agriculture, Agricultural Research Service, Pesticides Regulation Division, March 17, 1964), document no. 805, SHLF.

172. *Notice of Registration under the Federal Insecticide, Fungicide, and Rodenticide Act: Registration Number 201–140, Nemagon C Soil Fumigant.*

CHAPTER TWO

1. *Crop Recommendations and General Information on Dow Soil Fumigants* (Midland, MI: Dow Chemical Company, 1966), document no. 145, SHLF; *Nemagon Soil Fumigant Emulsible Concentrate* [Label] (New York: Shell Chemical Company, 1966), document no. 1931, SHLF.

2. Vernon White, Memo, August 28, 1969, document no. 2887, SHLF; E. L. Hobson to William Miller, Registration Division, Environmental Protection Agency, July 9, 1975, document no. 2934, SHLF.

3. Dow Chemical Company and Best Fertilizers Company, "Agreement" (n.p. : November 9, 1964), document no. 380, SHLF.

4. *Chemistry in the Economy* (Washington DC: American Chemical Society, 1973).

5. Eula Bingham, "Occupational Exposure to 1,2-Dibromo-3-Chloropropane (DBCP)," 43 Federal Register 11514–11533 (March 17, 1978).

6. R. W. Morgan, to C. Kenneth Bjork, December 27, 1974, document no. 3359, SHLF; "Fumazone," (n.p.: Dow Chemical Company, November, 1975), document no. 619, SHLF.

7. Manuel García y Griego, "The Importation of Mexican Contract Laborers to the United States, 1942–1964," in *Between Two Worlds: Mexican Immigrants in the United States,* ed. David G. Gutiérrez (Wilmington, DE: Scholarly Resources, 1996); Ernesto Galarza, *Merchants of Labor: The Mexican Bracero Story* (Charlotte, NC and Santa Barbara, CA: NcNally & Loftin, 1964).

8. For explorations of racial and nationalist thought on the experience of farmworkers and Latino immigrants more generally in the United States during the twentieth century, see Mae M. Ngai, *Impossible Subjects: Illegal Aliens and the Making of Modern America,* Politics and Society in Twentieth-Century America (Princeton, N.J.: Princeton University Press, 2004); Kathleen Mapes, *Sweet Tyranny: Migrant Labor, Industrial Agriculture, and Imperial Politics* (Urbana: University of Illinois Press, 2009); Laura Pulido, *Environmentalism and Economic Justice* (Tucson: University of Arizona Press, 1996); Leo R. Chavez, *The Latino Threat: Constructing Immigrants, Citizens, and the Nation* (Stanford, California: Stanford University Press, 2008); Aviva Chomsky, *Undocumented: How Immigration Became Illegal* (Boston: Beacon Press, 2014).

9. Sean A. Andrade, "Biting the Hand that Feeds You: How Federal Law has Permitted Employers to Violate the Basic Rights of Farmworkers and How This has Begun to Impact Other Industries," *University of Pennsylvania Journal of Labor*

& *Employment Law* 4 (2002), 609. Varden Fuller and Bert Mason, "Farm Labor," *Annals of the American Academy of Political and Social Science* 429 (1977): 76.

10. Varden Fuller and Bert Mason, "Farm Labor," 74–76; Guadalupe T. Luna, "An Infinite Distance? Agricultural Exceptionalism and Agricultural Labor," *University of Pennsylvania Journal of Labor & Employment Law* 1 (1998); "Labor Laws for Farmworkers," accessed June 4 2011, Farmworker Justice, http://www.fwjustice. org/labor-law; Linda C. Majka and Theo J. Majka, *Farm Workers, Agribusiness, and the State* (Philadelphia: Temple University Press, 1982).

11. Irma West, "Occupational Diseases of Farmworkers," *Archives of Environmental Health* 9 (1964): 92.

12. Douglas Costle, "Notice of Intent to Suspend and Conditionaly Suspend Registrations of Pesticide Products Containing Dibromochloropropane," 42 Federal Register 48915 (September 8, 1977), 48919

13. Jeff Kempter and Mitchell H. Bernstein, *Dibromochloropropane: Final Position Document* (Washington, DC: Special Pesticide Review Division, Office of Pesticide Programs, U.S. Environmental Protection Agency, September 6, 1978), Table 4.

14. Ralph F. Glasser, memo, 1968, document no. 3780, SHLF.

15. Richard Hawkins, *A Pacific Industry: The History of Pineapple Canning in Hawaii* (London, New York: I.B. Tauris, 2011), 143–164.

16. Earl J. Anderson, "Toxicity of BBC—a Warning," *PRI News* 11, no. 5 (1963): 150–51, SHLF

17. For accounts of the development of the banana industry in Central America see John Soluri, *Banana Cultures: Agriculture, Consumption, and Environmental Change in Honduras and the United States* (Austin: University of Texas Press, 2005); Lara Putnam, *The Company They Kept: Migrants and the Politics of Gender in Caribbean Costa Rica, 1870–1960* (Chapel Hill: University of North Carolina Press, 2002); Charles D. Kepner and Jay H. Soothill, *The Banana Empire: A Case Study of Economic Imperialism* (New York: Vanguard Press, 1935); Charles D. Kepner, *Social Aspects of the Banana Industry* (New York: AMS Press, 1967 [1936]); Stacy May and Galo Plaza, *The United Fruit Company in Latin America* (Washington, DC: National Planning Association, 1958); Thomas Lindas Karnes, *Tropical Enterprise: The Standard Fruit and Steamship Company in Latin America* (Baton Rouge: Louisiana State University Press, 1978); Thomas David Schoonover, *The United States in Central America, 1860–1911: Episodes of Social Imperialism and Imperial Rivalry in the World System* (Durham, NC: Duke University Press, 1991); Edelberto Torres-Rivas, *History and Society in Central America,* Translations from Latin America Series (Austin: University of Texas Press, 1993); Daniel J. Faber, *Environment under Fire: Imperialism and the Ecological Crisis in Central America* (New York: Monthly Review Press, 1993); Marcelo Bucheli, *Bananas and Business: The United Fruit Company in Colombia, 1899–2000* (New York: New York University Press, 2005); Marcelo Bucheli, "Multinational Corporations, Totalitarian Regimes, and Economic Nationalism: United Fruit Company in Central America, 1899–1975," *Business History Review* 50, no. 4 (2008): 433–54; Marcelo Bucheli, "Multinational

Corporations and the Politics of Vertical Integration: The Case of the Central American Banana Industry in the Twentieth Century (n.p.: n.d.), http://www. economia.unam.mx/amhe/pdfs/banano_bucheli.pdf.; and many of the excellent essays in Mark Moberg and Steve Striffler, eds., *Banana Wars: Power, Production, and History in the Americas* (Durham, NC and London: Duke University Press, 2003). For more general accounts of Central American history and U.S.-Central American relations, see Torres-Rivas; Schoonover; Walter LaFeber, *Inevitable Revolutions: The United States in Central America,* 2nd ed. (New York: W.W. Norton, 1993); John A. Booth, Christine J. Wade, and Thomas W. Walker, *Understanding Central America: Global Forces, Rebellion, and Change,* 5th ed. (Boulder, CO: Westview Press, 2010); Thomas W. Walker and Christine J. Wade, *Nicaragua: Living in the Shadow of the Eagle,* 5th ed. (Boulder, CO: Westview Press, 2011). Recent scholarship showing how banana companies exercised political and economic power in Central America include Philippe Bourgois's and Mark Moberg's contributions to Moberg and Striffler, eds.

18. Soluri, *Banana Cultures,* especially chapters 2 and 4; Steve Marquardt, "'Green Havoc': Panama Disease, Environmental Change, and Labor Process in the Central American Banana Industry," *American Historical Review* 106 no. 1 (2001). Steve Marquardt describes how a group of Costa Rican "organized workers" at the United Fruit Company made the health effects of Bordeaux-mixture exposure central to their organizing and grievances in his article "Pesticides, Parakeets and Unions in the Costa Rican Banana Industry, 1938–1962," *Latin American Research Review* 37, no. 2 (2001).

19. United Fruit Company Department of Research, *Problems and Progress in Banana Disease Research* (Boston: United Fruit Company, 1958), iv.

20. On how racial difference and racism shaped workers' experiences, see Putnam, *The Company They Kept;* Philippe Bourgois, *Ethnicity at Work: Divided Labor on a Central American Banana Plantation* (Baltimore: Johns Hopkins University Press, 1989).

21. For more on the labor carried out by banana workers, see Marquardt, "Pesticides, Parakeets and Unions"; Marquardt, "Green Havoc'"; Soluri, *Banana Cultures;* "Salud ocupacional en el sector bananero centroamericano: Salud ocupacional en trabajadores del Valle de la Estrella-Standard Fruit C. Costa Rica," *Revista centroamericana de ciencias de la salud* 4, no. 9 (1978): 18–23. Also see banana worker protest novels including Carlos Luis Fallas, *Mamita Yunai* (San José: Editorial Costa Rica, 2010); Ramón Amaya Amador, *Prisión Verde* (Buenos Aires: Ágape, 1957).

22. The history of the banana industry includes many examples of individual and organized resistance on the part of independent growers and workers. See Juan Pablo Pérez Sáinz, *From the Finca to the Maquila: Labor and Capitalist Development in Central America* (Boulder, CO: Westview Press, 1999), chapter 1; Soluri, *Banana Cultures;* Aviva Chomsky, *West Indian Workers and the United Fruit Company in Costa Rica, 1870–1940* (Baton Rouge: Louisiana State University Press, 1996); Avi Chomsky, "Afro-Jamaican Traditions and Labor Organizing on United Fruit Company Plantations in Costa Rica, 1910," *Journal of Social History* 28, no. 4 (1995):

837–55; Avi Chomsky, "West Indian Workers in Costa Rican Radical and National-ist Ideology 1900–1950," *Americas: A Quarterly Review of Inter-American Cultural History* 51, no. 1 (1994): 11–40; Darío Euraque, "The Threat of Blackness to the Mestizo Nation: Race and Ethnicity in the Honduran Banana Economy," in *Banana Wars: Power, Production and History in the Americas,* ed. Steve Striffler and Mark Moberg (Durham, NC and London: Duke University Press, 2003); Putnam, *The Company They Kept.*

23. For careful considerations of change in banana labor, production, and geog-raphy at around midcentury, see Soluri, *Banana Cultures*; Mario Bucheli, "United Fruit Company in Latin America," in *Banana Wars: Power, Production, and History in the Americas,* eds. Steve Striffler and Mark Moberg (Durham, NC and London: Duke University Press, 2003); Carolyn Hall, Héctor Pérez Brignoli, and John V. Cotter, *Historical Atlas of Central America* (Norman: University of Oklahoma Press, 2003), 4.20 and 4.21.

24. On the Central American Common Market (CACM) in historical eco-nomic context, see V. Bulmer-Thomas, "Economic Development over the Long Run: Central America since 1920," *Journal of Latin American Studies* 15, no. 2 (1983). Despite the successes of the CACM, Bulmer-Thomas argues that "the collapse of the CACM [by the early 1980s] . . . led to further intensification of the export-led model," 271.

25. In 1954, a strike of dock workers in Honduras's banana-growing North Coast led to another *gran huelga* in the region—a general strike that lasted over three months and ended with workers winning pay and benefit improvements, as well as the state's recognition of their right to organize and the development of a national labor code in 1959. This victory came with the involvement of the Inter-American Regional Labor Organization (Organización Regional Interamericana de Trabajadores, ORIT), which would remain involved in labor politics in the region over the following years, in an attempt to marginalize Communist unions and build support for U.S. policies. See Darío A. Euraque, *Reinterpreting the Banana Repub-lic: Region and State in Honduras, 1870–1972* (Chapel Hill: University of North Carolina Press, 1996); Andrew Herod, *Labor Geographies: Workers and the Land-scapes of Capitalism,* Perspectives on Economic Change (New York: Guilford Press, 2001); Soluri, *Banana Cultures.*

26. Marquardt, "Pesticides, Parakeets and Unions," 13. For an analysis of Costa Rican banana strikes from 1900 to 1955, see Carlos Hernández Rodríguez, "Del espontaneísmo a la acción concertada: Los trabajadores bananeros de Costa Rica, 1900–1955," *Revista de Historia,* no. 31 (1995).

27. The Central Intelligence Agency and other U.S. government actors success-fully plotted the overthrow of Guatemala's populist president Jacobo Árbenz in 1954. Árbenz's ouster was followed by state terror, antiunion activity, and the dis-mantling of populist gains won over the preceding decade. Banana company inves-tors nevertheless worried that the "political instability" signified by the defeated Guatemalan revolution would spread to other nations of the region and threaten the banana business.

28. A 1952 antitrust investigation of the company had required some divestment, but the mandate seemed to fit well with a broader shift that would see the company focus on shipping and marketing of bananas while retaining less land in direct production. See Mario Bucheli, "United Fruit Company in Latin America;" Soluri, *Banana Cultures*, 174–78.

29. James E. Austin, "Standard Fruit Co. In Nicaragua" (case study, Harvard Business School; Boston: Harvard University, 1994), 6.

30. While women had long worked in banana regions as cooks, launderers, and sex workers, packing-plant jobs were the first on-farm employment the companies offered them. Women's opportunities were circumscribed: they were only offered jobs in the new on-site packing plants. There, the *bananeras* cut bunches from the stem, then inspected the fruit, washed and treated it with a fungicide, then weighed, stickered, and packed the bunches in cardboard boxes. Women banana workers were exposed to new opportunities and new stresses, including unprotected exposure to chemicals and long hours standing on the concrete floors of the packing plants. Soluri, *Banana Cultures*; Dana Frank, *Bananeras: Women Transforming the Banana Unions of Latin America* (Cambridge, MA: South End Press, 2005); Putnam, *The Company They Kept*. For production changes in the early 1970s, see José Roberto López, *La economía del banano en Centroamérica*, 98–115.

31. José Roberto López, *La economía del banano en Centroamérica* (San José, Costa Rica: Editorial Departamento Ecuménico de Investigaciones, 1988).

32. Soluri, *Banana Cultures*, 207.

33. Q. L. Holdeman, *Annual Research Report* (United Fruit Company, 1959), cited in C. A. Loos, "Eradication of the Burrowing Nematode, Radopholus Similis, from Bananas," *Plant Disease Reporter* 45, no. 6 (1961): 457–58.

34. Fred E. Brooks, "Burrowing Nematode Disease," accessed 29 December, 2011, American Phytopathological Society, http://www.apsnet.org/edcenter/intropp/lessons/Nematodes/Pages/Burrowingnematode.aspx.

35. A. V. Adam and A. Rodriguez, "'Clean Seed' and 'Certified Seed' Programmes for Bananas in Mexico," *FAO Plant Protection Bulletin* 18, no. 3 (1970): 57.

36. N. M. Nayar, "The Bananas: Botany, Origin, Dispersal," in *Horticultural Reviews* 36, ed, J. Janick (New York: John Wiley & Sons, 2010), doi: 10.1002/9780470527238.ch2.

37. R. Leach, "Blackhead Toppling Disease of Bananas," *Nature* 181, no. 4602 (1958): 204–205; H. O'Bannon, "Worldwide Dissemination of Radopholus Similis and Its Importance in Crop Production," *Journal of Nematology* 9, no. 1 (1977).

38. Loos, "Eradication of Burrowing Nematode," 457; O'Bannon, "Worldwide Dissemination of Radopholus Similis," 17–18.

39. C. W. McBeth, Memorandum of Discussion: Observations on Nematode Problems in Jamaica, Bananas in Particular, July 21, 1959, document no. 3161, SHLF.

40. O'Bannon, "Worldwide Dissemination of Radopholus Similis," 17–18.

41. Robert Langdon, "The Banana as a Key to Early American and Polynesian History," *Journal of Pacific History* 28, no. 1 (1993): 16–17.

42. Woodville K Marshall, "Provision Ground and Plantation Labor in Four Windward Islands: Competition for Resources During Slavery," in *Cultivation and Culture: Labor and the Shaping of Black Life in the Americas,* eds. Ira Berlin and Philip Morgan (Charlottesville: University of Virginia, 1993); Dale Tomich, "'Une Petite Guinée': Provision Ground and Plantation in Martinique, 1830–1843," in *Cultivation and Culture: Labor and the Shaping of Black Life in the Americas,* eds. Ira Berlin and Philip Morgan (Charlottesville: University of Virginia, 1993), cited in Soluri, *Banana Cultures,* 6.

43. Soluri, *Banana Cultures,* 212–13.

44. E. J. Wehunt, D. J. Hutchison, and D. I. Edwards, "Reaction of Musa Acuminate to Radopholus Similis [Abstract]," *Phytopathology* 55, no. 10 (1965): 1082.

45. Leach, "Blackhead Toppling Disease of Bananas," 204.

46. "El banano: Sus plagas, enfermedades y malezas" (n.p.: Shell, 1959), 4.

47. H. C. Thornton to M. J. Sloan, April 21, 1961, document no. 17075, SHLF.

48. United Fruit Company Department of Research, iv–v.

49. Ibid., 29.

50. Ibid.

51. Wehunt, Hutchison, and Edwards, "Reaction to Radopholus Similis," 1082.

52. H. C. Thornton to M. J. Sloan, April 21, 1961.

53. K. E. Marple to Dr. J. L. Nickel, February 26, 1960, document no. 3163, SHLF.

54. H. C. Thornton to M. J. Sloan, April 21, 1961.

55. *Federal Insecticide, Rodenticide, and Fungicide Act,* section 3(b).

56. Shell Chemical Company and Dow Chemical Company, "Petition Proposing Tolerances for the Pesticide Chemical 1,2-Dibromo-3-Chloropropane in or on Certain Raw Agricultural Commodities" (February 6, 1961), document no. 16455, SHLF.

57. S. H. McAllister to Shell International Chemical Company Limited, March 12, 1962, document no. 2982, SHLF.

58. Shell Chemical Company and Dow Chemical Company, G-3.

59. It is worth noting that subsequent research has shown that children should receive special consideration when evaluating the health effects of pesticide consumption. Children are not only vulnerable due to their stage of development, but may eat large quantities of some items (such as bananas) in relation to their entire diet. See Philip J. Landrigan et al., "The Unique Vulnerability of Infants and Children to Pesticides," *Environmental Health Perspectives* 107, Supp 3 (1999).

60. Shell Chemical Company and Dow Chemical Company, F-1.

61. G. E. Lynn, Memorandum to Action File—Fumazone, March 7, 1962, document no. 4092, SHLF.

62. S. H. McAllister to Shell International Chemical Company Limited, August 16, 1962, document no. 2982, SHLF.

63. W. H. Stokes to S. H. McAllister and Calvin A. Campbell, March 6, 1963 document no. 4053, SHLF.

64. Commissioner of Food and Drugs, "Tolerances and Exemptions from Tolerances for Pesticide Chemicals in or on Raw Agricultural Commodities," 28 Fed. Reg. 3021 (May 18, 1963): 3021

65. D. J. Miner, memo, March 26, 1968, document no. 1417, SHLF.

66. "Cable to SICC London" (New York: Shell Chemical, Nov 16, 1967), document no. 1375, SHLF.

67. D. J. Miner to Sales Manager, Agricultural Chemicals Division, March 26, 1968, document no. 17684, SHLF.

68. Miner to Sales Manager, March 26, 1968.

69. Ibid.

70. N. L. Gianàkos, memo to Supervisor, New York Data Service Center, March 19, 1969, document no. 1441, SHLF; "Chemical Products Contract," in *Shell Chemical Company and Standard Fruit and Steamship Company* (August 12, 1968), document no. DBCP000907–913, SHLF. The dollar amount is equivalent to almost $5 million in 2013 dollars after adjustment for inflation.

71. "History: 1930–1969," last modified 2005, accessed December 29, 2005, Dole Food Company, http://www.dole.com/CompanyInformation/AboutDolee/History/History_19301969/tabid/1291/Default.aspx.

72. Hall, Pérez, and Cotter, *Historical Atlas of Central America*, 206; Austin, "Standard Fruit Co. in Nicaragua," 1.

73. Miner to Sales Manager, March 26, 1968.

74. Ibid.

75. D. J. Miner, memo, April 11, 1968, document no. 1350, SHLF.

76. D. J. Miner, memo, May 9, 1969, document no. 1445, SHLF.

77. D. J. Miner to Sales Manager, Agricultural Chemicals Division, March 26, 1968.

78. Miner, memo, May 9, 1969.

79. Jack DeMent to L. V. White, June 12, 1970, document no. 6457, SHLF.

80. "El banano: Sus plagas," 2,5.

81. T. Easterling to J. Neafcy, September 17, 1970, document no. 5641, SHLF.

82. "Estimate of DBCP Usage (Gallons)" (n.p.: [Standard Fruit Company], n.d., n.d. [1972]), SHLF.

83. J. B. Smith to C. E. Mumaw, October 18, 1972, document no. 10252, SHLF.

84. "Company Info History," accessed February 24, 2011, Chiquita Brands International, http://www.chiquitabrands.com/CompanyInfo/History.aspx.

85. J. D. Dickson and H. E. Ostmark, memo, October 9, 1969, document no. C1 001845–51, SHLF; *Annual Report 1970* (United Fruit Company Department of Research, 1970), document no. 13854, SHLF, 25, 28–36; J. J. Lawler, memo to San Ramon—Agricultural Division Marketing Manager, April 29, 1971, document no. 17722, SHLF.

86. W. C. Moye, memo to Walnut Creek Agricultural Divison Manager Regulatory Affairs, July 16, 1970, document no. 1390, SHLF; W.C. Moye, memo to Walnut Creek Agricultural Division Regulatory Affairs Manager, September 14, 1970, document no. 1388, SHLF; W.G.C.Forsyth to Mr. W. T. Van Diepen, May 22, 1970, document no. 9229, SHLF.

87. *Annual Report 1970*.

88. *Annual Report 1972* (United Brands Company Division of Tropical Research, 1972), document no. 13855, SHLF, 369.

89. *Annual Report 1973* (United Brands Company Division of Tropical Research, 1973), 7, document no. 12761, SHLF, 7.

90. *Annual Report 1974* (United Fruit Company Division of Tropical Research, 1974), 15, document no. 13858, SHLF, 15.

91. H. E. Ostmark to P. C. Butler, October 26, 1978, document no. 13726, SHLF.

92. Reynolds, "The Global Banana Trade," 26–27.

93. W. G. C. Forsyth to W. T. Van Diepen, July 12, 1973, document no. 9249, SHLF.

94. E. V. Daley to Paul Sink, July 1, 1976, document no. 615, SHLF.

95. M. D. Sullivan to P. G. Sink, November 13, 1975, document no. 5571, SGLF.

96. *Benefits of DBCP Use on Bananas* (Castle & Cooke, November 29, 1977), document no. DC0019642–59, SHLF, 5.

97. Sandra Piszk, "Expediente No. 250–23–98: Informe Final con Recomendaciones" (San José: Defensoría de los Habitantes de la República de Costa Rica, 1998), 16–17, CQLF; "Reglamento de Seguridad sobre Empleo de Sustancias Tóxicas en la Agricultura [Decreto No. 6]," La Gaceta [Costa Rica] 215 (21 December, 1968); "Ley General de Salud [Ley No. 5395]," La Gaceta [Costa Rica] 222 (November 24, 1973); Lori Ann Thrupp, "Pesticides and Policies: Approaches to Pest-Control Dilemmas in Nicaragua and Costa Rica," *Latin American Perspectives* 15, no. 4 (2004): 46–48, 54–58; "Reglamento sobre Importación, Distribución y Uso de Productos Químicos y Químico Biológicos para la Industria Agropecuaria," La Gaceta [Nicaragua] 104 (May 12, 1960); "Ley Reguladora sobre Importación, Elaboración, Almacenamiento, Transporte, Venta y Uso de Pesticidas [Decreto Numero 43–74]," Diario de Centro América [Guatemala] 93 (5 June, 1974); "Decreto No. 384," Gaceta Oficial [Panamá] 16.016 (December 11, 1967).

98. Political scientist Lori Ann Thrupp describes how, even in Costa Rica—where, in contrast to other Central American nations, a democratic welfare state had been established—the "lack of economic, institutional, educational, and political support" rendered many of the laws poorly enforced and, at worst, "purely rhetorical." Thrupp, "Pesticides and Policies," 51.

99. "Salud ocupacional en el sector bananero centroamericano, 9–67; Soluri, *Banana Cultures*. Soluri writes of Honduras in the 1950s–70s that the health effects of pesticides were usually only attended to by researchers when they threatened productivity; see his chapter "La Química," especially page 208.

100. Jorge Colindres Cárcamo, "Deposition," in *Jorge Colindres Carcamo et al. vs. Shell Oil Company et al. Civil Action No. H-94–2337* (District Court of Brazoria County, Texas, January 31, 1994), document no. 9596, SHLF.

101. Ibid., 15–20.

102. Germán Muñoz Moncada, "Deposition," in *Jorge Colindres Carcamo et al. vs. Shell Oil Company et al.* (District Court of Brazoria County, Texas; 23rd Judicial District; No. 93-C-2290, March 3, 1994), document no. 8429, SHLF.

103. Ibid., 13–15.

104. Ibid., 23–24.

105. Manuel Antonio Balderamos Erazo, "Deposition," in *Manuel Antonio Balderamos Erazo vs. Shell Oil Company; No. C-3665-93-D* (District Court of Hildalgo County, Texas, 206th Judicial District, July 12, 1996), document no. 9667, SHLF, 75–76.

106. Ernesto Garbanzo García, "Deposition," in *Jorge Colindres Carcamo et al. vs. Shell Oil Company et al.* (District Court of Brazoria County, Texas; 23rd Judicial District, March 17, 1994), document no. 9626, SHLF, 17–18; Jairo Elías Ramírez Contreras, "Deposition," in *Narciso Borja et al. vs. The Dow Chemical Company, et al.* (District Court of Dallas County, Texas, 116th Judicial District, November 12, 1996), document no. 9417, SHLF, 109–110.

107. Garbanzo, "Deposition," 31.

108. Ramírez, "Deposition", 110–112; Garbanzo, "Deposition," 62.

109. Hermenegildo Salazar Marroquín, "Deposition," in *Jorge Colindres Carcamo et al. vs. Shell Oil Company et al.; No. 93-C-2290* (District Court of Brazoria County, Texas, 23rd Judicial District, February 8, 1994), document no. 8514, SHLF, 154–160.

110. Garbanzo, "Deposition," 103.

111. Garbanzo, "Deposition," 18–20; Salazar, "Deposition, 56–57."

112. Salazar, "Deposition," 193–195.

113. Garbanzo, "Deposition," 55.

114. Ibid.

115. Johannes W. Klink to Jack DeMent, January 12, 1971, document no. 10304, SHLF.

116. "DBCP en la producción bananera," last modified 1999, accessed June 6, 2001, Foro Emaús, http://www.members.tripod.com/foro_emaus/dbcp.htm; Salazar, "Deposition," 131.

117. "DBCP en la producción bananera."

118. D.J. Miner memo to New York—Agricultural Chemicals Division Pesticide Regulations Department—Manager, March 30, 1970, document no. 6569, SHLF.

119. Ramírez, "Deposition," 127.

120. Garbanzo, "Deposition," 163–64.

121. Gerardo Dennis Patrickson, interview with the author, December 5, 2004, Siquirres, Costa Rica.

122. Garbanzo, "Deposition," 26–27.

123. W.F. Mai, "Appraisal of the Nematode Situation in the Bocas, Armuelles, and Golfito Divisions of the United Brands Company, March 28-April 3, 1976 " (n.p.: [United Brands Company], [1976]), document no. 7254, SHLF, 4.

124. Robert Dunn to Jack DeMent, July 15, 1974, document no. 10275, SHLF.

125. Rafael Guilbert, "Deposition," in *Franklin Rodriguez Delgado and Jorge Colindres Carcamo et al. vs. Shell Oil Company et al.; Civil Action No. H-94-2337.* (United States District Court for the District of Southern Texas, Houston Division, October 3, 1995), document no. 9646, SHLF, 81, 95.

126. Linda Nash, *Inescapable Ecologies: A History of Environment, Disease, and Knowledge* (Berkeley: University of California Press, 2003), chapter 4. Linda Nash ascribes the faith in protective equipment to a broader modern understanding of disease and ill health that depended on laboratory models to study chemicals in a carefully controlled context, looked at single chemicals in isolation, and saw the human body and the environment in which it existed as separate and discrete.

127. Richard F. Mountfort to Shell Chemical Company, January 29, 1974, document no. 2779, SHLF.

128. Ibid.

129. Guilbert, "Deposition," 69. In fact, Panama had required Spanish-language labels beginning in 1967, and Guatemala in 1974. "Ley Reguladora sobre Importación, Elaboración, Almacenamiento, Transporte, Venta y Uso de Pesticidas"; "Decreto No. 384."

130. Edwin L. Johnson, "Global Impacts of U.S. Pesticide Regulations" (paper presented at the Proceedings of the U.S. Strategy Conference on Pesticide Management, Washington DC, July 7–8 1979). For more on the 1978 amendments, see chapter 3.

131. Garbanzo, "Deposition," 110; Ramírez, "Deposition," 218.

132. Balderamos, "Deposition," 34.

133. Ramírez, "Deposition," 218; Dennis, interview with author, December 6, 2004.

134. Salazar, "Deposition," 32.

135. Garbanzo, "Deposition," 30–31.

136. Dennis, interview with author.

137. Garbanzo, "Deposition," 31.

138. Dennis, interview with author.

139. Ramírez, "Deposition,"111.

140. Garbanzo, "Deposition," 171.

141. W. S. Black to Jack DeMent, November 19, 1971, document no. 10289, SHLF; Jack DeMent to W. S. Black, November 22, 1971, document no. 10290, SHLF; Jack DeMent memo to J. W. Klink, October 28, 1971, document no. 10291, SHLF; Johannes W. Klink to Jack DeMent, October 18, 1971, document no. SFC0000239–240, SHLF.

142. Johannes W. Klink to Jack DeMent, January 12, 1971.

143. P. L. Taylor to W. Wade, December 19, 1973, document no. 9268, SHLF.

144. V. C. Heyl to W. T. Van Diepen, September 7, 1973, document no. 10291, SHLF.

145. Robert Dunn to M. Morillo, J. Kidier, F. Amaya, December 19, 1974, document no. 11287, SHLF.

146. T. Tobin and P. Bennett to Productores Asociados, December 9, 1969, document no. SFC50000494, SHLF.

147. Johannes W. Klink to Jack DeMent, January 22, 1971, document no. 10302, SHLF.

148. Johannes W. Klink to Jack DeMent, February 4, 1971, document no. 10292, SHLF.

149. Norma Stoltz Chinchilla, "Class Struggle in Central America: Background and Overview," *Latin American Perspectives* 7, no. 2/3 (1980): 8–9; Juan Manuel Sepúlveda Malbrán, ed. *Las organizaciones sindicales centroamericanas como actores del sistema de relaciones laborales* (San José, Costa Rica: Oficina Internacional de Trabajo [International Labor Organization], 2003); Jeffrey L. Gould, *To Lead as Equals: Rural Protest and Political Consciousness in Chinandega, Nicaragua, 1912–1979* (Chapel Hill: University of North Carolina Press, 1990), part 3; Walker and Wade, *Nicaragua: Living in the Shadow of the Eagle*; Thomas W. Walker, *Revolution & Counterrevolution in Nicaragua* (Boulder, CO: Westview Press, 1991), 31–43.

150. Chinchilla, "Class Struggle in Central America."

151. López, *La economía del banano*, 26.

152. P. L. Taylor to R. H. Stover, November 1, 1976, document no. C1 004163–64, SHLF.

153. Johannes W. Klink to O. N. Bellavita, January 22, 1971, document no. 10303, SHLF.

154. Johannes W. Klink to Jack DeMent, October 18, 1971.

155. Steve Johnson to W. T. Van Diepen, February 6, 1974, document no. C1 003010–11,

156. *Annual Report 1974*, 292–297; *Annual Report, Volume 1: Bananas and Plantains, 1975* (United Brands Company Division of Tropical Research,1975), document no. 13859, SHLF, 207–208.

157. J. Jeyaratman, "Acute Pesticide Poisoning: A Major Global Health Problem," *World Health Statistics Quarterly* 43 (1990): 139–44.

158. Robert Dunn to Jack DeMent, July 15, 1974.

159. Salazar, "Deposition," 141.

160. *Annual Report, Volume 1: Bananas and Plantains, 1975*, 209.

161. Ibid.

162. Ibid., 206.

163. R. H. Stover and P. L. Taylor to W. G. C. Forsyth, April 5, 1976, document no. C1 004104, SHLF.

164. Ralph L. Renzi to Herb Hamer, June 11, 1976, document no. 7252, SHLF. I was not able to locate the underlying report to which this memo referred.

165. P. L. Taylor to R. H. Stover, November 1, 1976, document number C1002320, CHLF.

166. Jack DeMent to P. W. Cox, April 13, 1977, document no. 291, SHLP.

CHAPTER THREE

1. Josh Hanig and David Davis, *Song of the Canary* (Wayne, NJ: New Day Films, 1978).

2. Angus Lindsay Wright, "Rethinking the Circle of Poison: The Politics of Pesticide Poisoning among Mexican Farm Workers," *Latin American Perspectives* 13 (1990): 26–59.

3. For an investigative journalist's account of Occidental workers' experience, see Cathy Trost, *Elements of Risk: The Chemical Industry and Its Threat to America* (New York: Times Books, 1984).

4. Ibid., 218.

5. Quoted in Hanig and Davis, *Song of the Canary*.

6. Ibid.

7. Ibid.

8. Rachel Carson's *Silent Spring* brought increased attention to the dangers of pesticides upon its publication in 1962. New environmentalist groups such as the Environmental Defense Fund used lawsuits and other tactics to force government response, and the United Farm Workers (UFW) brought attention to the dangers pesticides posed to farmworkers, with local and national actions ranging from strikes to popular theater to their famous grape boycotts. Laura Pulido and Linda Nash each provide nuanced discussions of the challenges and failures of UFW strategy, especially in relation to linking consumer concerns about pesticides in food to worker concerns about exposures in the field. Rachel Carson, *Silent Spring* (Cambridge, Massachusetts: The Riverside Press, 1962); Robert Gottlieb, *Forcing the Spring: The Transformation of the American Environmental Movement* (Washington, DC: Island Press, 2005), 121–126, 243–248; Laura Pulido, *Environmentalism and Economic Justice* (Tucson: University of Arizona Press, 1996), chapter 3; Linda Nash, *Inescapable Ecologies: A History of Environment, Disease, and Knowledge* (Berkeley: University of California Press, 2003), 161–166.

9. See "Earth Day: The History of a Movement," Earth Day Network, accessed October 2, 2013, http://www.earthday.org/earth-day-history-movement. For a helpful consideration of how Earth Day imagery defused critiques linking environmental damage to corporate and racialized power, see Finis Dunaway, "Gas Masks, Pogo, and the Ecological Indian: Earth Day and the Visual Politics of American Environmentalism," *American Quarterly* 60 (2008): 67–98.

10. William D. Ruckelshaus, *EPA Press Release: First Administrator on Establishment of EPA* (Washington, DC: Environmental Protection Agency, December 16, 1970), http://www2.epa.gov/aboutepa/first-administrator-establishment-epa.

11. See Pete Daniel, *Toxic Drift: Pesticides and Health in the Post–World War II South* (Baton Rouge: Louisiana State University Press in association with Smithsonian Institution, Washington, DC, 2005), chapter 8; David Zierler, *The Invention of Ecocide: Agent Orange, Vietnam, and the Scientists Who Changed the Way We Think about the Environment* (Athens: University of Georgia Press, 2011).

12. See Gottlieb, *Forcing the Spring*, 175–86.

13. The Federal Environmental Pesticide Control Act (FEPCA), as the 1972 changes to FIFRA were known, also required for the first time that "when used in accordance with widespread and commonly accepted practice [a pesticide] will not cause unreasonable adverse effects on the environment." *Federal Environmental Pesticide Control Act* P.L. 92–516, 7 USC 136–136y (1972), sec. 3(c)(5)(D). On DDT, see Thomas R. Dunlap, *DDT: Scientists, Citizens, and Public Policy* (Princeton, NJ: Princeton University Press, 1981).

14. By 1972, the EPA had failed to set "reentry intervals," minimum periods of time after pesticide application during which workers could not be required to enter recently treated fields. After industry pressure dissuaded EPA from setting these limits, a farmworker group, Migrant Legal Action Program, pressured OSHA to pass an "emergency temporary standard" to protect workers from pesticide exposures. OSHA passed the standard and then suspended it, provoking a series of conflicts and litigation involving agricultural interests and farmworker groups. Ultimately, EPA passed its own reentry standards in 1974, beating OSHA to the punch, and effectively claiming this regulatory ground as its own. See Nicholas A. Ashford, *Crisis in the Workplace: Occupational Disease and Injury* (Cambridge, MA: MIT Press, 1976), 183–184.

15. Robert Gordon, "Shell No! OCAW and the Labor-Environmental Alliance," *Environmental History* 3, no. 4 (1998): 460–87.

16. Beth Rosenberg and Charles Levenstein, "Unintended Consequences: Impacts of Pesticide Bans on Industry, Workers, the Public, and the Environment," Methods and Policy Report No. 13 (Lowell, MA: Toxics Use Reduction Institute; University of Massachusetts, Lowell, 1995), 36–37.

17. Trost, *Elements of Risk*, 216–17.

18. Ibid.

19. Douglas L. Murray, "The Politics of Pesticides: Corporate Power and Popular Struggle over the Regulatory Process" (Ph.D diss., University of California Santa Cruz, 1983), 106–08.

20. Murray, "The Politics of Pesticides," 108.

21. Donald Whorton, Thomas H. Milby, and Harrison A. Stubbs, "Health Hazard Evaluation: Occidental Chemical Company, Lathrop, California; August–October, 1977," document no. 717, SHLF.

22. D.L. Baeder memo to J.E. Baird et al., July 22, 1977, document no. DBCP003870, SHLF.

23. Frances C. O'Melia memo to D. Buchner, et al., July 22, 1977, document no. 648, SHLF.

24. Frances C. O'Melia letter to Mr. A.B. Horner, July 21, 1977, document no. 72, SHLF.

25. D.L. Baeder memo to J.E. Baird et al., July 22, 1977.

26. Trost, *Elements of Risk*, 225.

27. Murray, "The Politics of Pesticides," 142

28. Mitchell Zavon to James Galvin, August 1, 1977 ("Exhibit 10"), SHLF. By December 1977, the examining doctor would report that 14 of 25 nonvasectomized men at Oxy had azoospermia or oligospermia. See D. Whorton et al., "Infertility in Male Pesticide Workers," *Lancet* 2, no. 8051 (1977): 1259–61.

29. "Chemical Plant Workers Found Sterile," *LA Times*, August 24, 1977, 1.

30. Ibid.

31. H.C. Scharnweber, *Status Report: DBCP Issue* (Corporate Medical Department [Dow Chemical], August 11, 1977), document no. DC0042455, SHLF; W.E. Hall letter to Customer, August 24, 1977, document no. 4187, SHLF; Benjamin

B. Holder to Eula Bingham, Director, Occupational Safety and Health Administration, September 1, 1977, document no. 515, SHLF.

32. *Shell Chemical Announces Status Report on Fertility Testing* (Houston, TX: Shell Oil Company, August 26, 1977), document no. 6262, SHLF; C. E. Whitney letter to Shell Nemagon Soil Fumigant Customer, August 24, 1977, document no. 6157, SHLF; D. E. F. Gollin letter to Finance Manager Chemical Products, August 31, 1977, document no. 6159, SHLF.

33. "Order Adopting Regulations of the Department of Food and Agriculture Pertaining to Economic Poisons" (California Department of Food and Agriculture, August 12, 1977), document no. 4143, SHLF.

34. Murray, "The Politics of Pesticides," 145.

35. P. C. Butler to F. S. Mentz, August 25, 1977, document no. C1001488, SHLF.

36. Henry Cassity Telex to Bud Daley, August 18, 1977, document no. 5667, SHLF.

37. Jack DeMent to Distribution, August 16, 1977, document no. 5827, SHLF.

38. Dale G. Bottrell, "Government Influence on Pesticide Use in Developing Countries," *Insect Science and its Applications* 5 no. 3 (1984): 151–55.

39. Henry Cassity Telex to Bascom, Lasky, Fong, Gray, Guilbert, August 23, 1977, document no. SFC20000658, SHLF.

40. Henry Cassity Telex to Bascom et al., August 23, 1977; Henry Cassity to Harry Endsley, August 23, 1977, document no. 5633, SHLF.

41. Henry Cassity Telex to Bascom et al., August 23, 1977.

42. Generally, see Douglas L. Murray, *Cultivating Crisis: The Human Cost of Pesticides in Latin America* (Austin: University of Texas Press, 1994), 62–63. Lori Ann Thrupp places dilemmas of policy and enforcement in political-economic context, arguing that Costa Rica "has been relatively ineffective in carrying out laws and in undertaking badly needed actions," but finds that Nicaragua under the Sandinista government in the 1980s did a better job. Lori Ann Thrupp, "Pesticides and Policies: Approaches to Pest-Control Dilemmas in Nicaragua and Costa Rica," *Latin American Perspectives* 15, no. 4 (2004): 37–70. For a more recent discussion of problems with pesticide enforcement in Costa Rica see R. Sass, "Agricultural 'Killing Fields': The Poisoning of Costa Rican Banana Workers," *International Journal of Health Services* 30, no. 3 (2000): 491–514.

43. Henry Cassity Telex to Bud Daley, August 18, 1977.

44. V. K. Rowe to Mitchell Zavon, October 4, 1960, document no. DC0023934–36, SHLF; Ken Olson, *Results of Range Finding Toxicological Tests on M-1678 (Fumazone 70E)* (Biochemical Research Laboratory, The Dow Chemical Company, 1961), document no. 16451, SHLF; H. R. Hoyle, *Vapor Exposures Encountered by Workmen While Fumigating Soil at Wellston, Michigan* (Biochemical Research Laboratory, The Dow Chemical Company, 1958), document no. 4222, p. 12–20, SHLF; *Technical Bulletin: Summary of Basic Data for Nemagon Soil Fumigant* (New York: Shell Chemical Company, Agricultural Chemicals Division, May, 1960), CHLF.

45. "Occupational Safety and Health Administration List of Products Believed to Contain DBCP," *Occupational Safety and Health Reporter* (1977): 412–16.

46. Douglas Costle, "Notice of Intent to Suspend and Conditionaly Suspend Registrations of Pesticide Products Containing Dibromochloropropane," 42 Federal Register 48915 (September 8, 1977), 48919.

47. Harold W. Lembright to F. C. O'Melia et al., August 3, 1977, document no. 1227, SHLF.

48. Costle, "Notice of Intent to Suspend Dibromochloropropane," 48919.

49. Murray, "The Politics of Pesticides," 146–47.

50. Occupational Safety and Health Administration, "Emergency Temporary Standard for Exposure to 1,2-Dibromo-3-Chloropropane," 45 Federal Register 45535 (September 9, 1977).

51. Murray, "The Politics of Pesticides," 142–144.

52. Costle, "Notice of Intent to Suspend Dibromochloropropane," 48918.

53. Ibid., 48920.

54. Ibid., 48918–21.

55. Cora Roelofs, *Preventing Hazards at the Source* (Fairfax, VA: American Industrial Hygiene Association, 2007), 13–14.

56. W. D. Brink, Memo to Agricultural Chemicals Manager Sales, District Managers, October 13, 1977, document no. 6172, SHLF.

57. R. Hansen to J. M. Flynn, September 15, 1977, document no. 16350, SHLF.

58. Ibid.

59. *Addendum to Contract and Letter Agreement Concerning the Sale of Dow's Fumazone Inventory to Caste and Cooke Inc.* (Coral Gables, FL: Dow Chemical Latin America, February 9, 1978), document no. 553, SHLF; E. V. Daley to Paul Sink, February 23, 1978, document no. 554, SHLF.

60. E. V. Daley to Paul Sink, February 23, 1978. I was not able to locate a copy of the release itself.

61. *Dow Chemical Company and Standard Fruit Company Letter of Agreement* (Midland, MI: Dow Chemical Company, January-February, 1978), document no. 396, SHLF.

62. *Requirements for the Safe Handling and Application of Fumazone 86E Soil Fumigant to Banana Plantations* (Midland, Michigan: Dow Chemical Company, December 21, 1977), document no. 1267, SHLF; Frances C. O'Melia to R. E. Hefner, September 23, 1977, document no. 1244, SHLF.

63. Frances C. O'Melia to R. E. Hefner, September 23, 1977; Frances C. O'Melia to R. E. Hefner, December 21, 1977, document no. 1267, SHLF. I was unable to locate a report of these test results. Jack DeMent reported in a deposition in September 1995 that Dow personnel reported by phone that exposures were within safe levels. Jack DeMent, "Deposition," in *Delgado vs. Shell Oil Company, 890 F. Supp. 1324 (S.D. Tex. 1995)* (September 29, 1995), document no. 9945, SHLF.

64. *Requirements for the Safe Handling and Application of Fumazone.*

65. Frances C. O'Melia to Kenneth Moll, December 15, 1977, document no. 1266, SHLF.

66. *Requirements for the Safe Handling and Application of Fumazone*, 3.

67. *Results of Repeated Exposure of Laboratory Animals to Various Concentrations of 1,2-Dibromo-3-Chloropropane* (Biochemical Research Laboratory, The Dow Chemical Company: July 23, 1958), document no. 1106, SHLF.

68. E. R. Laning, Jr. to B. Daley, April 26, 1978, document no. 5681, SHLF; E. V. Daley to Paul Sink, May 10, 1978.

69. Jack DeMent to Distribution, March 3, 1978, document no. 5793, SHLF.

70. Ibid.

71. Ibid.

72. Among major producers, Occidental's production had been halted by Cal-OSHA, while Shell and Dow voluntarily stopped production. W. E. Hall letter to Customer, August 24, 1977; C. E.Whitney letter to Shell Nemagon Soil Fumigant Customer, August 24, 1977.

73. Eula Bingham, "Occupational Exposure to 1,2-Dibromo-3-Chloropropane (DBCP)," 43 Federal Register 11514–11533 (March 17, 1978).

74. Quoted in Rosenberg and Levenstein, "Unintended Consequences," 41.

75. *Shell to Sell DBCP Inventories* [News Release] (Houston, TX: Shell Chemical Company, April 12, 1978), document no. 6205, SHLF.

76. Quoted in David Weir and Mark Schapiro, *Circle of Poison: Pesticides and People in a Hungry World* (San Francisco: Institute for Food and Development Policy, 1981), 21.

77. Glenn. A. Wintemute to A. B. Horner, December 19, 1977, document no. 2691, SHLF.

78. Ronald B. Taylor, "Production of Highly Toxic Pesticide Shifts to Mexico," *Los Angeles Times,* September 9, 1978; Frances C. O'Melia Telex to A. Clark, July 20, 1978, document no. 556, SHLF.

79. Taylor, "Production of Pesticide Shifts to Mexico."

80. Jeff Kempter and Mitchell H. Bernstein, *Dibromochloropropane: Final Position Document* (Washington, DC: Special Pesticide Review Division, Office of Pesticide Programs, U.S. Environmental Protection Agency, September 6, 1978).

81. Ibid., 29.

82. Ibid., 56.

83. The "Final Position Document" had its origins in a process initiated a year before, in September 1977, when the EPA had issued a "Notice of Rebuttable Presumption against Registration" (RPAR) for DBCP. An RPAR was required when new evidence emerged that a pesticide might pose particularly serious risks. As the name suggested, it did not automatically cancel a chemical's registration, but gave "registrants and other interested persons" a chance to respond to official concerns before the EPA made a final decision. Office of Pesticide Programs Environmental Protection Agency, "Notice of Rebuttable Presumption against Registration and Continued Registration of Pesticide Products Containing Dibromochloropropane (DBCP)," 42 Federal Register 48026 (September 22, 1977). The chemical companies had expected an RPAR to be issued for DBCP as early as 1973, when National Cancer Institute studies had found that DBCP caused stomach and mammary cancer in mice and called DBCP "very carcinogenic." See William A. Olson et al., "Brief

Communication: Induction of Stomach Cancer in Rats and Mice by Halogenated Aliphatic Fumigants," *Journal of the National Cancer Institute*, no. 51 (1973): 1993; Jerrold M. Ward and Robert T. Habermann, "Pathology of Stomach Cancer in Rats and Mice Induced with the Agricultural Chemicals Ethylene Dibromide and Dibromochloropropane," *Bulletin of the Society of Pharmacological and Environmental Pathologists* 2, no. 2 (1974); Joanne E. Betso to Frances O'Melia et al., November 8, 1976, document no. 1194, SHLF; G. W. Paytt memo to R. G. Hansen, September 16, 1976, document no. 1556, SHLF.

84. Harold W. Lembright to Paige Taylor, October 5, 1977, document no. 13490, SHLF.

85. *Benefits of DBCP Use on Bananas* (Castle & Cooke, November 29, 1977), document no. DC0019642–59, SHLF; A. E. Schober to Jeff Kempter, December 6, 1977, document no. 15087, SHLF.

86. Jack DeMent to J. V. Scibetta et al., August 16, 1977, document no. 5827, SHLF.

87. *Benefits of DBCP Use on Bananas*, 9.

88. Ibid., 7.

89. Kempter and Bernstein, *Dibromochloropropane: Final Position Document*, 158.

90. *Benefits of DBCP Use on Bananas*, Figure A.

91. Kempter and Bernstein, *Dibromochloropropane: Final Position Document*, 156.

92. Ibid., Table 33.

93. Edwin L. Johnson, "Global Impacts of U.S. Pesticide Regulations" (paper presented at the Proceedings of the U.S. Strategy Conference on Pesticide Management, Washington, DC, July 7–8 1979).

94. In any case, EPA's performance in following legal notification procedures over the next decade would be far less than satisfactory. See United States General Accounting Office, "Report to the Chairman, Environment, Energy, and Natural Resources Subcommittee, Committee on Government Operations, House of Representatives: Export of Unregistered Pesticides Is Not Adequately Monitored by EPA" (April, 1989).

95. Robert A. Reeves to Worker Health and Safety Group, January 24, 1979, SHLF.

96. Steven D. Jellinek, "Amended Notice of Intent to Cancel the Registrations of Pesticide Products Containing Dibromochloropropane (DBCP), and Statement of Reasons; in Re: Shell Oil Company et al., Petitioners; FIFRA Docket Nos. 401, 402, 403" (U.S. Environmental Protection Agency, September 6, 1978).

97. D.J. Miner for J.L. Reed memo to P.F. Deisler et al., September 12, 1978, document no. 6417, SHLF.

98. Carlos E. Domínguez Vargas, interview with the author, February 14, 2005, San José, Costa Rica.

99. Lori Ann Thrupp, "Sterilization of Workers from Pesticide Exposure: The Causes and Consequences of DBCP-Induced Damage in Costa Rica and Beyond," *International Journal of Health Services* 21 no. 4 (1991): 747.

100. Domínguez Vargas.

101. Thrupp, "Sterilization of Workers," 748.

102. Fernando Urbina Salazar to Roberto Galva Jiménez, November 29, 1978, LATF.

103. Fernando Urbina Salazar to Vice President Rodrigo Altmann, December [illeg.], 1978, LATF.

104. Henry Nanne, *Draft: Meeting with Dr. Carmelo Calvosa, Minister of Health* (December 1, 1978), document no. 12, LATF.

105. Ibid., 2.

106. Ibid.

107. Ibid.

108. Ibid., 3.

109. Ibid.

110. U.K. Gray Telex to Rafael Guilbert, November 30, 1978, document no. SFC6–0000214, SHLF.

111. *Rough Draft: Considerations Involved in Protections for Banana Workers Who Apply DBCP* (Standard Fruit Company, December 28, 1978), document no. 11034, SHLF; *Synthesis of Report About Intoxications with Nemagon or DBCP* [Translation from the Spanish] (San José, Costa Rica: Medical Department, Unit of Occupational Helath, Instituto Nacional de Seguros, April 5, 1982), LATF.

112. *Rough Draft: Considerations Involved in Protections for Banana Workers Who Apply DBCP.*

113. Thrupp, "Sterilization of Workers," 749; Rafael Pagán Telex to Muñoz et al., January 25, 1979, document no. 5836, SHLF.

114. Kenneth Moll, *Review of Pesticide Uses in Costa Rica* (San Francisco: Castle & Cooke Foods, January 12, 1979), document no. 13, LATF.

115. Ibid.

116. Thrupp, "Sterilization of Workers."

117. Sandra Piszk, "Expediente No. 250–23–98: Informe Final con Recomendaciones" (San José: Defensoría de los Habitantes de la República de Costa Rica, 1998), CQLF.

118. *Synthesis of Report About Intoxications with Nemagon.*

119. Thrupp, "Sterilization of Workers," 478.

120. *Contract for Sale of Goods* (Amvac Chemical Corporation and Castle & Cooke Foods, June 15, 1979), document no. 10183, SHLF. Although Amvac was required to meet OSHA safety standards for production workers, by January 1979 the company was cited for multiple violations of occupational safety law. See Ronald B. Taylor, "Firm Cited for Exposing Workers to Carcinogen," *Los Angeles Times,* January 26, 1979.

121. Rafael Guilbert to Case Stek, May 4, 1979, document no. SFC6–0000012–13, SHLF.

122. H.E. Ostmark to P.C. Butler, October 26, 1978, document no. C1002064, SHLF.

123. Federal Insecticide, Fungicide, and Rodenticide Act Scientific Advisory Panel, "Transcript" (Arlington, VA: U.S. Environmental Protection Agency, Acme Reporting, April 26, 1979), document no. 730, SHLF.

124. J. J. Tolan to M. G. Rice, May 25, 1978, document no. 8494, SHLF.

125. Steven D. Jellinek, "Intent to Hold a Hearing to Determine Whether or Not the Registrations of Certain Uses of Pesticide Products Containig Dibromochloropropane (DBCP) Should Be Cancelled, and Statement of Issues," 44 Federal Register 11822 (March 2, 1979).

126. Ibid. The hearing, convened under statutory grounds different from those in the Amaya group's original request, would be folded into ongoing DBCP proceedings, which would also be accompanied by comment from the USDA and the EPA Scientific Advisory Panel.

127. FIFRA Scientific Advisory Panel, 73.

128. Ibid., 77–78.

129. "DBCP—Why Fungicide [sic] May Be Legalized Again," *Stockton Record,* July 27, 1978.

130. Ronald B. Taylor, "Growers, Industry Figures Defend Pesticide Controls," *Los Angeles Times,* October 6, 1978.

131. Jellinek, "Intent to Hold a Hearing," 11827.

132. Douglas Costle, "Intent to Suspend Registrations of Pesticide Products Containing Dibromochloropropane (DBCP)," 44 Federal Register 43335 (July 24, 1979), 43337–38.

133. Graham Tweedy memo to Mel Rice, May 2, 1979, document no. 17784, SHLF.

134. Ibid.

135. Ibid.

136. "DBCP," *Associated Press,* May 17, 1979; "Traces of Pesticide DBCP Found in 36 California Wells," *New York Times,* May 18, 1979.

137. "Dibromochloropropane (DBCP): Suspension Order and Notice of Intent to Cancel," 44 Federal Register 65135 (November 9, 1979), 65153.

138. Trost, *Elements of Risk,* 260.

139. "Dibromochloropropane (DBCP): Suspension Order and Notice of Intent to Cancel," 65153.

140. FIFRA Scientific Advisory Panel, "Review of a Notice of Intent to Hold a FIFRA Section 6(B)(2) Hearing on Dibromochloropropane (DBCP)" June 29, 1979.

141. "E.P.A. Prepares to Ban Pesticide Tied to Cancer," *New York Times,* July 20, 1979.

142. Costle, "Dibromochloropropane (DBCP): Suspension Order and Notice of Intent to Cancel," 65147–49, 65162–64.

143. Ibid., 65153–54.

144. "FIFRA Docket No. 485; in Re: Notice of Intent to Suspend Registrations of Pesticide Products Containing Dibromochloropropane" (Washington, DC: U.S. Environmental Protection Agency, [1979]).

145. Costle, "Dibromochloropropane (DBCP): Suspension Order and Notice of Intent to Cancel," 65169.

146. Ibid.; "EPA Administrator Costle Suspends All but Hawaiian Pineapple Uses of DBCP," *Pesticide and Toxic Chemical News,* October 31, 1979.

147. Costle, "Dibromochloropropane (DBCP): Suspension Order and Notice of Intent to Cancel," 65165–66, 65168.

148. Ibid., 65169.

149. Walter Wright, "Isle Firm Halts Use of DBCP Overseas," *Honolulu Advertiser,* November 16, 1979.

150. Edward C. Gray, Michael S. Winer, and Marcia E. Mulkey, "Memo to the Administrator: Resolution of the Dibromochloropropane (DBCP) Cancellation Proceeding (FIFRA Docket No. 402, et al.)—Action Memorandum" (Washington DC: US Environmental Protection Agency, January 23, 1981), document no. C1002139–45, SHLF.

151. Murray, "The Politics of Pesticides," 169. Murray tells how the EPA granted a request for an "emergency" exception for DBCP use on peaches in 1982. After protest by a local environmental group, use of DBCP was blocked by a judicial restraining order. The exemption would not have been possible with a nonvoluntary cancellation, 169–175.

152. *Nematocide Crop Guide: Bananas* (Amvac Chemical Corporation, 1980), document no. 13850, SHLF.

153. Glenn A. Wintemute to Jack DeMent, December 29, 1980, document no. 358, SHLF.

154. Ibid.

155. *Proceedings of the U.S. Strategy Conference on Pesticide Management.* July 7–8 (Washington, DC: U.S. Department of State and U.S. National Committee for Man and the Biosphere,1979.) The proceedings from the State Department conference show how information exchange was privileged over barriers to export by invoking national sovereignty and a slippery notion of national culture that in some conference participants' discourse verged into a suggestion that illness or injury caused by pesticides might not be such a big deal in a developing nation. For example, one speaker suggested that "less industrial countries may consider a few cases of cancer in older people a small price to pay for increased yields of food crops," anticipating the logic of Larry Summers's notorious 1991 World Bank memo "encouraging MORE migration of the dirty industries to the LDCs [less developed countries]." A more robust proposal, reported but not endorsed by the GAO, came from the House Committee on Government Operations, which proposed that "No product which is banned from the domestic market should be allowed to be exported" without either a U.S. regulatory determination that the intended use was safe or the importing government's explicit request. For this and President Carter's task force, see Comptroller General of the United States, "Report to the Congress of the United States; Better Regulation of Pesticide Exports and Pesticide Residues in Imported Foods Is Essential" (Washington, DC: United States General Accounting Office, 1979).

156. President Jimmy Carter, "Executive Order 12264: Export of Banned or Significantly Restricted Substances" (January 15, 1981).

157. David Weir, Mark Schapiro, and Terry Jacobs, "The Boomerang Crime," *Mother Jones,* November 1979. Concern with the "export of hazard" was already in the

public eye after 1978 reporting that children's clothes treated with the carcinogenic fire retardant Tris were being exported to the developing world following their being banned in the United States the previous year.

158. R. I. Glass et al., "Sperm Count Depression in Pesticide Applicators Exposed to Dibromochloropropane," *American Journal of Epidemiology* 109, no. 3 (1979): 346.

159. S. H. Sandifer et al., "Spermatogenesis in Agricultural Workers Exposed to Dibromochloropropane (DBCP)," *Bulletin of Environmental Contamination and Toxicology* 23, no. 4–5 (1979): 705.

160. A. L. Ramírez and C. M. Ramírez, "Esterilidad masculina causada por la exposición laboral al nematicida 1,2-Dibromo-3-Cloropropano" [Male Sterility Caused by Occupational Exposure to the Nematocide 1,2-Dibromo-3-Chlorpropane], *Acta médica costarricense* 23, no. 2 (1980): 219–22.

161. President Ronald Reagan, "Executive Order 12290—Federal Exports and Excessive Regulation" (February 17, 1981).

162. Steve Jellinek served as Acting Administrator for five days beginning January 21, and Walter Barber held the same position from January 26 until Anne Gorsuch was appointed on a permanent basis on May 20. "Chronology of EPA Administrators," United States Environmental Protection Agency, accessed August 30 http://www.epa.gov/aboutepa/history/admin/agency/index.html.

163. Gray, Winer, and Mulkey, "Memo to the Administrator: Resolution of DBCP Cancellation."

164. Ibid.

165. Walter C. Barber, "Dibromochloropropane: Withdrawal of Intent to Cancel Registrations for Use on Pineapples in Hawaii," 46 FR 19592 (March 31, 1981). The pineapple decision came despite opposition from a Hawaiian citizen group, water samples showing DBCP in low concentrations (on the order of parts per trillion) in some wells, and Castle & Cooke's finding of fumigator exposures reaching 30 ppb over a five-hour period. While it is not clear whether Castle & Cooke shared the worker exposure data with the EPA, the agency detailed the well data in its decision, ultimately finding that it didn't require cancellations because "exposure to DBCP can be maintained at extremely low levels." Barbara Hastings, "DBCP: Approval Would Keep This State as Nation's Only User of Pesticide," *Honolulu Advertiser*, March 6, 1981; E. D. Pattimore, memo to S. Ogata: DBCP Air Analysis, Subsoiler, (Lanai, Hawaii: Castle & Cooke Foods, February 1, 1980).

166. Sam Schneider Memorandum to Joe Scibetta, November 16, 1979, document no. 11978, SHLF.

167. Leonard R. Marks to the Honorable John Bingham, September 26, 1980, document no. SFC3–0000112–13, SHLF.

168. Marco A. Gurdián memo to Junta Directiva de Mariana de Gasteazoro y Cía. Ltda., Informe Agrícola al Período 11, 1980 [Teresa], November 12, 1980, document no. SFC9–001927–28, VBF; Marco A. Gurdián memo to Junta Directiva de Mariana de Gasteazoro y Cía. Ltda., Informe Agrícola al Período 11, 1980 [Coquimba], November 12, 1980, document no. SFC9–001924–26, VBF; Patricio Ballesteros S. memo to Lic. M. A. Gurdián, February 16, 1981, document no. SFC9–001921–23, VBF.

169. Weir and Schapiro, *Circle of Poison*, 20.

170. Leonard R. Marks to John Bingham, September 26, 1980; C. A. Campion memo to Distribution, December 4, 1980, document no. 5807, SHLF.

171. Randy Fleming to Jack DeMent, September 13, 1984, document no. SFC3–0000153, SHLF.

172. Robert Loe to Jack DeMent, April 10, 1984, document no. SFC30000149–51, SHLF.

173. Jack DeMent to Dave Green, January 22, 1985, document no. SFC3–0000159, SHLF.

174. Jack DeMent to R. Fleming, September 25, 1984, document no. SFC3–0000152, SHLF.

175. Jack DeMent to Dave Green, January 22, 1985.

176. This dynamic was at the center of "Boomerang Crime" and *Circle of Poison*, which drew much of the force of their argument from a critique of the reimportation of banned pesticides on foods eaten within the United States. DeMent's comment that DBCP use was "unofficially blessed" by the EPA adds to the evidence that Standard Fruit considered U.S. regulator opinion important in relationship to their Central American operations.

177. John A. Moore, "Dibromochloropropane: Intent to Cancel Registrations of Pesticide Products Containing Dibromochloropropane (DBCP)," 50 Federal Register 1122 (January 9, 1985), section (II)(B)(1).

178. Ibid, 1122.

179. Jack DeMent to Dave Green, January 22, 1985.

180. Lori Ann Thrupp reports that Nicaragua had ceased imports of DBCP in 1981 as part of a strengthening of pesticide registration under the Sandinista government; post-revolution, the National Pesticides Commission instituted a ban in 1993. Panama banned the use of DBCP in 1997. Thrupp, "Pesticides and Policies," 56; Ilsa Zapata, "Plaguicidas prohibidos en Panamá y que deben ser retirados del mercado," República de Panamá, Ministerio de Salud, Dirección Nacional de Farmacia y Drogas, accessed June 16, 2007, http://www.authorstream.com/Presentation/Belly-21401-Resumen-del-de-Plaguicidas-prohibidos-Cerro-Punta-REP-BLICA-PANAM-Ministerio-SaludDirecci-n-Nacional-Farmacia-Drogas-Antecedentes-Justificaci-31122004-as-ppt-powerpoint/.

181. "DBCP: Identification, Toxicity, Use, Water Pollution Potential, Ecological Toxicity, and Regulatory Information," Pesticide Action Network North America, accessed October 24, 2006, www.pesticideinfo.org/Detail_Chemical.jsp?Rec_Id = PC33459.

CHAPTER FOUR

1. *Saul Muñoz Sibaja et al. v. Dow Chemical Company, et al.*, 757 F.2d 1215 (11th Cir., 1985); Charles S. Siegel, Interview with author, November 21, 2011. Siegel remembered that *Sibaja* may have included some claims relating to fear of

cancer. I was unable to obtain any of the case materials from the court or lawyers involved.

2. For a discussion the challenges of transnational litigation and labor solidarity in the context of Colombian Coca-Cola workers' struggle for justice, see Leslie Gill, "The Limits of Solidarity: Labor and Transnational Organizing against Coca-Cola," *American Ethnologist* 36, no. 4 (2009): 667–80.

3. In August 2008, Law 8653 "Ley Reguladora del Mercado de Seguros," permitted private insurers to do business in Costa Rica. See "Historia del INS," Instituto Nacional de Seguros, accessed January 23, 2013, https://portal.ins-cr.com/portal. ins-cr.com/Institucional/Historia/.

4. "Salud ocupacional en el sector bananero centroamericano: Salud ocupacional en trabajadores del Valle de la Estrella-Standard Fruit C. Costa Rica," *Revista centroamericana de ciencias de la salud* 4, no. 9 (1978): 31–32; "Código de Trabajo" (San José: Congreso Constitucional de la República de Costa Rica, 1944), article 217.

5. "Código de Trabajo," table 3.

6. In 1986, *La Nación* reported that 300 men had received compensation between 11,000 and 15,000 *colones* (the Costa Rican unit of currency) since 1979; based on exchange rates for the years concerned, these values represented between US$220 and US$1,500 (assuming that the lowest compensation amount took place at the lowest value of the *colón* and the highest amount at the highest rate). In 1991, the *Los Angeles Times* reported that compensation ranged up to US$1,800. Barry Siegel, "Going an Extra Mile for Justice," *Los Angeles Times,* March 23, 1991; Edgar Espinoza and Gina Polini, "Pelea por obreros estériles se dará en Estados Unidos," *La Nación,* September 16, 1986.

7. Cathy Trost, *Elements of Risk: The Chemical Industry and Its Threat to America* (New York: Times Books, 1984), 288–89.

8. Erika Rosenthal, "The DBCP Pesticide Cases: Seeking Access to Justice to Make Agribusiness Accountable in the Global Economy," in *Agribusiness and Society: Corporate Responses to Environmentalism, Market Opportunities and Public Regulation,* eds. Kees Jansen and Sietze Vellema (London; New York: Zed Books, 2004), 181; Espinoza and Polini, "Pelea por obreros estériles."

9. Sandra Piszk, "Expediente No. 250–23–98: Informe Final con Recomendaciones," (San José: Defensoría de los Habitantes de la República de Costa Rica, 1998), CQLF, 22. I was unable to contact any of the U.S. lawyers who originated these cases to obtain a first-person account of the circumstances or motivation for the initiation of this litigation.

10. This conclusion is based primarily on my general experience working with or observing plaintiff attorneys in my role as a researcher at a litigation consulting firm in 2002–2007. Despite the common perception of plaintiff lawyers as "ambulance chasers," Anne Bloom argues that they may usefully be understood as "cause lawyers" in some cases, including in the DBCP litigation. Bloom used DBCP attorney Charles Siegel as an example of a plaintiff lawyer who was motivated by a "high degree of commitment to both the clients in the case and the broader issues at stake."

I would agree that such commitment was evident among many of the attorneys with whom I spoke in the course of my research. See Anne Bloom, "Cause Lawyering in the Shadow of the State: Why Personal Injury Litigation May Represent the Future of Transnational Cause Lawyering," in *Cause Lawyering and the State in a Global Era*, eds. Austin Sarat and Stuart A. Scheingold (Oxford; New York: Oxford University Press, 2001), 105.

11. Scott L. Cummings, "The Internationalization of Public Interest Law," *Duke Law Journal* 57, no. 4 (2008).

12. In *Filártiga v. Peña-Irala*, the United States Court of Appeals for the Second Circuit decided that the Alien Tort Statute provided grounds for the family of a Paraguayan torture victim to bring suit in the United States against a Paraguayan police inspector, for claims of torture in violation of international law, despite the fact that neither party was a U.S. citizen or resident. For a longer discussion of ATCA and human-rights litigation in the United States, see Beth Stephens, *International Human Rights Litigation in U.S. Courts* (Boston; Leiden: Martinus Nijhoff Publishers, 2008).

13. Marc Edelman, *Peasants against Globalization: Rural Social Movements in Costa Rica* (Stanford, CA: Stanford University Press, 1999), 75–81.

14. Ibid., 78.

15. For a consideration and defense of the contingency-fee arrangement, see Elihu Inselbuch, "Contingent Fees and Tort Reform: A Reassessment and Reality Check," *Law & Contemporary Problems* 64, nos. 2 & 3 (2001).

16. Marcy Strauss, "Toward a Revised Model of Attorney-Client Relationship: The Argument for Autonomy," *North Carolina Law Review* 65 (1987): 318.

17. Center for Professional Responsibility, *A Legislative History: The Development of the ABA Model Rules of Professional Conduct, 1982–2005* (Chicago: Center for Professional Responsibility, American Bar Association, 2006), 44.

18. See Strauss, "Revised Model of Attorney-Client Relationship."

19. Beau Baez, *Tort Law in the USA* (Alphen aan den Rijn, The Netherlands; Frederick, MD: Kluwer Law International; sold and distributed in North, Central, and South America by Aspen Publishers, 2010), 39–40, 86–87.

20. Ibid.

21. *Sibaja v. Dow*, 1216–1217; Barry Siegel, "Going an Extra Mile for Justice."

22. Federal courts have jurisdiction over cases that address issues of constitutionality, federal laws and treaties, disputes between two or more states or governments, and other cases that have particular resonance at the national level.

23. *Diversity of Citizenship; Amount in Controversy; Costs,* 28 U.S.C.1332 (1994).

24. In Florida, the doctrine could only be invoked under strict conditions. See Laurel E. Miller, "Forum Non Conveniens and State Control of Foreign Plaintiff Access to U.S. Courts in International Tort Actions," *University of Chicago Law Review* 58 (1991): 1373, 1376.

25. Don Mayer and Kyle Sable, "Yes! We Have No Bananas: Forum Non Conveniens and Corporate Evasion," *International Business Law Review* 4 (2004), 139–149.

26. *Gulf Oil v. Gilbert,* 330 U.S. 501 (1947).

27. Ibid., 509.

28. Ibid.

29. *Sibaja v. Dow,* 1218.

30. Ibid., 1217.

31. Roger A. Petersen, *The Legal Guide to Costa Rica* (Miami: Centro Legal R & M, S.A., 1994), 33–44; Henry Saint Dahl, "Forum Non Conveniens, Latin America and Blocking Statutes," *University of Miami Inter-American Law Review* 35, no. 1 (2003/2004): 39.

32. Petersen, *The Legal Guide to Costa Rica,* 38.

33. Emily Yozell, "The Castro Alfaro Case: Convenience and Justice—Lessons for Lawyers in Transcultural Litigation," in *Human Rights, Labor Rights, and International Trade,* eds. Lance A. Compa and Stephen F. Diamond (Philadelphia: University of Pennsylvania Press, 1996), 277.

34. These included *Perfecto Barrantes Cabalceta, et al. v. Standard Fruit Company, et al.,* 667 833 (United States District Court for the Southern District of Florida, 1987); *Aquilar v. Dow Chemical,* No. 86–4753 JGD (Superior Court for Los Angeles County, California, 1985).

35. "Víctimas de sustancia tóxica: 500 trabajadores quedan estériles en zona atlántica," *La Nación,* September 14, 1986.

36. Ibid.

37. Ibid.

38. Ibid.

39. Ibid.

40. Ibid.

41. Espinoza and Polini, "Pelea por obreros estériles."

42. *Dow Chemical, et al. v. Domingo Castro Alfaro, et al.,* 786 S.W.2d 674 (Tex., 1990).

43. Emily Yozell, "The Castro Alfaro Case: Convenience and Justice," 281.

44. *Act or Omission out of State,* Texas Statutes 71.031 (1986).

45. Charles S. Siegel to R.B. Ballanfant, October 20, 1988, document no. 11875, SHLF; Charles Szalk, e-mail to author, December 2, 2011.

46. *Domingo Castro Alfaro, et al. v. Dow Chemical, et al.,* 751 S.W.2d 208 (Tex. App., 1988), 208.

47. Ibid., 209.

48. Ibid.

49. Ibid., 211–12.

50. *Aquilar v. Dow Chemical.*

51. *Cabalceta v. Standard Fruit.*

52. I have been unable to determine exactly when new plaintiffs were added to the *Castro Alfaro* group, but a 1988 letter from Charles Siegel, of Baron & Budd, to Shell Oil counsel shows that the plaintiffs were considering adding 92 plaintiffs to the *Castro Alfaro* group, five of whom had been named as plaintiffs in the *Cabalceta* case. Charles S. Siegel to R.B. Ballanfant, October 20, 1988.

53. Yozell, "The Castro Alfaro Case."

54. Ibid., 282.

55. Ibid.

56. Charles S. Siegel, Interview with Author, March 30, 2006.

57. Siegel, "Going an Extra Mile for Justice."

58. *Dow Chemical v. Castro Alfaro*, 675. The majority opinion, written by Justice Ray, addressed both the legislative intent and the judicial history of *forum non conveniens* in Texas. To establish that the 1913 Texas legislature had indeed intended to abolish FNC, the justices needed to show that FNC existed at that point in time. Ray's opinion explains that FNC had indeed existed in Texas prior to 1913, when the predecessor to Section 71.031 was written. Even though the term "forum non conveniens" had only been "brought ... into American law" in 1929, the opinion explains, the substance of the doctrine was already in use in U.S. courts. According to Ray, by 1919 it was possible to cite hundreds of cases that were dismissed for reasons equivalent to the considerations dictated by FNC. Ray goes on to enumerate Texas cases dismissed for similar reasons before the passage of the predecessor of 71.031: Article 4678. To make their argument that FNC dismissal was no longer an option in Texas, the justices referred to *Allen v. Bass*, a 1932 case with a plaintiff from the state of New Mexico. In *Allen*, the Supreme Court decided that, while prior to 1913 Texas courts had "a discretion in the matter of exercising jurisdiction" in cases with nonresident parties, the passage of Article 4678 meant that "this discretion no longer existed and that it was now obligatory on the district courts to accept jurisdiction and try these cases."

59. Ibid., 708.

60. See Saskia Sassen, *Losing Control? Sovereignty in an Age of Globalization* (New York: Columbia University Press, 1996), chapter 3. For a consideration of the role of states and transnational capital in Central American migration, see Nora Hamilton and Norma Stoltz Chinchilla, "Central American Migration: A Framework for Analysis," *Latin American Research Review* 26, no. 1 (1991): 75–110, http://www.jstor.org/stable/2503765.

61. *Dow Chemical v. Castro Alfaro*, 698.

62. "Wildcatter" can also mean a worker who strikes without official union authorization, a meaning that is clearly not indicated in this usage but nevertheless underscores the sense of outlawry or lack of authorization that clings to the word.

63. Edward Broughton, "The Bhopal Disaster and Its Aftermath: A Review," *Environmental Health Perspectives* 4, no. 1 (2005): 6, http://www.ehjournal.net/content/4/1/6.

64. Monroe Leigh, "Decision: Forum Non Conveniens—Conditional Dismissal of Tort Claim by Foreign Plaintiffs," *American Journal of International Law* 80 (1986): 964–67.

65. *Dow Chemical v. Castro Alfaro*, 691.

66. Ibid., 707.

67. Ibid., 681.

68. Ibid., 682.

69. Ibid., 690.

70. Charles S. Siegel and David S. Siegel, "The History of DBCP from a Judicial Perspective," *International Journal of Occupational and Environmental Health* 5, no. 2 (1999): 132.

71. "Payment to Sterile Workers," *The Advertiser,* August 13, 1992.

72. Charles S. Siegel, interview with the author; Yozell, "The Castro Alfaro Case," 278.

73. Malcolm J. Rogge, "Towards Transnational Corporate Accountability in the Global Economy: Challenging the Doctrine of Forum Non Conveniens in Re: Union Carbide, Alfaro, Sequihua, and Aguinda," *Texas International Law Journal* 36 (2001): 300.

74. Piszk, "Expediente No. 250–23–98," 22–23; Cameron Barr, "Shootout at Texas' Legal Corral," *The Christian Science Monitor,* May 7, 1997.

75. Bill Lambrecht, "One Worker's Conclusion: Settling Case Was Mistake," *St. Louis Post-Dispatch,* November 14, 1993.

76. "DBCP en la producción bananera," Foro Emaús, last modified 1999, accessed June 6, 2001, http://www.members.tripod.com/foro_emaus/dbcp.htm.

77. Lambrecht, "One Worker's Conclusion."

78. Bill Lambrecht, "Farm Chemical Robs Couples of Dreams: Thousands Sterile after Using DBCP," *St. Louis Post-Dispatch,* November 14, 1993; "DBCP en la producción bananera."

79. Lambrecht, "Farm Chemical Robs Couples."

80. Piszk, "Expediente No. 250–23–98," 23.

81. "DBCP en la producción bananera"; Piszk, "Expediente No. 250–23–98," 23.

82. *Dow Chemical v. Castro Alfaro,* 681.

83. Walter Borges, "Forum Bill Backers Plot Senate Strategy; after House Passage, Fores May Have to Seek Richards' Veto," *Texas Lawyer,* January 14, 1991; "Forum Non Conveniens Implementation Act of 1993 (Senate Bill 2)," *Texas Lawyer,* January 25, 1993; *Forum Non Conveniens,* Texas Statutes 71.051 (1993). Anne Bloom explains how some plaintiff lawyers participated in drafting the legislation, in essence signing on to FNC codification in exchange for assurances that they would be able to continue to bring lucrative out-of-state (but not international) asbestos (and some other) cases in Texas courts. Bloom, "Cause Lawyering in the Shadow of the State."

84. Charles B. Camp, "Dow Tries to Limit Scope of Pesticide Suit; Pre-Emptive Litigation Aimed at Claims by Banana Workers," *Dallas Morning News,* August 24, 1993; *Forum Non Conveniens,* Texas Statutes 71.051 (1993).

85. *Franklin Delgado Rodriguez v. Shell Oil Company,* 890 F. Supp. 1324 (S.D. Tex., 1995).

86. While a foreign sovereign would normally be immune to suit in the United States, Dead Sea Bromine eventually waived its immunity for the purposes of the case—although it would also sign an agreement with the other defendants limiting its liability to 2.5% of any total award to the plaintiffs. For a detailed description and critique of Dead Sea's impleading, see Siegel and Siegel, "DBCP from a Judicial Perspective."

87. *Franklin Delgado Rodriguez v. Shell Oil Company,* 1337–40.

88. "Biographical Directory of Federal Judges: Lake, Simeon Timothy III," Federal Judicial Center, accessed 2 December 2011, http://www.fjc.gov/servlet /nGetInfo?jid = 1329&cid = 999&ctype = na&instate = na.

89. *Delgado Rodriguez v. Shell Oil.* Siegel and Siegel provide a lucid description of these and other legal issues regarding the presence and legal status of Dead Sea Bromine in the suit in their article, "DBCP from a Judicial Perspective."

90. As legal scholar Austin Sarat has argued, "for the United States, no set of conceptual boundaries is more important, or more in need of critical examination, than those associated with the idea of the rule of law. Today as in the past, Americans pride themselves on their commitment to the rule of law.... Invocations of the rule of law as a constitutive boundary separating this country from the rest of the world are pervasive." Austin Sarat, "At the Boundaries of Law: Executive Clemency, Sovereign Prerogative, and the Dilemma of American Legality," *American Quarterly* 57, no. 3 (2005): 611.

91. My analysis here draws on Sarat's treatment of executive clemency.

92. Gary B. Born, *International Civil Litigation in United States Courts* (New York: Kluwer Law International, 1996), 341.

93. *Delgado Rodriguez v. Shell Oil,* 1359.

94. Ibid.

95. "Forum non conveniens, Honduras," *McGraw-Hill's Spanish and English Legal Dictionary,* comp. Henry Saint Dahl (New York: McGraw-Hill, 2004), 121.

96. "Forum non conveniens, Guatemala," *McGraw-Hill's Spanish and English Legal Dictionary,* comp. Henry Saint Dahl (New York: McGraw-Hill, 2004), 119.

97. *Delgado Rodriguez v. Shell Oil,* 1359–1365.

98. Ibid., 1357.

99. A number of legal scholars, notably Henry Saint Dahl and Alejandro Garro, have detailed a number of discrepancies between Latin American and U.S. courts that support the notion that the former are either unavailable or inadequate. These reasons include the absence of jury trials, strict rules on evidence and witnesses, and conflicts of jurisdiction, among others. Saint Dahl points out that, where foreign court law conflicts with U.S. FNC practice, the true test of availability is whether a Latin American court would accept a case pursuant to an FNC dismissal (rather than whether it was originally filed in that court).

100. *In re Union Carbide Corp. Gas Plant Disaster at Bhopal,* 634 F. Supp. 842 (United States District Court for the Southern District of New York, 1986), 867.

101. These figures are based on a 6-day work week, with a day's labor commonly exceeding 8 hours. Foro Emaús Coordinating Committee, "Bananas for the World—and the Negative Consequences for Costa Rica?" (San José, Costa Rica: Foro Emaús, 1998), http://members.tripod.com/foro_emaus/2ing.html.

102. *Gulf Oil v. Gilbert,* 509.

103. *Delgado Rodriguez v. Shell Oil,* 1368.

104. Ibid., 1367.

105. Saint Dahl, "Latin America and Blocking Statutes," 38.

106. *Delgado Rodriguez v. Shell Oil,* 1371.

107. Ibid.

108. Ibid.

109. *Delgado Rodriguez v. Shell Oil,* 1372.

110. Ibid., 1376.

111. Saint Dahl, *Spanish and English Legal Dictionary,* 115–133; Pérez Vargas, "Los inconvenientes."

112. Pérez Vargas, "Los inconvenientes," 80.

113. Petersen, *The Legal Guide to Costa Rica,* 19.

114. Saint Dahl, *Spanish and English Legal Dictionary,* 115. For a brief history and contextualization of the Bustamante Code, see Alejandro M. Garro, "Unification and Harmonization of Private Law in Latin America," *American Journal of Comparative Law* 4 (1992).

115. Saint Dahl, *Spanish and English Legal Dictionary,* 115.

116. *Abarca v. Shell Oil Co.,* Expediente No. 1011–95, 15:05, 5 de Septiembre de 1995, (Costa Rica).

117. Pérez Vargas, "Los inconvenientes," 80; Translation in Saint Dahl, *Spanish and English Legal Dictionary,* 115.

118. Translation in Saint Dahl, *Spanish and English Legal Dictionary,* 115.

119. Ibid.

120. Later Costa Rican critiques of the FNC doctrine would expand on this analysis, one arguing that "the *FNC* doctrine is unacceptable in Costa Rica as it is a doctrine that conflicts with and is disrespectful of our country's sovereignty, as it attempts to reduce the application of national procedural principles of public order and imposes jurisdiction on the National Judiciary, against the wish of the plaintiff." Pérez Vargas, "Los inconvenientes," 71.

121. The defendants appealed the decision as well; but their appeal was rejected because it was filed after the appropriate time period. Ibid.

122. Ibid., 84.

123. Saint Dahl, *Spanish and English Legal Dictionary,* 119–20, 25, 29–30.

124. Ibid., 130.

125. Vicent Boix Bornay, *El parque de las hamacas: El químico que golpeó a los pobres* (Barcelona: Icaria Editorial, 2008), 195; Saint Dahl, *Spanish and English Legal Dictionary,* 121–24.

126. Fred Misko, Response to Petition for Writ of Mandamus, in re Standard Fruit Company, Standard Fruit and Steamship Co., Dole Food Company, Inc., and Dole Fresh Fruit Company (Tex. April 22, 2006), 3–4.

127. Charles S. Siegel, Interview with the author.

128. Misko, Interview with the author, April 12, 2006.

129. Charles S. Siegel, Interview with the author.

CHAPTER FIVE

1. Sandra Piszk, "Expediente No. 250–23–98: Informe Final con Recomendaciones" (San José: Defensoría de los Habitantes de la República de Costa Rica, 1998), CQLF, 14, 23.

2. "Real Historical Exchange Rates for Baseline Countries/Regions (2005 Base Year), 1970–2030," Economic Research Service, United States Department of Agriculture, http://www.ers.usda.gov/data/macroeconomics/#HistoricalMacroTables.

3. Piszk, "Expediente No 250–23–98," 23.

4. Charles S. Siegel and David S. Siegel, "The History of DBCP from a Judicial Perspective," *International Journal of Occupational and Environmental Health* 5, no. 2 (1999): 132–35; Charles S. Siegel, interview with the author, March 30, 2006.

5. Piszk, "Expediente No 250–23–98," 23–24. In 2010 the World Health Organization's lower limit for a "normal" sperm count was decreased to 15 million/milliliter of semen; see T. G. Cooper et al., "World Health Organization Reference Values for Human Semen Characteristics," *Human Reproduction Update* 16, no. 3 (2010): 5, http://www.ncbi.nlm.nih.gov/pubmed/19934213.

6. Cited in Piszk, "Expediente No 250–23–98," 24.

7. Carlos A. Villalobos, "Abogados recomiendan aceptación: Proponen arreglo a esterilizados," *La Nación*, June 16, 1997, www.nacion.com/ln_ee/1997/junio/15/proponen.html.

8. Marvin Barquero S., "Fustigan arreglo con esterilizados," *La Nación*, July 27, 1997, http://www.nacion.com/ln_ee/1997/junio/27/fustigan.html; Ángela Ávalos Rodríguez, "Pugna por trato con bananeros," *La Nación*, July 28, 1997, http://www.nacion.com/ln_ee/1997/julio/28/pais3.html.

9. Barquero S., "Fustigan arreglo."

10. Ávalos Rodríguez, "Pugna por trato."

11. M. Slutsky, J. L. Levin, and B. S. Levy, "Azoospermia and Oligospermia among a Large Cohort of DBCP Applicators in Twelve Countries," *International Journal of Occupational and Environmental Health* 5 (1999): 119.

12. Néfer Muñoz, "El ogroquímico de los 70," *La Nación* February 2, 1997, www.nacion.com/dominical/1997/febrero/02/dom_pagina08.html.

13. Avalos Rodríguez, "Pugna por trato."

14. Muñoz, "El ogroquímico de los 70."

15. Barquero S., "Fustigan arreglo."

16. See Daniel Teitelbaum, "The Toxicology of 1,2-Dibromo-3-Chloropropane (DBCP)," *International Journal of Occupational and Environmental Health* 5, no. 2 (1999).

17. "DBCP: The Legacy," 26.

18. Of course, it was possible that researchers had not found these problems simply because they hadn't looked for them.

19. For research on health problems of banana workers more generally, we Catharina Wesseling et al., "Cancer in Banana Plantation Workers in Costa Rica," *Internationanl Journal of Epidemiology* 25, no. 6 (1996); Catharina Wesseling, Berna van Wendel de Joode, and Patricia Monge, "Pesticide-Related Illness and Injuries among Banana Workers in Costa Rica: A Comparison between 1993 and 1996," *International Journal of Occupational and Environmental Health* 7 (2001); Catharina Wesseling et al., "Long-Term Neurobehavioral Effects of Mild Poisonings with Organophosphate and N-Methyl Carbamate Pesticides among Banana Workers,"

International Journal of Occupational and Environmental Health 8 no. 1 (2002); Jonathan Hofmann et al., "Mortality among a Cohort of Banana Plantation Workers in Costa Rica," *International Journal of Occupational and Environmental Health* 12, no. 4 (2006); B. N. Van Wendel de Joode et al., "Paraquat Exposure of Knapsack Spray Operators on Banana Plantations in Costa Rica," *International Journal of Occupational and Environmental Health* 2, no. 4 (1996); C. Wesseling et al., "Symptoms of Psychological Distress and Suicidal Ideation among Banana Workers with a History of Poisoning by Organophosphate or N-Methyl Carbamate Pesticides," *Occupational and Environmental Medicine* 67, no. 11 (2010); H. G. Penagos, "Contact Dermatitis Caused by Pesticides among Banana Plantation Workers in Panama," *International Journal of Occupational and Environmental Health* 8, no. 1 (2002).

20. Ávalos Rodríguez, "Pugna por trato"; Piszk, "Expediente No 250–23–98," 8.

21. Marc Edelman, *Peasants against Globalization: Rural Social Movements in Costa Rica* (Stanford, CA: Stanford University Press, 1999), 51–56.

22. Ibid., 56–73; Vladimir de la Cruz, "Características y rasgos históricos del movimiento sindical en Costa Rica," in *El sindicalismo frente al cambio: Entre la pasividad y el protagonismo,* ed. Jorge Nowalski (San José, Costa Rica: FES/DEI, 1997).

23. Edelman, *Peasants against Globalization,* 73.

24. Sindy Mora Solano, "Costa Rica en la década de 1980: Estrategias de negociación política en tiempos de crisis: ¿Qué pasó después de la protesta?," *Intercambio* 4, no. 5 (2007): 165; José Roberto López, *La economía del banano en Centroamérica* (San José, Costa Rica: Editorial Departamento Ecuménico de Investigaciones, 1988); Vladimir de la Cruz, "Características y rasgos históricos."

25. Robert G. Williams, *Export Agriculture and the Crisis in Central America* (Chapel Hill: University of North Carolina Press, 1986), 183–86.

26. Carazo would later become a leading critic of IMF policies and a leading opponent of CAFTA.

27. Edelman, *Peasants against Globalization,* 75–83. Arias served a second term from 2006 to 2010.

28. Mora Solano, "Costa Rica en la década de 1980," 170–71.

29. Antoni Royo, "La ocupación del Pacífico Sur costarricense por parte de la compañía bananera (1938–1984)," *Diálogos* 4, no. 2 (2003), http://historia.fcs.ucr.ac.cr/articulos/2003/zonasur.htm.

30. Leda Abdallah Arrieta, "Solidarismo: Nuevo referente 'laboral' del libre comercio" (San José, Costa Rica: ASEPROLA and Coordinadora Latinoamericana de Sindicatos Bananeros, n.d.), http://aseprola.net/media_files/download/Solidarismo2.pdf.

31. Foro Emaús Coordinating Committee, "Bananas for the World—and the Negative Consequences for Costa Rica?" (San José, Costa Rica: Foro Emaús, 1998), http://members.tripod.com/foro_emaus/2ing.html.

32. Engström, "Economic Globalization," 23–24.

33. Dana Frank, *Bananeras: Women Transforming the Banana Unions of Latin America* (Cambridge, MA: South End Press, 2005), 61–65; Henry J. Frundt, *Fair*

Bananas: Farmers, Workers, and Consumers Strive to Change an Industry (Tucson, AZ: University of Arizona Press, 2009), passim; Eduardo Mora Castellanos, "Obreros, pesticidas, salud y relaciones de fuerza en los bananales del Caribe costarricense," *Ambien-Tico* 33–34 (1995), http://www.una.ac.cr/ambi/ambientico/amb33–34.html. See Frundt for a broader discussion of the effects of neoliberalism on Central American union organizing in this period, and union responses including greater engagement in social and economic policy debates and changed organizing strategies.

34. Allen Cordero Ulate, "Nuevas desigualdades; nuevas resistencias: El caso de los ex trabajadores bananeros costarricenses afectados por los agroquímicos" (paper presented at the XXVIII International Congress of the Latin American Studies Association, Rio de Janeiro, Brazil, June 11–14 2009), http://lasa.international.pitt.edu/members/congress-papers/lasa2009/files/CorderoUlate Allen.pdf, 9.

35. See Alan Scott, *Ideology and the New Social Movements* (London: Unwin Hyman, 1990); Alberto Melucci, "The New Social Movements: A Theoretical Approach," *Social Science Information* 19 (1980).

36. For a thoughtful exploration of the character and aims of new social movements across lines of identity, class, and geography, see Robert Fisher and Joseph M. Kling, *Mobilizing the Community: Local Politics in the Era of the Global City* (Newbury Park; London: Sage, 1993).

37. Cordero Ulate, "Nuevas desigualdades; nuevas resistencias," 10.

38. Ibid.

39. Ibid., 12.

40. It is also worth noting that CONATRAB's unique character differed from "social movement unionism," defined as trade unions organizing around issues that affect a working-class community beyond those represented by the union, including through organizing in conjunction with social movement organizations. In CONATRAB's case, the group called itself a union, but its strategies, membership, and activities aligned it more closely with social movement groups than with traditional trade unions. See Kim Moody, *Workers in a Lean World: Unions in the International Economy* (London; New York: Verso, 1997); Ian Robinson, "Does Neoliberal Restructing Promote Social Movement Unionism? U.S. Developments in Comparative Perspective," in *Unions in a Globalized Environment: Changing Borders, Organizational Boundaries, and Social Roles,* ed. Bruce Nissen (Armonk, NY: M. E. Sharpe, 2002).

41. Henry Frundt describes CONATRAB as a "small, Trotskyist" union and notes that CONATRAB "often criticized [COSIBA] to the detriment of union and Foro activity." H.J. Frundt, "Sustaining Labor-Environmental Coalitions: Banana Allies in Costa Rica," *Latin American Politics and Society* 52, no. 3 (2010). By the time of my first visit to Costa Rica in 2004, CONATRAB was no longer active in Foro Emaús due to intergroup conflicts.

42. Marc Edelman, *Peasants against Globalization.*

43. Ibid., 185.

44. At the same time, Barrantes would keep CONATRAB connected to far-left union and party politics. Well known and widely respected by the Costa Rican left, Barrantes was a leader not only of CONATRAB, but also of the Trotskyite *Movimiento de Trabajadores y Campesinos* (Workers' and Peasants' Movement, which would gain official political party status in Costa Rica in 2005) and of the *Central General de Trabajadores*, or CGT (a coalition of union and popular organizations).

45. Orlando Barrantes Cartín, "Las secuelas del DBCP," (San José: Foro Emaús, n.d.), http://www.foroemaus.org/espanol/ambiental/04_04.html, 2.

46. Accordingly, I use the term *afectado* to denote a person who self-identifies as DBCP-affected or participates in DBCP-related political action, not as an attribution of any particular physical state.

47. Edelman, *Peasants against Globalization*, 187.

48. Steve Marquardt, "Pesticides, Parakeets, and Unions in the Costa Rican Banana Industry, 1938–1962," *Latin American Research Review* 37, no. 2 (2001): 3.

49. "Acerca de la Defensoría," accessed December 19, 2006, http://www.dhr. go.cr/acerca-respon.html; Piszk, "Expediente No. 250–23–98."

50. Saskia Sassen, *Losing Control? Sovereignty in an Age of Globalization* (New York: Columbia University Press, 1996).

51. "Acerca de la Defensoría"; Amaru Barahona, "Costa Rican Democracy on the Edge," *Revista Envío,* May 2002.

52. Piszk, "Expediente No. 250–23–98," 5–8.

53. Ibid., 1.

54. Ibid.

55. Ibid.

56. Ibid., 27.

57. Ibid., 26–27. Some *afectados* reported that "they were presented with a series of documents stapled with one sheet left blank where they had to place their signature, fingerprint or initial."

58. Ibid., 25.

59. Of course, prohibitions against bringing any suits on such grounds would not apply to those who were not included in the settlement, but the companies' unwillingness to make deals on claims of this type indicated that they were likely unafraid that such claims would hold much water in a U.S. legal setting.

60. Piszk, "Expediente No. 250–23–98," 27–28.

61. The report does not specify whether the atypical sex ratio was observed when male, female, or both parents had been exposed. The papers cited record altered sex ratios among children of exposed fathers. J.R. Goldsmith, G. Potashnik, and R. Israeli, "Reproductive Outcomes in Families of DBCP Exposed Men," *Archives of Environmental Health* 39, no. 85–89 (1984).

62. Piszk, "Expediente No. 250–23–98," 32.

63. Ibid., 33.

64. Ibid., 33.

65. Ibid., 32.

66. Ibid., 33–35.

67. "Constitución Política de Costa Rica," (November 7, 1949), http://www. apse.or.cr/webapse/docum/docu12.htm.

68. Piszk, "Expediente No. 250–23–98," 35.

69. Ibid., 39.

70. Piszk, Sandra. "Informe Defensoría de los Habitantes 99–2.000." San José, Costa Rica: Defensoría de los Habitantes, 2000.

71. Alexánder Ramírez S, "Estado apoyará a esterilizados," *La Nación*, November 20, 1998; Piszk, "Expediente No. 250–23–98," 38.

72. Patricia Monge, Sonia Román, and Catharina Wesseling, "Informe de la Comisión Médica INS/CONATRAB sobre los efectos adversos del DBCP y los criterios para indemnización de trabajadores y trabajadoras afectado/as," (n.p.: Instituto Nacional de Seguros, Consejo Nacional de Trabajadores Bananeros, 2002), CWF, 2.

73. Ramírez, "Estado apoyará a esterilizados."

74. Régulo Sánchez Barrantes, interview with the author, February 8, 2005, Heredia, Costa Rica.

75. Ramírez, "Estado apoyará a esterilizados."

76. Ibid.; "Diputados Dan Apoyo a Bananeros Esterilizados," Assamblea Nacional de Costa Rica, www.asamblea.go.cr/actual/boletin/1998/nov98/17nov98. htm.

77. "Diputados Dan Apoyo a Bananeros."

78. Barrantes Cartín, "Las secuelas del DBCP." My analysis of the import of the Interinstitutional Commission's report is based on an article by Orlando Barrantes that was written after an oral presentation of the report's findings to CONATRAB, but before the January 2000 release of the report. In *El parque de las hamacas,* Vicent Boix characterizes the Commission's report as "disappointing to the workers [because it] criticized and dismissed as inaccurate the Defensoría's report" (222). I was unable to obtain the Interinstitutional Commission's report itself, so could not compare the claims in "Las secuelas del DBCP" with that document or form an opinion on the various perspectives offered on the impact of the Interinstitional Commission's conclusions.

79. Acta de la Sesión Plenaria No. 079, Segundo Período de Sesiones Ordinarias, Primera Legislatura, Asamblea Legislativa de la República de Costa Rica (San José: September 26, 2006), http://www.asamblea.go.cr/Centro_de_informacion/biblioteca/Investigaciones%20Realizadas/ral2012/ral_portal/actas_presidencia/Acta%2018–10–07.pdf.

80. "Síntesis Nacional," *La Nación,* July 29, 2000.

81. Vanessa Loaiza Naranjo, "Bananeros bloquearon vías," *La Nación,* July 30, 2000.

82. Ibid.

83. Ibid.

84. Alexánder Ramírez S, "Caliente protesta en Limón," *La Nación,* December 13, 2000, http://www.nacion.com/ln_ee/2000/diciembre/13/pais1.html.

85. "Enfrentamientos entre policía y manifestantes deja al menos 50 afectados en bloqueo de carretera en Limón," *La Nación,* December 12, 2000, http://wvw

.nacion.com/ln_ee/2000/diciembre/12/ultima1.html; Ramírez, "Caliente protesta"; "Disturbios en Costa Rica," *La Prensa,* December 13, 2000.

86. Ramírez, "Caliente protesta en Limón." Based on this incident, Orlando Barrantes was later charged with "extortive abduction," of the police officers in question, a charge resulting in a number of suspended trials and annulled sentences, but apparently no lasting verdict against Barrantes. See Vicent Boix Bornay, "Suspendido temporalmente el juicio a Orlando Barrantes," accessed June 26, Lahaine.org, http://www.lahaine.org/index.php?p=14198&more=1&c=1.

87. Régulo Sánchez Barrantes, interview with the author.

88. Álvaro Murillo, "Terminó conflicto en Limón," *La Nación,* December 16, 2000, http://www.nacion.com/ln_ee/2000/diciembre/16/pais2.html.

89. Ibid.

90. Marc Edelman has described how Costa Rican officials reacted to *campesino* protests in the 1980s and 1990s, arguing that "the state had played its hand well, reacting to the agriculturalists' mobilization with a classically Costa Rican combination of prolonged negotiations, minor concessions, vague promises, and sudden—yet basically sporadic and mild—repression." This characterization rings true in the DBCP case as well. Edelman, *Peasants against Globalization,* 150.

91. Ana Lorena Brenes Esquivel, "Ficha del Pronunciamiento: Opinión Jurídica: 169—J del 11/09/2003," (San José, Costa Rica: Procuradora de Costa Rica, 2003),

92. *Determinación de Beneficios Sociales y Económicos para la Población Afectada por el DBCP,* Expediente No. 14.357, Asamblea Legislativa de la República de Costa Rica (May 23, 2001).

93. Ibid., article 3.

94. Ibid., article 13.

95. Ibid., article 14.

96. "Determinación de Beneficios Sociales y Económicos para la Población Afectada por el 'DBCP' [Ley 8130]", La Gaceta [Costa Rica] 181 (San José: September 20, 2001).

97. "Instituto Regional de Estudios en Sustancias Tóxicas (IRET): Quiénes Somos," Universidad de Costa Rica, accessed May 22, 2012, http://www.una.ac.cr /iret/.

98. For example, IRET researchers and colleagues argued in 2001 that "the use of hazardous pesticides persists through deficiencies in government-driven assessment and risk management; excessive focus on regional harmonization; short-term economic interests; strong links between industry and governments; aggressive marketing; weak trade unions; and failure of universities to reach decision makers." Catharina Wesseling, Aurora Aragón, L. Castillo, et al., "Hazardous Pesticides in Central America," *International Journal of Occupational and Environmental Health* 7, no. 4 (2001): 287.

99. Monge, Román, and Wesseling, "Informe de la Comisión Médica." I was unable to determine why the INS appointed only one physician to the Commission.

100. Ibid., 6 (citations not reproduced).

101. Monge, Román, and Wesseling, "Informe de la Comisión Médica," 10.

102. Ibid., 15.

103. Ibid., 14.

104. Patricia Monge, interview with the author, February 15, 2005, Heredia, Costa Rica.

105. Carlos Arguedas Mora to Dr. Abel Pacheco de la Espriella, Presidente de la República de Costa Rica, June 14, 2002, CAMF.

106. *Asociación Pro Defensa de los Trabajadores Agrícolas y del Medio Ambiente et al. v. Presidente Ejecutivo del Instituto Nacional de Seguros et al.*, Expediente No. 02–007789–0007-CO, 14:55, 5 de Noviembre de 2002 (Costa Rica).

107. Ibid., 2.

108. Ibid.

109. Just ten days after the courts' November 5 opinion, *La Nación* reported that INS had paid 1.5 billion *colones* to 2,212 workers, a figure that, if correct, would seem to mean either that the bulk of them would have to have been paid under either the Memorandum of Understanding or through Workers' Compensation claims filed before the agreement between CONATRAB and INS, or that INS had increased the rate of processing claims substantially since reporting the 976 figure to the *Sala Constitucional*. The number of workers reported as compensated may have been erroneous, however, as the newspaper later reported a smaller number. Vanessa Loaiza Naranjo, "Apuros con riesgos de trabajo," *La Nación*, November 15, 2002, www.nacion.com/ln_ee/2002/noviembre/15/pais1.html.

110. Vanessa Loaiza Naranjo, "INS tuvo un déficit de 1.194 millones en el 2001," *La Nación*, November 15, 2002.

111. Berlioth Herrera and Freddy Parrales, "Protesta de bananeros por indemnización," *La Nación*, November 24, 2002, http://wvw.nacion.com/ln_ee/2002/noviembre/24/pais10.html.

112. *Lucas Pastor Canales Martinez, et al. v. Dow Chemical Company, et al.*, 219 F. Supp. 2d 719 (E.D. La., 2002).

113. John W Joyce, "Forum Non Conveniens in Louisiana," *Louisiana Law Review* 60 (1999).

114. *Canales Martinez v. Dow Chemical.*, 725–41.

115. Here the judge referred to *Abarca* as well as to *Aguilar*, a similar but less commented upon case from 1996 that Judge Javier Víquez Herrera of the Second Civil and Labor Court of Limón held could not be brought in Costa Rica. Ibid., 726–728.

116. Also notable in Judge Barbier's opinion was his response to defendants' nationalist appeals that "Costa Rican law should not be able to dictate to American judges what cases they must hear" (726) by noting that the cases in question had clearly-established jurisdiction in the United States and were properly brought there. In another departure from the exclusionary logic of FNC, the judge found that despite defendants' "insinuation that plaintiffs' effort to secure the most favorable forum is somehow unscrupulous or unsporting, the Court finds instead that it is consistent with the usual workings of our adversary system." (732)

117. Scott Hendler, "Bend It Like Beckham: Forum Manipulation and Abuse of the Foreign Sovereign Immunities Act by Multinational Corporations" (paper presented at the Inter-American Bar Association/Inter-American Federation of Lawyers Annual Conference, Madrid, 2004).

118. *Dole Food Company, et al. v. Gerardo Dennis Patrickson, et al.; Dead Sea Bromine Co., Ltd., and Bromine Compounds. Ltd. v. Gerardo Dennis Patrickson, et al.,* 538 U.S. 468 (2003); Hendler, "Bend It Like Beckham."

119. Hendler, "Bend It Like Beckham."

120. Vanessa Loaiza Naranjo, "Afectadas por agroquímico tomaron el INS," *La Nación,* June 12, 2003, http://wvw.nacion.com/ln_ee/2003/junio/12/pais3 .html.

121. Ana Lorena Brenes Esquivel, "Ficha del Pronunciamiento: Dictamen 213 del 14/07/2003," (San José, Costa Rica: Procuradora de Costa Rica, July 14, 2003).

122. Loaiza Naranjo, "Afectadas por agroquímico."

123. Ana Lorena Brenes Esquivel, "Ficha del Pronunciamiento: Dictamen 181 del 16/06/2003," (San José, Costa Rica: Procuradora de Costa Rica, June 6, 2003).

124. Brenes Esquivel, "Ficha del Pronunciamiento: Opinión Jurídica."

125. Ibid. A letter identical to Serrano's had been sent to the attorney general from an administrator in INS's Workers' Compensation Division but had been returned, in part because regulations required that requests for the attorney general's opinion originate from institution heads. Ana Lorena Brenes Esquivel, "Ficha del Pronunciamiento: Opinión Jurídica."

126. Jairo Villegas S. and Freddy Parrales, "Exbananeros reclaman por pago," *La Nación,* July 13, 2003, http://wvw.nacion.com/ln_ee/2003/julio/13/pais2.html; Vanessa Loaiza Naranjo, "Buscan salida a quejas limonenses," *La Nación,* September 3, 2003, http://wvw.nacion.com/ln_ee/2003/septiembre/03/pais10.html.

127. Rocío Pérez Sáenz, "Acuerdo por Nemagón sujeto a dictamen de Procuraduría," *Diario Extra,* September 9, 2003, http://www.diarioextra.com/2003/setiembre/09/nacionales08.shtml.

128. "Seguro por Nemagón," *La Nación,* October 3, 2003, http://wvw.nacion. com/ln_ee/2003/octubre/03/pais5.html.

129. Vanessa Loaiza Naranjo, "2.800 exigen indemnización por Nemagón," *La Nación,* May 5, 2004, http://wvw.nacion.com/ln_ee/2004/mayo/05/pais6.html.

130. "Protestan ex trabajadores bananeros," *Al Día,* May 4, 2004, http://wvw. aldia.cr/ad_ee/2004/mayo/04/ultimahora3.html; "Ex trabajadores bananeros reclaman indemnización a gobierno," *La Nación,* May 5, 2004, http://www.nacion.com /ln_ee/2004/mayo/05/ultima-cr6.html.

131. "Ex trabajadores bananeros reclaman indemnización."

132. Shirley Sandí, "Afectados por Nemagón depusieron huelga de hambre," *Diario Extra Online,* May 6, 2004, www.diarioextra.com/2004/06/nacionales08. shtml; "Afectados por plaguicida consiguen acuerdo de pago," *La Nación,* May 6, 2004, www.nacion.com/ln_ee/2004/06/ultima-cr7.html.

133. Sandí, "Afectados por Nemagón depusieron huelga."

134. Ibid.

135. "Costa Rica: DBCP Victims Organise for Justice," *Banana Trade News Bulletin*, no. 31 (2004): 4, http://www.bananalink.org.uk/webfm_send/36.

136. María de la Cruz Naranjo Pérez, "Indemnización por Nemagón," *La Nación*, September 3, 2004, http://wvw.nacion.com/ln_ee/2004/septiembre/03 /opinion7.html.

137. "Comunicado de Prensa" (Asamblea Legislativa de Costa Rica; Diputado José Merino del Río; Partido Frente Amplio, September 21, 2004), http://www. mail-archive.com/cr-denuncia@gruposyahoo.com/msg04979.html.

138. "Ex Trabajadores Bananeros Reclaman Indemnización"; Coordinadora de Sindicatos Bananeros de Costa Rica, "Víctimas del Nemagón-DBCP, sin indemnizar," Regional Latinoamericana de la Unión Internacional de Trabajadores de la Alimentación, Agrícolas, Hoteles, Restaurantes, Tabaco y Afines (Rel-UITA), accessed April 20, 2012, http://www.rel-uita.org/agricultura/agrotoxicos/nemagon /costa-rica-sin-indemnizacion.htm.

139. Acta de la Sesión Plenaria No. 079.

140. Reforma del Inciso C) del Artículo 2, el Artículo 9, el Artículo 13, el Párrafo Final del Artículo 14, el Artículo 15 y el Artículo 17 y Adición de un Transitorio a la Ley No. 8130, La Gaceta [Costa Rica] 224 (San José: October 19, 2006).

141. "Registro legislativo: Otros temas," *La Nación*, September 27, 2006, www. nacion.com/ln_ee/2006/septiembre/27/pais840857.html; José Enrique Rojas, "Afectados por Nemagón serán indemnizados," *La Nación*, September 22, 2006, 2006, www.nacion.com/ln_ee/2006/septiembre/22/pais835966.html; "Rápidas: Ayuda por Nemagón," *Al Día*, September 27, 2006, http://wvw.aldia.cr /ad_ee/2006/septiembre/27/nacionales840242.html.

142. On the anti-CAFTA fight in Costa Rica, see J. Cupples and I. Larios, "A Functional Anarchy: Love, Patriotism, and Resistance to Free Trade in Costa Rica," *Latin American Perspectives* 37, no. 6 (2010).

143. Carlos Arguedas Mora, e-mail to author, February 1, 2007.

144. "Trabajadores afectados por Nemagón anuncian medidas en defensa de sus derechos y contra el TLC," Cumbre Continental de Pueblos y Nacionalidades Indígenas de Abya Yala, http://www.movimientos.org/enlacei/cumbre-abyayala /show_text.php3?key = 9567; Rodolfo Ulloa, "[CR-Denuncia] Trabajadores afectados por el Nemagón se movilizarán contra el TLC," (February 24, 2007), http:// www.mail-archive.com/cr-denuncia@gruposyahoo.com/msg06231.html.

145. Juan Luis Acuña Vargas, "Cartas a la Columna: Víctimas del Nemagón," *La Nación*, August 25, 2007, http://wvw.nacion.com/ln_ee/2007/agosto/25/opinion1215815.html.

146. Lino Carmona López, "Cartas a la Columna: Víctimas del Nemagón," *La Nación*, August 3, 2007, http://wvw.nacion.com/ln_ee/2007/agosto/03/opinion1190596.html.

147. Alonso Mata B., "INS indemnizará en Junio a 900 afectados por el Nemagón," *La Nación*, May 11, 2008, http://wvw.nacion.com/ln_ee/2008/mayo/11 /pais1532276.html.

148. Ibid.

149. Ibid.

150. G. Potashnik and I. Yanai-Inbar, "Dibromochloropropane (DBCP): An 8-Year Reevaluation of Testicular Function and Reproductive Performance," *Fertility and Sterility* 47, no. 2 (1987): 317–23.

151. "Actas de Votación," (Sala Constitucional de la Corte Suprema de Justicia, July 30, 2010), http://www.docstoc.com/docs/104538622/San-Jos%EF% BF%BD-30-de-julio-de-2010#.

152. Ibid.

153. Luis Eduardo Díaz, "INS deberá realizar exámenes médicos a afectados por pesticida Nemagón " *La Nación,* September 22, 2010, http://www.nacion. com/2010–09–22/ElPais/FotoVideoDestacado/ElPais2529998.aspx.

154. Ximena Alfaro M, "Afectados por Nemagón consiguen pago del INS," *La Nación,* October 9, 2010, http://www.nacion.com/2010–10–09/ElPais/NotasSe-cundarias/ElPais2549464.aspx.

155. Ibid.

CHAPTER SIX

1. Moisés Castillo Zeas, "5 marchas a Managua," *El Nuevo Diario,* November 23, 1999, http://archivo.elnuevodiario.com.ni/1999/noviembre/23-noviembre-1999 /nacional/nacional14.html; Moisés Castillo Zeas, "Viacrucis bananero," *El Nuevo Diario,* November 24, 1999, http://archivo.elnuevodiario.com.ni/1999 /noviembre/24-noviembre-1999/nacional/nacional2.html.

2. James E. Austin, *Standard Fruit Company In Nicaragua* (Boston: Harvard University, 1994); Jacinto Obregón Sánchez, "Research Summary: Application of Nemagon and Fumazone in Banana Farms in Nicaragua," (Managua, Nicaragua: 2004), JOSF.

3. Jeffrey L. Gould, *To Lead as Equals: Rural Protest and Political Consciousness in Chinandega, Nicaragua, 1912–1979* (Chapel Hill: University of North Carolina Press, 1990); Michelle Dospital, *Siempre Más Allá . . . : el Movimiento Sandinista en Nicaragua, 1927–1934* (Managua, Nicaragua: Centro Francés de Estudios Mexicanos y Centroamericanos; Instituto de Historia de Nicaragua, 1996).

4. The others were *Asociación de Mujeres Nicaragüenses Luisa Amanda Espinoza* (AMNLAE, Nicaraguan Women's Association), *Confederación Sandinista de Trabajadores* (CST, Sandinista Workers' Confederation) and the *Comités de Defensa Sandinista* (CDS, Sandinista Defense Committees).

5. For an analysis of how conflicting understandings of the revolutionary role of the peasantry shaped agrarian-reform policies under FSLN leadership, see María Josefina Saldaña-Portillo, "Irresistible Seduction: Rural Subjectivity under Sandinista Agricultural Policy," in *The Revolutionary Imagination in the Americas and the Age of Development* (Durham, NC; London: Duke University Press, 2003).

6. Erica Polakoff and Pierre LaRamée, "Grass-Roots Organizations," in *Nicaragua without Illusions: Regime Transition and Structural Adjustment in the 1990s,* ed. Thomas W. Walker (Wilmington, DE: SR Books, 1997), 193.

7. Richard Stahler-Sholk, "Review: Sandinista Economic and Social Policy: The Mixed Blessings of Hindsight," *Latin American Research Review* 30, no. 2 (1995): 250.

8. Polakoff and LaRamée, "Grass-Roots Organizations," 186; Phillip J. Williams, "Dual Transitions from Authoritarian Rule: Popular and Electoral Democracy in Nicaragua," *Comparative Politics* 26, no. 2 (1994).

9. Victorino Espinales Reyes, interview with the author, November 13, 2006, Chinandega, Nicaragua.

10. Austin, *Standard Fruit in Nicaragua*, 8.

11. José Adán Silva, *El Nemagón en Nicaragua: Génesis de una pesadilla* (Managua: Instituto de Historia de Nicaragua y Centroamérica, Diario La Prensa, 2007), 69–70; "El Nemagón en el banquillo: Acusan los bananeros," *Revista Envío*, March, 1998, http://www.envio.org.ni/articulo/109.

12. Scott Hendler, telephone call with author, April 17, 2013.

13. Henry Saint Dahl, *McGraw-Hill's Spanish and English Legal Dictionary* (New York: McGraw-Hill, 2004), 115–133; Henry Saint Dahl, "*Forum Non Conveniens*, Latin America and Blocking Statutes," *University of Miami Inter-American Law Review* 35, no. 1 (2003/2004).

14. Castillo Zeas, "Viacrucis bananero"; Castillo Zeas, "5 marchas a Managua"; "Debates de Ley," Asamblea Nacional de Nicaragua, accessed December 29, 2012, http://legislacion.asamblea.gob.ni/Diariodebate.nsf/1e91f0054ac77a85062572e500 67fde4/5534f759ed5adfe60625792f007746a0?OpenDocument.

15. Adán Silva, *El Nemagón en Nicaragua*, 60.

16. Giorgio Trucchi, "Intervista a Victorino Espinales Reyes" (October, n.d.) http://www.itanica.org/itanica/campagne/bananeras/Intervista_a_Victorino_Espinales_Reyes.htm.

17. Scott Hendler, telephone call with the author, April 17, 2013.

18. Thomas W. Walker and Christine J. Wade, *Nicaragua : Living in the Shadow of the Eagle,* 5th ed. (Boulder, CO: Westview Press, 2011), 72–76.

19. These were *Diputados* (Representatives) Dámaso Vargas (FSLN), Orlando Mayorga Sánchez (Camino Cristiano), and Rosa Argentina Mayorga (the alternate for Camino Cristiano *Diputado* Marco Antonio Castillo Ortiz).

20. Trucchi, "Intervista a Victorino Espinales."

21. Humberto Meza, "Afectados por Nemagón emplazan al Parlamento " *La Prensa,* September 6, 2000, http://archivo.laprensa.com.ni/archivo/2000/septiembre/06/nacionales/nacionales-20000906–08.html.

22. Benjamín Chávez and Félix Thomas, "Conmoción social en Chinandega," *El Nuevo Diario,* July 17, 2000, http://achivo.elnuevodiario.com.ni/2000/julio/17-julio-2000/departamentos/departamentos.

23. Iván Castro, "Afectados por pesticidas siguen pidiendo justicia," *El Nuevo Diario,* September 10, 2000, http://archivo.elnuevodiario.com.ni/2000/septiembre /10-septiembre-2000/nacional/nacional14.html.

24. Edgard Barberena S., "'Marcharemos desnudos,'" *El Nuevo Diario,* September 6, 2000, http://archivo.elnuevodiario.com.ni/2000/septiembre/06-septiembre-2000/nacional/nacional2.html.

25. Trucchi, "Intervista a Victorino Espinales."

26. Vicent Boix Bornay, "DBCP: Un artefacto químico que sigue estallando," (unpublished manuscript, 2005), MS Word.

27. Trucchi, "Intervista a Victorino Espinales."

28. Edgard Barberena S., "Se toman Asamblea Nacional," *El Nuevo Diario*, September 8, 2000, http://archivo.elnuevodiario.com.ni/2000/septiembre/08-septiembre-2000/nacional/nacional13.html.

29. "Ley 364: Ley Especial para la Tramitación de Juicios Promovidos para las Personas Afectadas por el Uso de Pesticidas Fabricados a Base de DBCP," La Gaceta, Diario Oficial [Nicaragua] 12 (October 5, 2001).

30. Oliver P. Garza cable to Secretary of State, May 17, 2001, FOIA; Trucchi, "Intervista a Victorino Espinales"; "Ley 364."

31. "Víctimas de Nemagón presentarán demanda," *El Nuevo Diario*, February 27, 2001, http://archivo.elnuevodiario.com.ni/2001/febrero/27-febrero-2001/departamentos/departamentos2.html.

32. Lizbeth García, "Más víctimas de Nemagón formalizan sus demandas," *El Nuevo Diario*, March 17, 2001, http://archivo.elnuevodiario.com.ni/2001/marzo/17-marzo-2001/nacional/nacional7.html.

33. Oliver Gómez, "Avanza demanda de los afectados por Nemagón," *El Nuevo Diario*, May 26, 2001, http://archivo.elnuevodiario.com.ni/2001/mayo/26-mayo-2001/nacional/nacional18.html.

34. Walker and Wade, *Shadow of the Eagle*, 69–70.

35. Joaquín Tórrez A., "Danzan millones en juicios Nemagón," *El Nuevo Diario*, August 30, 2001, http://archivo.elnuevodiario.com.ni/2001/agosto/30-agosto-2001/nacional/nacional15.html.

36. María Haydée Brenes Flores, "Inescrupulosos engañan a víctmas del Nemagón," *El Nuevo Diario*, October 9, 2001, http://archivo.elnuevodiario.com.ni/2001/octubre/09-octubre-2001/departamentos/departamentos1.html.

37. Joaquín Tórrez A., "Admiten demanda del Nemagón," *El Nuevo Diario*, March 12, 2002, http://archivo.elnuevodiario.com.ni/2002/marzo/12-marzo-2002/nacional/nacional7.html.

38. M. McBride cable to ARA Central American Collective, January 23, 2002, FOIA; V. Alvarado cable to American Embassy in Managua, February 1, 2002; E. Monster cable to American Embassy in Managua, February 2, 2002, FOIA. The Alvarado cable, received pursuant to a FOIA request, was heavily redacted.

39. E. Monster cable to American Embassy in Managua, February 2, 2002.

40. Oliver P. Garza cable to Secretary of State, March 8, 2002, FOIA.

41. "2002 National Trade Estimate Report on Foreign Trade Barriers," United States Trade Representative, accessed December 28, 2012, http://www.ustr.gov/archive/Document_Library/Reports_Publications/2002/2002_NTE_Report/Section_Index.html.

42. Oliver P. Garza cable to Secretary of State, March 8, 2002.

43. Ibid.

44. Ibid.

45. M. McBride cable to ARA Central American Collective, March 18, 2002, FOIA.

46. V. Alvarado cable to American Embassy in Managua, April 15, 2002, FOIA.

47. M. McBride cable to ARA Central American Collective, October 25, 2002, FOIA.

48. Marco Centeno Caffarena, "El Nemagón apesta," *El Nuevo Diario,* November 12, 2002, http://archivo.elnuevodiario.com.ni/2002/noviembre/12-noviembre-2002/opinion/opinion6.html.

49. Octavio Enríquez, "Insólito abuso del Procurador," *El Nuevo Diario,* October 8, 2002, http://archivo.elnuevodiario.com.ni/2002/octubre/08-octubre-2002/nacional/nacional2.html.

50. Ibid.

51. Vida Benavente, "Sentencia" [Transcription of the original] (Chinandega, Nicaragua: Juzgado Tercero Civil del Distrito, December 11, 2002), GTF, 8–10.

52. Dole Food Company, Inc., *Form 10Q for the Quarterly Period Ended June 15, 2002* (Washington, DC: Securities and Exchange Commission, 2003), 12.

53. McBride cable to Central American Collective, October 25, 2002.

54. Octavio Enríquez and Juan Carlos Bow, "EmbUSA intervino en caso Nemagón," *El Nuevo Diario,* October 9, 2002, http://archivo.elnuevodiario.com.ni/2002/octubre/09-octubre-2002/nacional/nacional2.html; Octavio Enríquez, "'Invitación a la impunidad,'" *El Nuevo Diario,* October 10, 2002, http://archivo.elnuevodiario.com.ni/2002/octubre/10-octubre-2002/nacional/nacional8.html.

55. Lisa Haugaard, "Nicaragua," *Foreign Policy in Focus* 2, no. 32 (1997), http://www.fpif.org/reports/nicaragua; Augusto Zamora, "Relaciones con USA: Camino de doble vía," *Revista Envío,* October, 1996, www.envio.org.ni/articulo/2972.

56. María Haydée Brenes Flores, "Marcharán víctimas del Nemagón," *El Nuevo Diario,* October 15, 2002, http://archivo.elnuevodiario.com.ni/2002/octubre/15-octubre-2002/nacional/nacional11.html; Roberto Collado y Octavio Enríquez, "'No a juego sucio,'" *El Nuevo Diario,* November 7, 2002, http://archivo.elnuevodiario.com.ni/2002/noviembre/07-noviembre-2002/nacional/nacional10.html; Rafael Lara, "'Defenderemos Ley 364,'" *El Nuevo Diario,* October 21, 2002, http://archivo.elnuevodiario.com.ni/2002/octubre/21-octubre-2002/nacional/nacional14.html.

57. Ramón Cruz Dolmus, "Avanza a Managua marcha de víctimas del Nemagón," *El Nuevo Diario,* November 16, 2002, http://archivo.elnuevodiario.com.ni/2002/noviembre/16-noviembre-2002/nacional/nacional18.html.

58. Ibid.

59. Ibid.

60. María Haydée Brenes Flores, "Logran apoyo marchistas del Nemagón," *El Nuevo Diario,* November 21, 2002, http://archivo.elnuevodiario.com.ni/2002/noviembre/21-noviembre-2002/nacional/nacional10.html.

61. *Certification* (Supreme Court of Justice of the Republic of Nicaragua, October 16, 2003), http://boudreaudahl.com/en_nicaragua_consultation.html.

62. Benavente, "Sentencia." Plaintiffs had dropped their cases against Occidental, Chiquita, and Del Monte. Confusion over the correct name of the Dole entities left the status of Dole or Standard Fruit–associated defendants unclear. See Dole Food Company Inc., *Form 10K.*

63. Benavente, "Sentencia," 11–13.

64. Dole did make a deposit, although the timing, method, and purpose of that deposit were contested by the parties. Ibid., 5–6.

65. *Shell Oil Company v. Sonia Eduarda Franco Franco et al.*, No. CV 03–8846 NM (PJWx) (C.D. Cal., 2004), 9–12.

66. Roberto Collado Narváez, "Bálsamo a Nemagonicidio," *El Nuevo Diario,* December 16, 2002, http://archivo.elnuevodiario.com.ni/2002/diciembre/16-diciembre-2002/nacional/nacional9.html.

67. Ibid.

68. *Sonia Eduarda Franco Franco et al. v. The Dow Chemical Company, Shell Chemical Company, and Dole Food Company, Inc.*, No. CV 03–5094 NM (PJWx) LEXIS 26639 (C.D. Cal., 2003).

69. David Gonzalez and Samuel Loewenberg, "Banana Workers Get Day in Court," *New York Times,* January 18, 2003, http://www.nytimes.com/2003/01/18/business/worldbusiness/18BANA.html.

70. *Franco Franco v. Dow, Shell, and Dole.*

71. Although both Shell entities were associated with Royal Dutch Shell (RDS), Shell Chemical was the petrochemicals arm of RDS, while Shell Oil Company was its U.S. affiliate. The Nicaraguan suit had also named "Dow Chemical, also known as Dow Agro Sciences" as one defendant, when in fact the agricultural science company was a wholly owned subsidiary of Dow Chemical.

72. Juan Carlos Tijerino A., "El veredicto en caso Nemagón," *La Prensa,* October 28, 2003, http://archivo.laprensa.com.ni/archivo/2003/octubre/28/opinion/opinion-20031028–04.html.

73. The matter of the technologist had been reported in the daily *La Prensa* as a convoluted multipart story in which the technologist said he had admitted to falsifying tests only under coercion and payment from Dole. Dole admitted to paying for his travel and hotel costs, but his story was made less believable by accusations (later retracted) that he had been kidnapped at gunpoint by false CIA and FBI agents. See Ary Neil Pantoja, "Controversia por veracidad de documentos en caso Nemagón," *La Prensa* November 5, 2003, http://archivo.laprensa.com.ni/archivo/2003/noviembre/05/nacionales/nacionales-20031105–04.html; Ary Neil Pantoja and José Adán Silva, "Mafia en caso Nemagón," *La Prensa,* November 8, 2003, http://archivo.laprensa.com.ni/archivo/2003/noviembre/08/nacionales/nacionales-20031108–09.html; Eduardo Marenco Tercero, "Tecnólogo fue con su gusto a la Dole," *La Prensa,* November 9, 2003, http://archivo.laprensa.com.ni/archivo/2003/noviembre/09/nacionales/nacionales-20031109–06.html.

74. *Dole Food Company, Inc. v. Walter Antonio Gutierrez, et al.*, No. CV 03–9416 NM (PJWx) LEXIS 28429 (C.D. Cal, 2004).

75. Rafael Lara, "'Nemagonicidas' contrademandan en EU," *El Nuevo Diario,* January 8, 2004, http://archivo.elnuevodiario.com.ni/2004/enero/08-enero-2004/nacional/nacional12.html.

76. In related litigation in 2005, Judge Manella would grant a summary judgment that the Nicaraguan suit against Shell Oil was unenforceable in the United States, based on her finding that Shell Oil did not market DBCP for use in Nicaragua. *Shell Oil v. Franco Franco et al.*

77. Carol Munguía, "Abogados citan a demandantes en caso del Nemagón," *La Prensa,* February 5, 2003, http://archivo.laprensa.com.ni/archivo/2003/febrero/05/regionales/regionales-20030205-04.html; Carol Munguía, "Claman unidad de afectados del Nemagón," *La Prensa,* February 11, 2003, 2003, http://archivo.laprensa.com.ni/archivo/2003/febrero/11/regionales/.

78. José Adán Silva and Carol Munguía, "Negocio redondo con víctimas de Nemagón " *La Prensa,* August 28, 2003, http://archivo.laprensa.com.ni/archivo/2003/agosto/28/nacionales/nacionales-20030828-01.html; Mirna Velásquez Sevilla, "Se disputan futura indemnización," *La Prensa,* February 26, 2003, http://archivo.laprensa.com.ni/archivo/2003/febrero/26/nacionales/nacionales-20030226-07.html.

79. Adán Silva and Munguía, "Negocio redondo."

80. Vicent Boix Bornay, *El parque de las hamacas: El químico que golpeó a los pobres* (Barcelona: Icaria Editorial, 2008), 243; Ethan Grundberg, "Confronting the Perils of Globalization: Nicaraguan Banana Workers' Struggle for Justice," *Iowa Historical Review* 1, no. 1 (2007): 110–11. Ethan Grundberg's account emphasizes a decline in OGESA's client load during this period, but attributes it to an active campaign by Provost Umphrey to attract clients, and to high fees charged by OGESA.

81. Octavio Enríquez, "Nemagón los mata y abandono los remata," *El Nuevo Diario,* February 14, 2004, http://archivo.elnuevodiario.com.ni/2004/febrero/14-febrero-2004/nacional/nacional15.html; Moisés González Silva, "Víctimas del Nemagón regresaron a Managua," *El Nuevo Diario,* February 11, 2004, http://archivo.elnuevodiario.com.ni/2004/febrero/11-febrero-2004/nacional/nacional11.html.

82. González Silva, "Víctimas del Nemagón regresaron a Managua."

83. M. McBridecable to ARA Central American Collective, October 25, 2002.

84. Erika Rosenthal, "Who's Afraid of National Laws? Pesticide Corporations Use Trade Negotiations to Avoid Bans and Undercut Public Health Protections in Central America," *International Journal of Occupational and Environmental Health* 11, no. 4 (2005); "North American Free Trade Agreement," (Ottawa; Washington D.C.; Mexico City: NAFTA Secretariat, 1994), http://www.nafta-sec-alena.org/DefaultSite/index_e.aspx?DetailID = 78.

85. *U.S. & Central American Countries Conclude Historic Free Trade Agreement* (Washington, DC: The Office of the United States Trade Representative, December 17, 2003), http://www.sice.oas.org/TPD/USA_CAFTA/Negotiations/US_CA_concludeagreemt_e.pdf.

86. Mark Smith, *U.S. Proposal to Initiate Free Trade Negotiations with the Dominican Republic: Statement on Behalf of the U.S. Chamber of Commerce and the Association of American Chambers of Commerce in Latin America* (Washington, DC: U.S. Chamber of Commerce, October 8, 2003), http://www.uschamber.com /issues/testimony/2003/031008msmith.htm.

87. Valeria Imhof, "Dispuestos a morir frente a Parlamento," *El Nuevo Diario,* February 24, 2004, http://archivo.elnuevodiario.com.ni/2004/febrero /24-febrero-2004/nacional/nacional19.html.

88. Francisco Javier Sancho Más, "El Nemagón y el distinguido visitante," *El Nuevo Diario,* February 21, 2004, http://archivo.elnuevodiario.com.ni/2004 /febrero/21-febrero-2004/opinion/opinion1.html; Paul Baker Hernández, "President Bush Announces Intention to Sign CAFTA: Jeb Bush Lobbies for Miami," *Nicaragua News Service,* February 16–22, 2004, http://www.tulane.edu/~libweb /RESTRICTED/NICANEWS/2004_0216.txt.

89. "Vía crucis de víctimas del Nemagón," *El Nuevo Diario,* March 6, 2004, http://archivo.elnuevodiario.com.ni/2004/febrero/17-febrero-2004/nacional /nacional17.html.

90. Edgard Barberena S., "Parlamento escuchará a víctimas del Nemagón," *El Nuevo Diario,* February 12, 2004, http://archivo.elnuevodiario.com.ni/2004 /febrero/12-febrero-2004/nacional/nacional14.html; Sergio Aguirre Aragón, "Esperan firma de respaldo de Bolaños," *El Nuevo Diario,* February 15, 2004, http:// archivo.elnuevodiario.com.ni/2004/febrero/15-febrero-2004/nacional/nacional4 .html.

91. "Nuevo grupo de ex bananeros llega a pedir apoyo a Managua," *La Nación,* February 24, 2004, http://www.nacion.com/ln_ee/2004/febrero/24/ueconomia-la12.html.

92. Ibid.

93. Edgard Barberena S., "Fallece frente al Parlamento," *El Nuevo Diario,* February 13, 2004, http://archivo.elnuevodiario.com.ni/2004/febrero/13-febrero-2004 /nacional/nacional16.html.

94. Karla Castillo, "No hay por qué mentir," *El Nuevo Diario,* February 23, 2004, http://archivo.elnuevodiario.com.ni/2004/febrero/23-febrero-2004/opin-ion/opinion2.html.

95. Ibid.

96. Eloísa Ibarra, "US$82 millones para 81 mujeres," *El Nuevo Diario,* March 4, 2004, http://archivo.elnuevodiario.com.ni/2004/marzo/04-marzo-2004/nacional /nacional34.html.

97. Paul Trivelli cable to Secretary of State, February 27, 2006, FOIA; Barbara Moore cable to Secretary of State, April 2, 2004, FOIA.

98. Jaoquín Tórrez A., "Ex bananeros logran Acuerdos El Raizón," *El Nuevo Diario,* March 21, 2004, http://archivo.elnuevodiario.com.ni/2004/marzo /21-marzo-2004/nacional/nacional7.html.

99. According to an American Embassy document, the Nicaragua Attorney General for Human Rights, Benjamín Pérez, refused to attend the annual United

Nations Human Rights Commission due to the fact that he only had one day's notice to travel to Geneva for the hearings. Barbara Moore cable to Secretary of State, April 2, 2004.

100. Paul Trivelli cable to Secretary of State, February 27, 2006.

101. Giorgio Trucchi, "09/05/2004: Continua la lotta . . ." Associazione Italia-Nicaragua Livorno, accessed February 21, 2013, http://www.nicalivo.com/nicaragua/bananeras_040509.htm.

102. Paul Trivelli cable to Secretary of State, February 27, 2006.

103. Trucchi, "Continua la lotta."

104. Silvia E. Carrillo, "Víctimas del Nemagón 'a la Buena de Dios,'" *El Nuevo Diario,* September 1, 2004, http://archivo.elnuevodiario.com.ni/2004/septiembre/01-septiembre-2004/nacional/nacional-20040901-08.html.

105. Denis H. Meléndez Aguirre, *El expediente de La Marcha Sin Retorno* (Managua: Centro Alexander Von Humboldt and Centro de Información y Servicios de Asesoría en Salud, 2006), http://www.cieets.org.ni/media/contenido/attachments/DMA_Expediente_Marcha_sin_Retorno_090311.pdf.

106. Giorgio Trucchi, "La lucha de los afectados por el Nemagón y la IRC en Nicaragua" [Powerpoint Presentation], (n.d.), GTF. IRC is the Spanish acronym for *Insuficiencia Renal Crónica,* or Chronic Kidney Disease.

107. "Victims of Nemagon Hit the Road," *Revista Envío* 287 June, 2005, www.envio.org.ni/articulo/2972; Doren Roa, "Variedades," *El Nuevo Diario,* May 27, 2005, archivo.elnuevodiario.com.ni/2005/mayo/27-mayo–2005/variedades/variedades-20050527.html.

108. "Nemagón," UITA-Secretaría Latinoamericana, accessed June 2, http://www.rel-uita.org/agricultura/agrotoxicos/nemagon/index.htm.

109. Boix Bornay, "Artefacto químico."

110. Ibid. The efforts of Italian activist GiorgioTrucchi, who lived in Nicaragua and accompanied the people of ASOTRAEXDAN in many of their struggles, led to a particularly strong role for the Italy-Nicaragua Solidarity Association.

111. Jason Glaser, interview with the author, April 3, 2013; Jason Glaser, "Untitled Character List and Timeline (2010)," JGF; Antonio Hernández, interview with the author, November 15, 2006, Chinandega, Nicaragua.

112. "Los del Nemagón regresan mañana a sus comunidades," *El Nuevo Diario,* October 8, 2005, http://impreso.elnuevodiario.com.ni/2005/10/08/nacionales/2918.

113. Meléndez Aguirre, *La Marcha Sin Retorno.*

114. Dole Food Company, Inc., *Form 10K for the Fiscal Year Ended January 1, 2005* (Washington, DC: Securities and Exchange Commission, 2005).

115. *Sánchez Osorio y Otros vs. Standard Fruti [sic] Company y Otros,* Sentencia #0271–2005; Demanda #0214–0425–02cv, 8 de Augusto de 2005, (Nicaragua), JOSF.

116. Dole Food Company, Inc., *Form 10K for the Fiscal Year Ended January 1, 2005,* 19.

117. Dole Food Company, Inc., *Form 10Q for the Quarterly Period Ended October 8, 2005* (Washington, DC: Securities and Exchange Commission, 2005), http://

www.rocketfinancial.com/FetchDoc.aspx?fid = 218006; Paul Trivelli cable to Ruehc/ Secstate Washdc, February 27, 2006, 15–16.

118. Valeria Imhof, "Juicios a transnacional Shell en diferentes países de América," *El Nuevo Diario,* January 18, 2006, http://www.elnuevodiario.com. ni/2006/01/18/nacionales/10569

119. *Dow Chemical Co. et al. v. Jeronimo Anibal Florian Calderon et al.,* 422 F.3d 827 (9th Cir., 2005).

120. Valeria Imhof and Carlos Salinas, "Histórica decisión judicial," *El Nuevo Diario,* January 13, 2006, http://elnuevodiario.com.ni/2006/01/13/nacionales /01099.

121. Ibid.

122. Paul Trivelli cable to Secretary of State, February 22, 2006.

123. Paul Trivelli cable to Secretary of State, February 27, 2006.

124. Damon Vis-Dunbar, "Shell Launches Claim against Nicaragua over Seizure of Intellectual Property," *Investment Treaty News,* October 13 2006, http:// www.iisd.org/pdf/2006/itn_oct13_2006.pdf.

125. "About ICSID," International Centre for Settlement of Investment Disputes, accessed January 13, 2006, https://icsid.worldbank.org/ICSID/ICSID /AboutICSID_Home.jsp.

126. Luis Galeano, "Shell nos clava en un arbitraje," *El Nuevo Diario* 2006, http://www.elnuevodiario.com.ni/2006/08/29/nacionales/27650; "Background Note: Nicaragua," U.S. Department of State, accessed November 25, 2007, http:// www.state.gov/r/pa/ei/bgn/1850.htm.

127. Royal Dutch Shell, PLC, *Annual Report and Form 20-F for the Year Ended December 31, 2006* (London: Royal Dutch Shell, 2007), 4.

128. Luis Galeano, "Juez suplente levanta el embargo a la Shell," *El Nuevo Diario* 2006, http://impreso.elnuevodiario.com.ni/2006/07/08/nacionales/23642. The substitute judge decided that the embargo unfairly affected SIPC, which was an entity distinct from the Shell Oil Company named in Judge Benavente's verdict. Ironically, SIPC was not a named party in the ICSID case; Shell's complicated corporate structure, rarely explained to the public or even investors, seemed to be working in its favor in the confusing debate over who owned its trademarks in Nicaragua.

129. Galeano, "Shell nos clava en arbitraje"; Luis Galeano, "'Shell ya se sometió a juicio en Nicaragua,'" *El Nuevo Diario,* August 30, 2006, http://impreso.elnuevo-diario.com.ni/2006/08/30/nacionales/27748.

130. Damon Vis-Dunbar and Luke Erik Peterson, "Shell Drops ICSID Suit against Nicaragua over Seizure of Trademarks," *Investment Treaty News* 2007, http://www.iisd.org/pdf/2007/itn_may9_2007.pdf; "List of Concluded Cases," International Centre for Settlement of Investment Disputes, accessed January 13, 2006, https://icsid.worldbank.org/ICSID/FrontServlet?requestType = GenCaseDtlsRH&actionVal = ListConcluded.

131. Róger Olivas and Valeria Imhof, "Otra victoria legal para afectados de Nemagón," *El Nuevo Diario,* December 5, 2006, http://impreso.elnuevodiario.com. ni/2006/12/05/nacionales/35648.

132. María Haydée Brenes Flores, "Afectados por Nemagón intentan acercamiento," *El Nuevo Diario,* October 19, 2005, http://impreso.elnuevodiario.com.ni/2005/10/19/nacionales/3714.

133. Valeria Imhof, "La Dole 'dobla el brazo' de algunos ex bananeros," *El Nuevo Diario, February 10, 2006,* http://www.elnuevodiario.com.ni/2006/02/10/nacionales/12367.

134. This agreement was brokered over the objection of many Honduran workers who preferred a Nicaraguan-style approach. Shell joined the program in 2007. "Dole Food Company Inc. Announces Signing of an Agreement Creating a Worker Program for Honduran Banana Workers Claiming Injuries as a Result of Exposure to DBCP," October 23, 2006, http://www.dole.com/Company-Info/Press-Releases/Press-Release-20061023; Boix Bornay, *El parque de las hamacas,* 235–7.

135. Valeria Imhof, "'Retiren demandas y deroguen Ley 364,'" *El Nuevo Diario,* March 22, 2006, http://impreso.elnuevodiario.com.ni/2006/03/22/nacionales/15530.

136. Ibid.

137. Valeria Imhof, "'Contubernio bendecido por gobierno,'" *El Nuevo Diario* 2006, http://impreso.elnuevodiario.com.ni/2006/03/29/nacionales/16082.

138. Valeria Imhof, "Bananeros contra los arreglos de Espinales," *El Nuevo Diario,* April 3, 2006, http://www.elnuevodiario.com.ni/2006/04/03/departamentales/16446.

139. Valeria Imhof, "'Ni entregado ni sometido, sigo peleando,'" *El Nuevo Diario,* April 2, 2006, http://impreso.elnuevodiario.com.ni/2006/04/02/nacionales/16392.

140. "Afectados por Nemagón marchan en Chinandega," *El Nuevo Diario,* April 23, 2006, http://impreso.elnuevodiario.com.ni/2006/04/23/nacionales/17815; Róger Olivas, "Concentración en estadio de víctimas del Nemagón," *El Nuevo Diario,* May 15, 2006, http://impreso.elnuevodiario.com.ni/2006/05/15/nacionales/19470.

141. Gary Sanderson, "Temporary restraining order and order setting hearing for preliminary injunction," *Provost & Umphrey Law Firm, LLP vs. Dole Food Company, Inc.,* No. E177–138 (Dist. Ct. Jefferson County, 172nd Jud. Dist., June 9, 2006), VBF.

142. Imhof, "'Retiren demandas y deroguen Ley 364.'"

143. For example, A. L. Ramírez and C. M. Ramírez, "Esterilidad masculina causada por la exposición laboral al nematicida 1,2-Dibromo-3-Cloropropano [Male sterility caused by occupational exposure to the nematicide 1,2-Dibromo-3-Chloropropane]," *Acta médica costarricense* 23, no. 2 (1980); S. H. Sandifer et al., "Spermatogenesis in Agricultural Workers Exposed to Dibromochloropropane (DBCP)," *Bulletin of Environmental Contamination and Toxicology* 23, no. 4–5 (1979); R. I. Glass et al., "Sperm Count Depression in Pesticide Applicators Exposed to Dibromochloropropane," *American Journal of Epidemiology* 109, no. 3 (1979); and other research reported in "Dibromochloropropane (DBCP): Suspension Order and Notice of Intent to Cancel," 44 Federal Register 65135 (November 9, 1979), 65141, 65143.

144. Gary Sanderson, "Temporary restraining order."

145. Giorgio Trucchi, "Esperamos que otros se sumen a la lucha," Rel-UITA, accessed April 10, 2013, http://www6.rel-uita.org/agricultura/agrotoxicos/nemagon/con_victorino_espinales-2.htm.

146. Walker and Wade, *Shadow of the Eagle*, 76.

147. Léster Juárez, "Alianza del MRS demanda unas elecciones justas," *La Prensa,* May 30, 2006, http::/archivo.laprensa.com.ni/archivo/2006/mayo/30/noticias/politica/ 2/.

148. Trucchi, "Esperamos que otros se sumen a la lucha."

149. Jason Glaser, a filmmaker who worked for several years as an investigator for Provost Umphrey in Nicaragua, e-mailed me (on January 3, 2013) a sound recording of a man he identified as ASOTRAEXDAN leader Hilario Centeno affirming that the group would take 10–15% of any settlement. Glaser reported the date of the recording was October 2011.

150. Michael Carter, Victorino Espinales Reyes, Jorge Ali Sánchez, Denis Zapata, Melba Poveda, Sergio García, Jaime González, and Manuel Hernández to Daniel Ortega Saavedra, Wilfredo Navarro, and Luis Callejas, June 28, 2007, VBF.

151. Paul Trivelli, cable to Secretary of State, Central American Collective, Departments of Justice, Treasury and Commerce, September 4, 2007, WLF.

152. Ibid.

153. Filadelfo Alemán, "Banana Workers to Negotiate in DBCP Case," *Associated Press,* July 11, 2007, http://www.abcmoney.co.uk/news/112007101460.htm.

154. Antonio Hernández, interview with the author, November 15, 2006.

155. Christian Miller, "Pesticide Company Settles Sterility Suit for $300,000," *Los Angeles Times,* April 16, 2007, http://articles.latimes.com/2007/apr/16/local /me-amvac16; Christian Miller, "Pesticide Trial Begins against Dole, Dow," *Los Angeles Times,* July 20, 2007, http://articles.latimes.com/2007/jul/20/local /me-pest20.

156. "Division One: Associate Justice Victoria Gerrard Chaney," Calfornia Courts: The Judicial Branch of California, http://www.courts.ca.gov/2384.htm.

157. John Spano, "Dole Must Pay Farmworkers $3.2 Million," *Los Angeles Times,* November 6, 2007, http://articles.latimes.com/2007/nov/06/local/me-dole6.

158. John Spano, "Dole Must Pay $2.5 Million to Banana Workers," *Los Angeles Times,* November 16, 2007, http://articles.latimes.com/2007/nov/16/local /me-dole16.

159. Ibid.

160. David Hechler, "The Kill Step," *Corporate Counsel,* October 1, 2009, http://www.law.com/corporatecounsel/PubArticleCC.jsp?id = 1202433709311& The__Kill_Step_&slreturn = 20130416154308.

161. Naoki Schwartz, "Calif. Jury Awards Nicaraguan Banana Workers $2.5 Million in Punitive Damages from Dole," *Associated Press,* November 16, 2007.

162. José Adán Silva and Róger Olivas, "Histórica victoria en caso del Nemagón," *El Nuevo Diario,* November 6, 2007, http://www.elnuevodiario.com.ni /nacionales/1748.

163. Mauricio Miranda, "Dejan champas y dicen que Ortega quiere 'terminar' con su líder, Víctor [Victorino] Espinales," *El Nuevo Diario,* January 1, 2008, http://www.elnuevodiario.com.ni/sucesos/4874; Paolo Lüers, "Haciendo política con los pobres," *elsalvador.com,* October 8, 2008, http://www.elsalvador.com/mwedh /nota/nota_completa.asp?idCat = 6351&idArt = 2896920.

164. Victoria G. Chaney, "Ruling on defendants' motion for JNOV or new trial on punitive damages," *Jose Adolfo Tellez, et al. v. Dole Food Company, Inc., et al.,* No. BC 312 852 (LASC, March 7, 2008), VBF, 4.

165. Hechler, "The Kill Step."

166. Gibson, Dunn & Crutcher attorneys have used similar tactics in litigation brought against Texaco Chevron by Ecuadorians for health and environmental problems they link to oil extraction. For example, in 2010, the firm's Andrea Neuman—who also worked on the DBCP cases—was sanctioned for "blatant intimidation tactics" in the Chevron case. "Sanctioned Chevron Lawyers Violating New Court Order in Ecuador Environmental Trial," *PR Newswire,* November 19, 2010, http://www.prnewswire.com/news-releases/amazon-defense-coalition-chevron-lawyers-at-gibson-dunn-sanctioned-by-federal-court-over-ecuador-case-109221244 .html.

167. Victoria G. Chaney, "Amended discovery and protective order regarding protected witness," *Jose Adolfo Tellez, et al. v. Dole Food Company, Inc., et al.,* No. BC 312 852 (LASC, February 8, 2008), VBF, 1.

168. Ibid.

169. Victoria G. Chaney, "Redacted version of statement of decision for order granting petitions for Writ of Error *Coram Vobis* vacating judgement and dismissing with prejudice," *Jose Adolfo Tellez, et al. v. Dole Food Company, Inc., et al.,* No. BC312852 (LASC, March 11, 2011), VBF, 5; Steve Condie, "Appellants' opening brief," *José Antonio Rojas Laguna, et al. v. Dole Food Company, Inc, et al.,* No. BC233497 (Cal. App., April 14, 2012), VBF, 51–53.

170. Victoria G. Chaney, "Redacted version of statement for granting petitions," 5.

171. Victoria G. Chaney, "Findings of fact and conclusions of law supporting order terminating Mejia and Rivera cases for fraud on the court; Redacted version of document filed under seal," *Rodolfo Mejia et al. vs. Dole Food Company, Inc. et al.; Hilario Valenzuela Rivera et al. v. Dole Food Company, Inc. et al.,* Nos. BC340049, BC379820 (LASC, June 17 2009), 29.

172. Victoria G. Chaney, "Findings of fact and conclusions of law."

173. Condie, "Appellant's opening brief, 59–64"; Theodore J. Boutros Jr. and Robert W. Loewen, "(Redacted version) Respondents' brief," *José Antonio Rojas Laguna, et al. v. Dole Food Company, Inc, et al.,* No. BC233497 (Cal. App., December 12, 2012), VBF, 30–31.

174. Cited in Condie, "Appellant's opening brief," 92–93.

175. Ibid., 93.

176. Ibid., 117–18; "US$ 50 Mil, Visa, Casa en EU y Dole Food 'Inocente,'" *El Nuevo Diario,* March 9, 2009, http://www.elnuevodiario.com.ni/opinion/42328.

177. "US$ 50 mil, visa, casa en EU y Dole Food 'Inocente.'"

178. Ibid.; "100 mil dólares era el precio para traicionar," *El Nuevo Diario,* March 10, 2009, http://www.elnuevodiario.com.ni/imprimir.php/42424.

179. Condie, "Appellant's opening brief," 90–93. The following month Hernández sent a letter on behalf of his clients dismissing Domínguez from his position as their counsel; see Chaney, "Findings of fact and conclusions of law," 54.

180. "Reporter's transcript of proceedings: Tuesday, April 21," *Rodolfo Mejia, et al. vs. Dole Food Company, Inc. et al.; Hilario Valenzuela Rivera, et al. v. Dole Food Company, Inc. et al.,* Nos. BC340049, BC379820 (LASC, April 21,, 2009), VBF, 26.

181. Ibid.; "Reporter's transcript of proceedings: Thursday, April 23," *Rodolfo Mejia, et al. vs. Dole Food Company, Inc. et al.; Hilario Valenzuela Rivera, et al. v. Dole Food Company, Inc. et al.,* Nos. BC340049, BC379820 (LASC, April 23, 2009), VBF.

182. Victoria G. Cheney, "Oral Ruling," *Rodolfo Mejia, et al. vs. Dole Food Company, Inc. et al.,* Nos. BC340049 (LASC, April 23, 2009), VBF.

183. Ibid., 3, 6–7.

184. Ibid., 3–4. On derogatory U.S. attitudes toward Latin America, see Lars Schoultz, *Beneath the United States: A History of U.S. Policy toward Latin America* (Cambridge, MA: Harvard University Press, 1998).

185. Cheney, "Oral Ruling," 16.

186. Ibid., 17.

187. Ibid.

188. Ibid., 19.

189. Paul Huck, "Order denying recognition of judgment," *Miguel Angel Sanchez Osorio, et al. v. Dole Food Company, et al.,* No.: 07–22693-CIV-HUCK (S.D.Fla, October 20, 2009), VBF.

190. Ibid. A number of lawyers for the *afectados* and their supporters attributed Huck's decision to a call he had received from Judge Chaney expressing her views on fraud in Nicaragua. (*Osorio* was one of the cases supposedly under negotiation at the Monserrat meeting). Huck noted he would leave aside considerations of fraud in his initial decision (2), but it is difficult to know with certainty how Chaney's call may have influenced his thinking.

191. To address the issue of irrefutable presumption of causation, Huck waded into a consideration of some medical evidence. He argued that the Nicaraguan trials did not provide enough evidence of causation, as they depended exclusively on sperm tests and limited diagnostic information. He also noted that Dole tried to introduce into evidence a number of birth certificates of children born to plaintiffs alleging a series of sperm pathologies, arguing that the judge's refusal to admit this evidence on causation was a violation of Dole's due-process rights. However, Huck's contention that "of the six types of sperm impairments listed in the Judgment, only azoospermia has been linked to DBCP exposure, and only in the factory setting— never to farm workers" suggests that consideration (or petitioner's presentation) of the medical issues was inadequate, as there is clear clinical and epidemiological evidence of DBCP health effects (specifically azoospermia and oligospermia) in

farm workers, including Kempter and Bernstein, *Dibromochloropropane: Final Position Document*; Ramírez and Ramírez, "Esterilidad masculina"; Glass et al., "Sperm Count Depression"; Sandifer et al., "Spermatogenesis in Agricultural Workers."

192. Huck, "Order denying recognition of judgment," 24.

193. Ibid., 60.

194. Steve Condie, interview with the author, August 11, 2010.

195. P. J. Boren and J. Chavez, "Order to show cause," *Dole Food Company, Inc. et al. v. The Superior Court of Los Angeles County, Respondent Jose Adolfo Tellez et al., Real Parties in Interest; Dow Chemical Company the Superior Court of Los Angeles County, Respondent Jose Adolfo Tellez et al., Real Parties in Interest,* No. B216182 (Cal. App., July 7, 2009), VBF.

196. At the time, *Tellez* was in the Court of Appeals, which in response to Dole's petition ordered that the case be revisited. In July 2009, Judge Cheney had been elevated to the Court of Appeals, but Dole successfully requested she be sent back to the Superior Court to preside over the latest developments in *Tellez*.

197. "Bananas!*" Accessed November 15, http://www.bananasthemovie.com/.

198. Steve Condie, interview with the author.

199. Róger Olivas, "Conspiración Dole, la fábrica infame de 'evidencias,'" *El Nuevo Diario,* May 12, 2010, http://www.elnuevodiario.com.ni/especiales/74226.

200. Condie, "Appellant's opening brief," 84; Boutros and Loewen, "(Redacted version) Respondents' brief," 218.

201. "Dole Food Inc. and the Dow Chemical Company's Motion of Entry of Protective Order Pursuant to FRCP 26(C) Plaintiffs' Response in Opposition to Dole Food Company," *Miguel Angel Sanchez Osorio et al., vs. Dole Food Company, Inc et al.,* (S.D. Fla., June 25, 2009), MSF, 13.

202. Victoria Kim and Alan Zarembo, "New Lawyer Aids Nicaraguan Farmworkers Who Sued Dole," *Los Angeles Times,* August 7, 2009, http://articles.latimes.com/2009/aug/07/local/me-dole7.

203. "For the Record," *Los Angeles Times,* August 15, 2009, http://articles.latimes.com/2009/aug/07/local/me-dole7.

204. Steve Condie, "The State Bar of California: California Attorney Complaint Form [Completed with Supporting Materials]," (October 4, 2010), 27, VBF.

205. Róger Olivas, "La Dole nos sobornó para declarar contra víctimas del Nemagón," *El Nuevo Diario,* May 10, 2010, http://elnuevodiario.com.ni/nacionales/74115#top; Fabiola de los Ángeles Dávila Gutiérrez, *Absolución de Posiciones,* 0101–0424–09CV (Juzgado Primero de Distrito de lo Civil, September 4, 2009), VBF; Bruno Filemón Herrera Jarquín, *Absolución de Posiciones,* 0096–0424–09CV (Juzgado Primero de Distrito de lo Civil, August 31, 2009), VBF; Juan José Herrera Jarquín, *Absolución de Posiciones,* 0093–0424–09CV (Juzgado Primero de Distrito de lo Civil, August 25, 2009); Rufino Hermógenes Pérez Vallestero, "Declaración Jurada," (December 11, 2009), VBF; Ernesto Aquiles Franco, "Declaración Jurada," *Miguel Angel Sanchez Osorio, et al. v. Dole Food Company, et al.,* Case 1:07-cv-22693-PCH December 15, 2009), VBF; Antenor Cano Centeno, "Declaración Jurada," Case 1:07-cv-22693-PCH January 11, 2010), VBF; Irving

Jacinto Castro Agüero, "Testimonio (Declaración Ante Notario) [Doc. 260–4]," *Miguel Angel Sanchez Osorio, et al. v. Dole Food Company, et al.,* (S.D.Fla, April 22, 2009), VBF.

206. Olivas, "La Dole nos sobornó."

207. José Adán Silva, "Conspiración: El otro veneno en la tragedia del Nemagón," *El Nuevo Diario,* July 4, 2010, http://elnuevodiario.com.ni/nacionales/74115#top.

208. Glaser, interview with the author, August 5, 2010; Condie, "The State Bar of California: California Attorney Complaint Form [Completed with Supporting Materials],"5–6.

209. Condie, "California Attorney Complaint Form," 6.

210. Michael Carter, Victorino Espinales Reyes, Jorge Ali Sánchez, Denis Zapata, Melba Poveda, Sergio García, Jaime González, and Manuel Hernández to Daniel Ortega Saavedra, Wilfredo Navarro, Luis Callejas, June 28, 2007.

211. Condie, "California Attorney Complaint Form," 42.

212. Ibid., 10, 42.

213. Steve Condie, *Amended Return to Order to Show Cause Re: Dole and Dow's Petitions for Writ of Error Corum [sic] Vobis,* No. BC312852 Jose Adolfo Tellez, et al. vs. Dole Food Company, Inc. et al. (LASC, December 17, 2009), VBF, 23.

214. Ibid., 11–12.

215. José Adán Silva, "Lecciones del Nemagón: de la dignidad al olvido," *El Nuevo Diario,* July 6, 2010, http://impreso.elnuevodiario.com.ni/2010/07/07/nacionales/127646; José Adán Silva, "Sangre envenenada: La herencia del Nemagón," *El Nuevo Diario,* July 5, 2010, http://www.elnuevodiario.com.ni/especiales/78246; José Adán Silva, "Nemagón: Una década de tragedia impune," *El Nuevo Diario,* July 4, 2010, http://www.elnuevodiario.com.ni/especiales/78064.

216. Adán Silva, "Lecciones del Nemagón."

217. Róger Olivas, "Fallo a favor de Dole 'descarado e injusto,'" *El Nuevo Diario,* July 15, 2010, http://www.elnuevodiario.com.ni/especiales/78064.

218. José Adán Silva, "Gobierno les hará casas a los del Nemagón," *El Nuevo Diario,* July 7, 2010, http://www.elnuevodiario.com.ni/nacionales/78380.

219. "Entregan viviendas a afectados por pesticida Nemagón en Nicaragua," *Deutsche Presse-Agentur,* February 25, 2011; "Ortega anuncia Congreso Sandinista que decidiría su postulación," *Deutsche Presse-Agentur,* February 26, 2011.

220. José Adán Silva, "Ya se ve la primera casita de los Nemagón," *El Nuevo Diario,* August 4, 2010, http://www.elnuevodiario.com.ni/nacionales/80422.

221. "Dole Food indemnizará a más de 5.000 ex bananeros en Centroamérica," *Agence France Press,* August 11, 2011.

222. Dole Food Company, Inc., "Dole Food Company Signs Definitive Settlement Agreement with Provost & Umphrey Regarding Banana Workers' Alleged Exposure to DBCP," news release, October 3, 2011, http://investors.dole.com/phoenix.zhtml?c = 231558&p = irol-newsArticle&id = 1613086.

223. Daniel Siegal, "Calif. Appeals Court Tosses $2.3m Dole Pesticide Verdict," *Law 360,* March 7, 2014, http://www.law360.com/articles/516638/calif-appeals-court-tosses-2–3m-dole-pesticide-verdict.

224. Steve Condie, e-mail to author, June 2, 2014.

225. Mark Sparks reported that Provost Umphrey is focusing on enforcement rather than filing new cases. Mark Sparks, interview with the author, May 1, 2013. Dole's annual filing for 2012 notes the latest verdict against it dates to 2009. Dole Food Company, Inc., *Form 10K for the Fiscal Year Ended December 29, 2012* (Washington, DC: Securities and Exchange Commission, 2013), http://investing.businessweek.com/research/stocks/financials/drawFiling.asp?docKey=136–00011931251310325 4–6SN7A4LP81N468832T2FI8BB14&docFormat = HTM&formType = 10-K.

226. Mark Sparks, interview with the author, May 1, 2013.

CONCLUSION

1. Jack DeMent to Distribution, March 3, 1978, document no. 5793, SHLF.

2. Jack DeMent to Dave Green, January 22, 1985, document no. SFC3–0000159, SHLF.

3. I say "so-called" here, and use quotations around this term elsewhere, because the scope of these agreements extends far beyond what has traditionally been defined as "trade."

4. Steve Condie, e-mail to author, June 2, 2014.

5. For example, a class action case was dismissed in May 2014 by a Delaware federal court over timing-related issues. Michael Lipkin, "DBCP Claims against Dole Tossed over Class Action Tolling," *Law 360*, May 28, 2014, http://www.law360.com/environmental/articles/542351/dbcp-claims-against-dole-tossed-over-class-action-tolling.

6. Mark Sparks, interview with the author, May 1, 2013; Scott Hendler, telephone call with the author, April 17, 2013.

7. Vicent Boix Bornay, *El parque de las hamacas: El químico que golpeó a los pobres* (Barcelona: Icaria Editorial, 2008), 236–37.

8. Jason Glaser, interview with the author, April 3, 2013.

9. Paul M. Barrett, "Chevron's RICO Victory Provides a Model for Other Companies," *Bloomberg Businessweek,* March 5, 2014, http://www.businessweek.com/articles/2014-03-05/chevrons-rico-victory-provides-a-model-for-other-companies. For a brief but sobering overview of the Chevron litigation, see Karen Hinton, "Chevron's Ecuador Plan B," *Huffington Post*, May 9, 2014, http://www.huffingtonpost.com/karen-hinton/chevrons-ecuador_b_5272031.html. Gibson Dunn has also been fingered as responsible for oil giant BP's efforts to block payments under a settlement they negotiated with oil spill claimants in 2012. Susan Beck, "How BP Decided to Fight the Deepwater Settlement," *American Lawyer*, March 5, 2014, http://www.americanlawyer.com/id=1202643826256/How-BP-Decided-to-Fight-the-Deepwater-Settlement#ixzz34ukBUYRX.

10. "Formación de un equipo interdisciplinario para la investigación de la enfermedad renal crónica en las regiones cañeras de Mesoamérica" (Heredia, Costa Rica: Instituto Regional de Estudios en Sustancias Tóxicas (IRET) de la Universidad

Nacional, Heredia, Costa Rica, November 13–14, 2009), http://www.saltra.una. ac.cr/index.php/descarga-memoria-taller-erc-nov-09; Catharina Wesseling et al., "The Epidemic of Chronic Kidney Disease of Unknown Etiology in Mesoamerica: A Call for Interdisciplinary Research and Action," *American Journal of Public Health* 103, no. 11 (2013).

11. S. Peraza et al., "Decreased Kidney Function among Agricultural Workers in El Salvador," *American Journal of Kidney Disease* 59, no. 4 (2012); J. Crowe et al., "Heat Exposure in Sugarcane Harvesters in Costa Rica," *American Journal of Industrial Medicine* 56, no. 10 (2013); C. Torres et al., "Decreased Kidney Function of Unknown Cause in Nicaragua: A Community-Based Survey," *American Journal of Kidney Disease* 55, no. 3 (2010); C. M. Orantes et al., "Chronic Kidney Disease and Associated Risk Factors in the Bajo Lempa Region of El Salvador: Nefrolempa Study, 2009," *MEDICC Review* 13, no. 4 (2011); "Formación de equipo interdisciplinario en regiones cañeras."

12. "Anatomy of a Riot," La Isla Foundation, accessed September 25, 2013, http://laislafoundation.org/epidemic/anatomy-riot-page/.

13. See Dana Frank, "Where Are the Workers in Consumer-Worker Alliances? Class Dynamics and the History of Consumer-Labor Campaigns," *Politics & Society* 31, no. 3 (2003): 363–79; G. Seidman, *Beyond the Boycott: Labor Rights, Human Rights, and Transnational Activism* (New York: Russell Sage Foundation, 2007).

14. During my field work, almost every banana worker I spoke with about organic production told me that bananas marketed as organic were just small or substandard conventionally-grown fruits. Apocryphal or not, this contention effectively conveys banana worker skepticism about the value of organic. On the business of organic production, see Julie Guthman, *Agrarian Dreams: The Paradox of Organic Farming in California* (Berkeley: University of California Press, 2004). On consumer perspectives on organic and local agriculture, see Sarah Wald, "Visible Farmers/Invisible Workers: Locating Immigrant Labor in Food Studies," *Food, Culture, and Society* 14, no. 4 (2011); Andrew Szasz, *Shopping Our Way to Safety: How We Changed from Protecting the Environment to Protecting Ourselves* (Minneapolis: University of Minnesota Press, 2007).

15. Saskia Sassen, *Losing Control? Sovereignty in an Age of Globalization* (New York: Columbia University Press, 1996), 56.

16. Carolyn Raffensperger and Joel A. Tickner, *Protecting Public Health and the Environment: Implementing the Precautionary Principle* (Washington, DC: Island Press, 1999), 1.

17. Ricardo Chavarría, "Diputados prohíben uso de 53 agroquímicos en el país," *La Prensa Gráfica*, October 2, 2013, http://www.laprensagrafica.com/2013/09/06/diputados-prohiben-uso-de-53-agroquimicos-en-el-pais.

18. "Importing and Exporting Pesticides and Devices," U.S. Environmental Protection Agency, accessed September 25, 2013, http://www.epa.gov/compliance/monitoring/programs/fifra/importexport.html; "Import and Export Trade Requirements," U.S. Environmental Protection Agency, accessed September 25, 2013, http://www.epa.gov/oppfead1/international/trade/.

19. Comptroller General of the United States, "Report to the Congress of the United States; Better Regulation of Pesticide Exports and Pesticide Residues in Imported Foods Is Essential," (Washington, DC: United States General Accounting Office, 1979); General Accounting Office, "Report to the Chairman, Subcommittee on Oversight and Investigations, Committee on Energy and Commerce, House of Representatives; Pesticides: Adulterated Imported Foods are Reaching U.S. Grocery Shelves," (Washignton, DC: Government Printing Office, 1992); C. Smith, K. Kerr, and A. Sadripour, "Pesticide Exports from U.S. Ports, 2001–2003," *International Journal of Occupational and Environmental Health* 14, no. 3 (2008).

20. The most notable attempt at changing U.S. policy in this regard was Senator Patrick Leahy's "Circle of Poison Prevention Act," which proposed banning the export of unregistered pesticides, but failed to pass in 1990, 1991, and 1998.

21. An imperfect example on banning trade in hazardous chemicals is the European Union's new chemical regulation, called REACH (Registration, Evaluation, and Authorization of Chemicals). A supranational regulation, REACH has been in force in all EU member states since 2007. REACH applies to chemicals exported in volumes of one ton or more, as well as those produced in or imported to the EU, but exempts most pesticides, whose trade is regulated under other EU directives that incorporate the Stockholm and Rotterdam conventions. REACH could be further leveraged as a tool for global chemical safety by including pesticides and removing exemptions for chemicals in transit through European ports. European Parliament and Council, "Regulation (EC) No 1907/2006 Concerning the Registration, Evaluation, Authorisation, and Restriction of Chemicals (REACH), Establishing a European Chemicals Agency," Official Journal of the European Union L396 (December 30, 2006).

22. Cassandra Burke Robertson, "Forum Non Conveniens on Appeal: The Case for Interlocutory Review," *Southwestern Journal of International Law* 18 (2011): 453.

23. Ibid.

24. *Kiobel et al. v. Royal Dutch Petroleum Co. et al.,* 133 US 1659 (2013).

25. In their 2007 decision in *Sinochem,* the U.S. Supreme Court held that federal courts did not have to make a determination on jurisdiction before dismissing under FNC. Mitchell Wong describes this as "fast track to dismissal" and discusses other impacts on litigation. *Sinochem International Co. v. Malaysia International Shipping Corp,* 549 US 422 (2007); Mitchell M. Wong, "Forum Non Conveniens: Circumstances after 'Sinochem,'" *New York Law Journal* 237, no. 58 (2007): 5, http://www.mofo.com/docs/pdf/forumnono32707.pdf.

26. Robertson, "Forum Non Conveniens on Appeal"; Mark E. Gray, "Don't Leave US Just Yet: Forum Non Conveniens and the Federal Court's Power to Stay and Monitor Actions in the 'Interest of Justice,'" *Loyola of Los Angeles Law Review* 46 (2012); Nicholas A. Fromherz, "A Call for Stricter Appellate Review of Decisions on Forum Non Conveniens," *Washington University Global Studies Law Review* 11 (2012); Erin Foley Smith, "Right to Remedies and the Inconvenience of *Forum Non Conveniens*: Opening U.S. Courts to Victims of Corporate Human Rights Abuses," *Columbia Journal of Law & Social Problems* 44 (2010).

27. On the treatment of evidence in the courtroom, see Sheila Jasanoff, "What Judges Should Know about the Sociology of Science," in *Science and Public Reason* (New York: Routledge, 2012); Carl F. Cranor, *Toxic Torts: Science, Law and the Possibility of Justice* (Cambridge; New York: Cambridge University Press, 2006); D. Egilman, J. Kim, and M. Biklen, "Proving Causation: The Use and Abuse of Medical and Scientific Evidence inside the Courtroom—an Epidemiologist's Critique of the Judicial Interpretation of the Daubert Ruling," *Food and Drug Law Journal* 58, no. 2 (2003).

28. This need has been sensitively articulated by at least one attorney involved in DBCP litigation. See Emily Yozell, "The Castro Alfaro Case: Convenience and Justice—Lessons for Lawyers in Transcultural Litigation," in *Human Rights, Labor Rights, and International Trade,* eds. Lance A. Compa and Stephen F. Diamond (Philadelphia: University of Pennsylvania Press, 1996). See also Leslie Gill, "The Limits of Solidarity: Labor and Transnational Organizing against Coca-cola," *American Ethnologist* 36, number 4 (2009).

29. Mary Jane Angelo et al., "Reclaiming Global Environmental Leadership: Why the United States Should Ratify Ten Pending Environmental Treaties" (Washington DC: Center for Progressive Reform, 2012), accessed August 12, 2013, http://www.progressivereform.org/articles/International_Environmental_Treaties_1201.pdf.

30. Trish Kelly, "The WTO, the Environment and Health and Safety Standards," *The World Economy* 26, no. 2 (2003).

31. Garrett Brown, "Protecting Workers' Health and Safety in the Globalizing Economy through International Trade Treaties," *International Journal of Occupational and Environmental Health* 11, no. 2 (2005); David A. Gantz, "Labor Rights and Environmental Protection under NAFTA and Other U.S. Free Trade Agreements," *The University of Miami Inter-American Law Review* 42 (2011); Linda Delp et al., "NAFTA's Labor Side Agreement: Fading into Oblivion? An Assessment of Workplace Health and Safety Cases" (Los Angeles: UCLA Labor Center, 2004), www.labor.ucla.edu/publications/reports/nafta.html.

32. "Advisory Committees," Office of the United States Trade Representative; Executive Office of the President, accessed October 2, 2013, http://www.ustr.gov/about-us/intergovernmental-affairs/advisory-committees.

33. "Campaign: Public Health Voice in Trade Policy," Center for Policy Analysis on Trade and Health—CPATH, http://www.cpath.org/id4.html; "Trade Advisory Committees December 2010," Center for Policy Analysis on Trade and Health—CPATH, http://www.cpath.org/id46.html.

34. Michael R. Bloomberg, "Why is Obama Caving on Tobacco?," *New York Times,* August 22, 2013, http://www.nytimes.com/2013/08/23/opinion/why-is-obama-caving-on-tobacco.html?_r = 0; Ellen Shaffer, "Stop TPP Protections for Big Tobacco," *Huffington Post,* September 9, 2013, http://www.huffingtonpost.com/ellen-r-shaffer/stop-tpp-protections-for-big-tobacco_b_3886771.html; Editorial Board, "The Hazard of Free-Trade Tobacco," *New York Times,* August 31, 2013, http://www.nytimes.com/2013/09/01/opinion/sunday/the-hazard-of-free-trade-tobacco.html.

35. See John Wargo, *Our Children's Toxic Legacy: How Science and Law Fail to Protect Us from Pesticides* (New Haven: Yale University Press, 1998); Jennifer Sass and Mae Wu, "Superficial Safeguards: Most Pesticides Are Approved by Flawed EPA Process" (Washington, DC: Natural Resources Defense Council, March, 2013).

36. Mary H. Graffam, "The Web of Tort 'Reform,'" *Trial* 48, no. 12 (2012), http://www.justice.org/cps/rde/xchg/justice/hs.xsl/19614.htm.

37. "Campaign: Public Health Voice in Trade Policy."

SELECTED BIBLIOGRAPHY

ARCHIVAL SOURCES

Litigation Files

Christian Hartley Litigation Files (CHLF)
Scott Hendler Litigation Files (SHLF)
Jacinto Obregón Litigation Files (JOSF)
Caroline Quintero Litigation Files (CQLF)
Mark Sparks Litigation Files (MSF)

Privately Held Files

Private papers of Carlos Arguedas Mora (CAMF)
Private papers of Vicent Boix (VBF)
Private papers of Jason Glaser (JGF)
Private papers of Lori Ann Thrupp (LATF)
Private papers of Giorgio Trucchi (GTF)
Private papers of Catharina Wesseling (CWF)

U.S. State Department Files

Produced in response to Freedom of Information Act request (FOIA)

Wikileaks Files

Obtained online (WLF)

INTERVIEWS BY THE AUTHOR

Arguedas Mora, Carlos. Siquierres, Costa Rica, 2007.
Condie, Steve. By telephone, 2010.
Domínguez Vargas, Carlos E. San José, Costa Rica, 2005.
Espinales Reyes, Victorino. Chinandega, Nicaragua, 2006.
Glaser, Jason. By Skype, 2010; 2013
Hendler, Scott. By telephone, 2013
Hernández Ordeñana, Antonio. Chinandega, Nicaragua, 2006.
Misko Jr, Fred. By telephone, 2006.
Monge, Patricia. Heredia, Costa Rica, 2005.
Patrickson, Gerardo Dennis. Siquirres, Costa Rica, 2004.
Sánchez Barrantes, Régulo. Heredia, Costa Rica, 2005; San José, Costa Rica, 2005
Siegel, Charles S. By telephone, 2006; 2011.
Sparks, Mark. By telephone, 2010; 2013.
Trucchi, Giorgio. Managua, Nicaragua, 2006.

TRADE MAGAZINES AND NEWSPAPERS

Agricultural Chemicals
American Lawyer
Banana Trade News Bulletin
Chemical Week
Corporate Counsel
Down to Earth
Farm Chemicals
Investment Treaty News
Law 360
New Jersey Farm and Garden
Pesticide and Toxic Chemical News
Plant Disease Reporter
Shell News
Texas Lawyer

NEWSPAPERS AND NEWSWIRES

Agence France Press
Al Día (Costa Rica)
Associated Press Wire
Bloomberg Businessweek
Christian Science Monitor

Dallas Morning News
Diario Extra (Costa Rica)
El Nuevo Diario (Nicaragua)
Honolulu Advertiser
Huffington Post
La Gaceta (Costa Rica)
La Nación (Costa Rica)
La Prensa (Nicaragua)
Los Angeles Times
New York Times
PR Newswire
St. Louis Post Dispatch
Stockton Record
Time Magazine
Upside Down World

PUBLISHED SOURCES

Abdallah Arrieta, Leda. "Solidarismo: Nuevo referente 'laboral' del libre comercio." San José, Costa Rica: ASEPROLA and Coordinadora Latinoamericana de Sindicatos Bananeros, n.d., http://aseprola.net/media_files/download/Solidarismo2.pdf. Reprint.

Adam, A. V., and A. Rodriguez. "'Clean Seed' and 'Certified Seed' Programmes for Bananas in Mexico." *FAO Plant Protection Bulletin* 18, no. 3 (1970): 57–63.

Amaya Amador, Ramón. *Prisión Verde.* Buenos Aires: Ágape, 1957.

Anderson, Earl J. "Toxicity of BBC—a Warning." *PRI News* 11, no. 5 (1963): 150–51.

Andrade, Sean A. "Biting the Hand that Feeds You: How Federal Law Has Permitted Employers to Violate the Basic Rights of Farmworkers and How This Has Begun to Impact Other Industries." *University of Pennsylvania Journal of Labor & Employment Law* 4 (2002): 601–621.

Angelo, Mary Jane, Rebecca Bratspies, David Hunter, John H. Knox, Noah Sachs, and Sandra Zellmer. "Reclaiming Global Environmental Leadership: Why the United States Should Ratify Ten Pending Environmental Treaties." Washington DC: Center for Progressive Reform, 2012, http://www.progressivereform.org/articles/International_Environmental_Treaties_1201.pdf. Reprint.

Ashford, Nicholas A. *Crisis in the Workplace: Occupational Disease and Injury.* Cambridge, MA: MIT Press, 1976.

Austin, James E. "Standard Fruit Co. In Nicaragua." Case study, Boston: Harvard University, 1994.

Baez, Beau. *Tort Law in the USA.* Alphen aan den Rijn, The Netherlands; Frederick, MD: Kluwer Law International, 2010.

Banuett Bourrillon, Marcelle. "Situación laboral en las zonas bananeras del Caribe costarricense." San José, Costa Rica: Asociación Servicios de Promoción Laboral; Foro Emaús, 2003.

Barahona, Amaru. "Costa Rican Democracy on the Edge." *Revista Envío,* May 2002.

Barrantes Cartín, Orlando. "Las secuelas del DBCP." San José: Foro Emaús, n.d., http://www.foroemaus.org/espanol/ambiental/04_04.html. Reprint.

Barraza, D., K. Jansen, B. van Wendel de Joode, and C. Wesseling. "Social Movements and Risk Perception: Unions, Churches, Pesticides, and Bananas in Costa Rica." *International Journal of Occupational and Environmental Health* 19, no. 1 (2013): 11–21.

Bishopp, F. C., G. J. Haeussler, H. L. Haller, W. L. Popham, B. A. Porter, E. R. Sasscer, J. S. Wade, Benjamin Schwartz, Karl S. Quisenberry, E. R. McGovran, and Alfred Stefferud. *Insects: The Yearbook of Agriculture.* Washington, DC: United States Department of Agriculture, 1952.

Bloom, Anne. "Cause Lawyering in the Shadow of the State: Why Personal Injury Litigation May Represent the Future of Transnational Cause Lawyering." In *Cause Lawyering and the State in a Global Era,* edited by Austin Sarat and Stuart A. Scheingold, xi, 417. Oxford; New York: Oxford University Press, 2001.

Boix Bornay, Vicent. *El parque de las hamacas: El químico que golpeó a los pobres:* Barcelona: Icaria Editorial, 2008.

Booth, John A., Christine J. Wade, and Thomas W. Walker. *Understanding Central America: Global Forces, Rebellion, and Change.* 5th ed. Boulder, CO: Westview Press, 2010.

Born, Gary B. *International Civil Litigation in United States Courts.* New York: Kluwer Law International, 1996.

Bottrell, Dale G. "Government Influence on Pesticide Use in Developing Countries." *Insect Science and its Applications* 5 no. 3 (1984): 151–55.

———. *Integrated Pest Management.* Washington, DC: Council on Environmental Quality, 1979.

Bourgois, Philippe I. *Ethnicity at Work: Divided Labor on a Central American Banana Plantation.* Johns Hopkins Studies in Atlantic History and Culture. Baltimore: Johns Hopkins University Press, 1989.

Brickman, Ronald, Sheila Jasanoff, and Thomas Ilgen. *Controlling Chemicals: The Politics of Regulation in Europe and the United States.* Ithaca, NY: Cornell University Press, 1985.

Broughton, E. "The Bhopal Disaster and Its Aftermath: A Review." *Environmental Health Perspectives* 4, no. 1 (2005): 6.

Brown, Garrett. "Protecting Workers' Health and Safety in the Globalizing Economy through International Trade Treaties." *International Journal of Occupational and Environmental Health* 11, no. 2 (2005): 207–09.

Brown, Phil. *Contested Illnesses: Citizens, Science, and Health Social Movements.* Berkeley: University of California Press, 2012.

Brown, Phil, Stephen Zavestoski, Sabrina McCormick, Brian Mayer, Rachel Morello-Frosch, and Rebecca Gasior Altman. "Embodied Health Movements: New Approaches to Social Movements in Health." *Sociology of Health & Illness* 26, no. 1 (2004): 50–80.

Bucheli, Marcelo. *Bananas and Business: The United Fruit Company in Colombia, 1899–2000:* New York: New York University Press, 2005.

———. "Multinational Corporations, Totalitarian Regimes, and Economic Nationalism: United Fruit Company in Central America, 1899–1975." *Business History Review* 50, no. 4 (2008): 433–54.

Bucheli, Mario. "United Fruit Company in Latin America." In *Banana Wars: Power, Production and History in the Americas,* edited by Steve Striffler and Mark Moberg, 80–100. Durham, NC, and London: Duke University Press, 2003.

Buhs, Joshua Blu. *The Fire Ant Wars: Nature, Science, and Public Policy in Twentieth-Century America.* Chicago: University of Chicago Press, 2004.

Bulmer-Thomas, V. "Economic Development over the Long Run: Central America since 1920." *Journal of Latin American Studies* 15, no. 2 (1983): 269–94.

Carey, Frank. "War Develops Powerful Attack against Insects." *Mobile Press Register,* 1943.

Carson, Rachel. *Silent Spring.* Cambridge, MA: The Riverside Press, 1962.

Caswell, E. P. and W. J. Apt, "Pineapple Nematode Research in Hawaii: Past, Present and Future," *Journal of Nematology* 21, no. 2 (1989): 147–157.

Center for Professional Responsibility. *A Legislative History: The Development of the ABA Model Rules of Professional Conduct, 1982–2005.* Chicago: Center for Professional Responsibility, American Bar Association, 2006.

Chavez, Leo R. *The Latino Threat: Constructing Immigrants, Citizens, and the Nation.* Stanford, CA: Stanford University Press, 2008.

Chemistry in the Economy. Washington, DC: American Chemical Society, 1973.

Chinchilla, Norma Stoltz. "Class Struggle in Central America: Background and Overview." *Latin American Perspectives* 7, no. 2/3 (1980): 2–23.

Chomsky, Aviva. "Afro-Jamaican Traditions and Labor Organizing on United Fruit Company Plantations in Costa Rica, 1910." *Journal of Social History* 28, no. 4 (1995): 837–55.

———. *West Indian Workers and the United Fruit Company in Costa Rica, 1870–1940.* Baton Rouge : Louisiana State University Press, 1996.

———. "West Indian Workers in Costa Rican Radical and Nationalist Ideology 1900–1950." *Americas: A Quarterly Review of Inter-American Cultural History* 51, no. 1 (1994): 11–40.

Clark, Claudia. *Radium Girls: Women and Industrial Health Reform, 1910–1935.* Chapel Hill, NC: University of North Carolina Press, 1997.

Comptroller General of the United States. "Report to the Congress of the United States: Better Regulation of Pesticide Exports and Pesticide Residues in Imported Foods Is Essential." Washington, DC: United States General Accounting Office, 1979.

Conner, John Davis, and George Aloysius Burroughs. *Manual of Chemical Products Liability: An Analysis of the Law Concerning Liability Arising from the Manufacture and Sale of Chemical Products.* Washington, DC: Manufacturing Chemists' Association and National Agricultural Chemical Association, 1952.

Cooper, T. G., E. Noonan, S. von Eckardstein, J. Auger, H. W. Baker, H. M. Behre, T. B. Haugen, T. Kruger, C. Wang, M. T. Mbizvo, and K. M. Vogelsong. "World Health Organization Reference Values for Human Semen Characteristics." *Human Reproduction Update* 16, no. 3 (2010): 231–45.

Cranor, Carl F. *Toxic Torts: Science, Law and the Possibility of Justice.* Cambridge; New York: Cambridge University Press, 2006.

Crowe, J., C. Wesseling, B. R. Solano, M. P. Umana, A. R. Ramirez, T. Kjellstrom, D. Morales, and M. Nilsson. "Heat Exposure in Sugarcane Harvesters in Costa Rica." *American Journal of Industrial Medicine* 56, no. 10 (2013): 1157–64.

Cumbre Continental de Pueblos y Nacionalidades Indígenas de Abya Yala. "Trabajadores afectados por Nemagón anuncian medidas en defensa de sus derechos y contra el TLC." http://www.movimientos.org/enlacei/cumbre-abyayala/show_text.php3?key = 9567. ·

Cummings, Scott L. "The Internationalization of Public Interest Law." *Duke Law Journal* 57, no. 4 (2008): 891–1036.

Cupples, J., and I. Larios. "A Functional Anarchy: Love, Patriotism, and Resistance to Free Trade in Costa Rica." *Latin American Perspectives* 37, no. 6 (2010): 93–108.

Dahlberg, Kenneth A. "Government Policies that Encourage Pesticide Use in the United States." In *The Pesticide Question: Environment, Economics, and Ethics,* edited by David Pimentel and Hugh Lehman. New York and London: Chapman & Hall, 1993.

Daniel, Pete. *Toxic Drift: Pesticides and Health in the Post-World War II South.* Baton Rouge: Louisiana State University Press in association with Smithsonian Institution, Washington, DC, 2005.

De la Cruz, Vladimir. "Características y rasgos históricos del movimiento sindical en Costa Rica." In *El sindicalismo frente al cambio: Entre la pasividad y el protagonismo,* edited by Jorge Nowalski. San José, Costa Rica: FES/DEI, 1997.

Dosal, Paul J. *Doing Business with the Dictator: A Political History of United Fruit in Guatemala, 1899–1944.* Wilmington, DE: SR Books, 1993.

Dospital, Michelle. *Siempre Más Allá . . . : El Movimiento Sandinista en Nicaragua, 1927–1934.* Managua, Nicaragua: Centro Francés de Estudios Mexicanos y Centroamericanos; Instituto de Historia de Nicaragua, 1996.

Dunaway, Finis. "Gas Masks, Pogo, and the Ecological Indian: Earth Day and the Visual Politics of American Environmentalism." *American Quarterly* 60 (2008): 67–98.

Dunbar, Paul. "The Food and Drug Administration Looks at Insecticides." *Food-Drug-Cosmetic Law Quarterly.* June (1949): 233–39.

Dunlap, Thomas R. *DDT: Scientists, Citizens, and Public Policy.* Princeton, NJ: Princeton University Press, 1981.

Edelman, Marc. *Peasants against Globalization: Rural Social Movements in Costa Rica.* Stanford, CA: Stanford University Press, 1999.

Egilman, David S., and Susanna Rankin Bohme. "A Brief History of Warnings." In *Handbook of Warnings,* edited by Michael Wogalter. Mahwah, NJ: Lawrence Erlbaum Associates, 2005.

Egilman, David S., J. Kim, and M. Biklen. "Proving Causation: The Use and Abuse of Medical and Scientific Evidence inside the Courtroom:An Epidemiologist's Critique of the Judicial Interpretation of the Daubert Ruling." *Food and Drug Law Journal* 58, no. 2 (2003): 223–50.

Engström, Johan. "Economic Globalization, Neo-Liberal Reforms, and Costa Rican Banana Unions' Struggles in a Context of Regulation." *Iberoamericana: Nordic Journal of Latin American and Caribbean Studies* 31, no. 2 (2001): 9–35.

Escobar, Arturo. *Encountering Development: The Making and Unmaking of the Third World.* Princeton Studies in Culture/Power/History. Princeton, NJ: Princeton University Press, 1995.

Euraque, Darío A. *Reinterpreting the Banana Republic : Region and State in Honduras, 1870–1972.* Chapel Hill: University of North Carolina Press, 1996.

———. "The Threat of Blackness to the Mestizo Nation: Race and Ethnicity in the Honduran Banana Economy." In *Banana Wars: Power, Production and History in the Americas,* edited by Steve Striffler and Mark Moberg, 23–47. Durham, NC, and London: Duke University Press, 2003.

Faber, Daniel J. *Environment under Fire: Imperialism and the Ecological Crisis in Central America.* New York: Monthly Review Press, 1993.

Fallas, Carlos Luis. *Mamita Yunai.* San José: Editorial Costa Rica, 2010.

Ferleger, Louis. "Arming American Agriculture for the Twentieth Century: How the USDA's Top Managers Promoted Agricultural Development." *Agricultural History* 74, no. 2 (2000).

———. "Uplifting American Agriculture: Experiment Station Scientists and the Office of Experiment Stations in the Early Years after the Hatch Act." *Agricultural History* 64, no. 2 (1990): 5–23.

Finley-Brook, Mary. "Geoeconomic Assumptions, Insecurity, and 'Free' Trade in Central America." *Geopolitics* 17, no. 3 (2012): 629–57.

Fisher, Robert, and Joseph M. Kling. *Mobilizing the Community: Local Politics in the Era of the Global City.* Newbury Park; London: Sage, 1993.

Foro Emaús. "DBCP en la producción bananera." Foro Emaús. http://www.members.tripod.com/foro_emaus/dbcp.htm.

Foro Emaús Coordinating Committee. "Bananas for the World—and the Negative Consequences for Costa Rica? The Social and Environmental Impacts of the Banana Industry in Costa Rica." San José, Costa Rica: Foro Emaús, 1998, http://members.tripod.com/foro_emaus/2ing.html. Reprint.

Fortun, Kim. *Advocacy after Bhopal : Environmentalism, Disaster, New Global Orders.* Chicago: University of Chicago Press, 2001.

Foster, Robert J. "Tracking Globalization: Commodities and Value in Motion." In *Sage Handbook of Material Culture.* London: Sage Publications, 2008.

Frank, Dana. *Bananeras: Women Transforming the Banana Unions of Latin America*. Cambridge, MA: South End Press, 2005.

———. "Where Are the Workers in Consumer-Worker Alliances? Class Dynamics and the History of Consumer-Labor Campaigns." *Politics & Society* 31, no. 3 (2003): 363–79.

Fromherz, Nicholas A. "A Call for Stricter Appellate Review of Decisions on Forum Non Conveniens." *Washington University Global Studies Law Review* 11 (2012): 527–64.

Frundt, Henry J. *Fair Bananas: Farmers, Workers, and Consumers Strive to Change an Industry*. Tucson: University of Arizona Press, 2009.

———. "Sustaining Labor-Environmental Coalitions: Banana Allies in Costa Rica." *Latin American Politics and Society* 52, no. 3 (2010): 99–129.

Fuller, Varden, and Bert Mason. "Farm Labor." *Annals of the American Academy of Political and Social Science* 429 (1977): 63–80.

Galarza, Ernesto. *Merchants of Labor: The Mexican Bracero Story*. Charlotte, NC, and Santa Barbara, CA: NcNally & Loftin, 1964.

Gantz, David A. "Labor Rights and Environmental Protection under NAFTA and Other U.S. Free Trade Agreements." *The University of Miami Inter-American Law Review* 42 (2011): 297–356

García y Griego, Manuel. "The Importation of Mexican Contract Laborers to the United States, 1942–1964." In *Between Two Worlds: Mexican Immigrants in the United States,* edited by David G. Gutiérrez, 45–85. Wilmington, DE: Scholarly Resources, 1996.

Garro, Alejandro M. "Forum Non Conveniens: 'Availability' and 'Adequacy' of Latin American Fora from a Comparative Perspective." *The University of Miami Inter-American Law Review* 35 (2003/04): 65–99.

———. "Unification and Harmonization of Private Law in Latin America." *American Journal of Comparative Law* 4 (1992): 587–616.

General Accounting Office. "Report to the Chairman, Subcommittee on Oversight and Investigations, Committee on Energy and Commerce, House of Representatives; Pesticides: Adulterated Imported Foods are Reaching U.S. Grocery Shelves." Washignton, DC: Government Printing Office, 1992.

Gianessi, Leonard P., Cressida S. Silvers, Sujatha Sankula, and Janet E. Carpenter. "Plant Biotechnology: Current and Potential Impact for Improving Pest Management in U.S. Agriculture; An Analysis of 40 Case Studies; Nematode Resistant Pineapple." Washington, DC: National Center for Food and Agricultural Policy, 2002.

Gill, Leslie. "The Limits of Solidarity: Labor and Transnational Organizing against Coca-Cola." *American Ethnologist* 36, no. 4 (2009): 667–80.

Glass, R. I., R. N. Lyness, D. C. Mengle, K. E. Powell, and E. Kahn. "Sperm Count Depression in Pesticide Applicators Exposed to Dibromochloropropane." *American Journal of Epidemiology* 109, no. 3 (1979): 346–51.

Goldsmith, J. R., G. Potashnik, and R. Israeli. "Reproductive Outcomes in Families of DBCP Exposed Men." *Archives of Environmental Health* 39, no. 85–89 (1984).

Gordon, Robert. "Shell No! OCAW and the Labor-Environmental Alliance." *Environmental History* 3, no. 4 (1998): 460–87.

Gordon, Todd, and Jeffery R. Webber. "Honduran Labyrinth." *Jacobin,* no. 10 (2013).

Gottlieb, Robert. *Forcing the Spring: The Transformation of the American Environmental Movement.* Washington, DC: Island Press, 2005.

Gould, Jeffrey L. *To Lead as Equals: Rural Protest and Political Consciousness in Chinandega, Nicaragua, 1912–1979.* Chapel Hill: University of North Carolina Press, 1990.

Graffam, Mary H. "The Web of Tort 'Reform.'" *Trial* 48, no. 12 (2012).

Grandin, Greg. *Empire's Workshop: Latin America, the United States, and the Rise of the New Imperialism.* New York: Metropolitan Books, 2006.

Gray, Mark E. "Don't Leave US Just Yet: Forum Non Conveniens and the Federal Court's Power to Stay and Monitor Actions in the 'Interest of Justice.'" *Loyola of Los Angeles Law Review* 46 (2012): 293–325.

Grundberg, Ethan. "Confronting the Perils of Globalization: Nicaraguan Banana Workers' Struggle for Justice." *Iowa Historical Review* 1, no. 1 (2007): 96–130.

Guidry, John A. "The Useful State? Social Movements and the Citizenship of Children in Brazil." In *Globalizations and Social Movements: Culture, Power and the Transnational Public Sphere,* edited by John A. Guidry, Michael D. Kennedy, and Mayer N. Zald, 1–32. Ann Arbor: University of Michigan Press, 2000.

Gunter, V. J., and C. K. Harris. "Noisy Winter: The DDT Controversy in the Years before Silent Spring." *Rural Sociology* 63, no. 2 (1998): 179–98.

Guthman, Julie. *Agrarian Dreams: The Paradox of Organic Farming in California.* Berkeley: University of California Press, 2004.

Haglund, LaDawn. "Hard-Pressed to Invest: The Political Economy of Public Sector Reform in Costa Rica." *Revista centroamericana de ciencias sociales* 1, no. 3 (2006): 5–46.

Hall, Carolyn, Héctor Pérez Brignoli, and John V. Cotter. *Historical Atlas of Central America.* Norman: University of Oklahoma Press, 2003.

Hamilton, Nora, and Norma Stoltz Chinchilla. "Central American Migration: A Framework for Analysis." *Latin American Research Review* 26, no. 1 (1991): 75–110.

Hanig, Josh, and David Davis. *Song of the Canary.* Wayne, NJ: New Day Films, 1978.

Harrington, Jerry. "The Midwest Agricultural Chemical Association: A Regional Study of an Industry on the Defensive." *Agricultural History* 70, no. 2 (1996): 415–38.

Harvey, David. *A Brief History of Neoliberalism.* New York: Oxford University Press, 2005.

———. *Spaces of Hope.* Berkeley: University of California Press, 2000.

Haugaard, Lisa. "Nicaragua." *Foreign Policy in Focus* 2, no. 32 (1997).

Hawkins, Richard. *A Pacific Industry: The History of Pineapple Canning in Hawaii.* London and New York: I. B. Tauris, 2011.

Hays, Samuel. *Beauty, Health, and Permanence: Environmental Politics in the United States: 1955–1985.* Cambrigde: Cambrigde University Press, 1987.

Henke, Christopher. *Cultivating Science, Harvesting Power: Science and Industrial Agriculture in California.* Cambridge, MA: MIT Press, 2008.

Hernández Rodríguez, Carlos. "Del espontaneísmo a la acción concertada: Los trabajadores bananeros de Costa Rica, 1900–1955." *Revista de Historia,* no. 31 (1995): 69–125.

Herod, Andrew. *Labor Geographies : Workers and the Landscapes of Capitalism.* Perspectives on Economic Change. New York: Guilford Press, 2001.

Hofmann, Jonathan, Jorge Guardado, Matthew Keifer, and Catharina Wesseling. "Mortality among a Cohort of Banana Plantation Workers in Costa Rica." *International Journal of Occupational and Environmental Health* 12, no. 4 (2006): 321–28.

Hooks, Gregory. "From an Autonomous to a Captured State Agency: The Decline of the New Deal in Agriculture." *American Sociological Review* 55, no. 1 (1990): 29–43.

Inselbuch, Elihu. "Contingent Fees and Tort Reform: A Reassessment and Reality Check." *Law & Contemporary Problems* 64, nos. 2 & 3 (2001): 175–95.

Jasanoff, Sheila. *States of Knowledge: The Co-Production of Science and Social Order.* New York and London: Routledge, 2004.

———. "What Judges Should Know about the Sociology of Science," in *Science and Public Reason.* New York and London: Routledge, 2012.

Jeyaratman, J. "Acute Pesticide Poisoning: A Major Global Health Problem." *World Health Statistics Quarterly* 43 (1990): 139–44.

Johnson, Edwin L. "Global Impacts of U.S. Pesticide Regulations." Paper presented at the Proceedings of the U.S. Strategy Conference on Pesticide Management, Washington, DC, July 7–8 1979.

Joyce, John W. "Forum Non Conveniens in Louisiana." *Louisiana Law Review* 60 (1999): 293–319.

Karnes, Thomas Lindas. *Tropical Enterprise: The Standard Fruit and Steamship Company in Latin America.* Baton Rouge: Louisiana State University Press, 1978.

Kelly, Trish. "The WTO, the Environment and Health and Safety Standards." *The World Economy* 26, no. 2 (2003): 131–51.

Kent, Noel J. *Hawaii, Islands under the Influence.* New York: Monthly Review Press, 1983.

Kepner, Charles D. *Social Aspects of the Banana Industry.* New York: AMS Press, 1967 [1936].

Kepner, Charles D., and Jay H. Soothill. *The Banana Empire: A Case Study of Economic Imperialism.* New York : Vanguard Press, 1935.

Kinkela, David. *DDT and the American Century: Global Health, Environmental Politics, and the Pesticide that Changed the World.* Chapel Hill: University of North Carolina Press, 2011.

Kroll-Smith, J. Stephen, Phil Brown, and Valerie J. Gunter. *Illness and the Environment: A Reader in Contested Medicine.* New York: New York University Press, 2000.

LaFeber, Walter. *Inevitable Revolutions: The United States in Central America.* 2nd ed. New York: W. W. Norton, 1993.

Landrigan, Philip J, Luz Claudio, Steven B. Markowitz, Gertrud S. Berkowitz, Barbara L. Brenner, Harry Romero, James G. Wetmur, Thomas D. Matte, Andrea C. Gore, James H. Godbold, and Mary S. Wolff. "The Unique Vulnerability of Infants and Children to Pesticides." *Environmental Health Perspectives* 107, no. Supp 3 (1999).

Langdon, Robert. "The Banana as a Key to Early American and Polynesian History." *Journal of Pacific History* 28, no. 1 (1993): 15–35.

Langston, Nancy. *Toxic Bodies: Hormone Disruptors and the Legacy of DES.* New Haven, CT: Yale University Press, 2010.

Leach, R. "Blackhead Toppling Disease of Bananas." *Nature* 181, no. 4602 (1958): 204–05.

Lear, Linda J. "Bombshell in Beltsville: The USDA and the Challenge of 'Silent Spring.'" *Agricultural History* 66, no. 2 (1992): 151–70.

———. *Rachel Carson: Witness for Nature.* New York: Henry Holt, 1997.

Leigh, Monroe. "Decision: Forum Non Conveniens—Conditional Dismissal of Tort Claim by Foreign Plaintiffs." *American Journal of International Law* 80 (1986): 964–67.

López, José Roberto. *La economía del banano en Centroamérica.* San José, Costa Rica: Editorial Departamento Ecuménico de Investigaciones, 1988.

Luna, Guadalupe T. "An Infinite Distance? Agricultural Exceptionalism and Agricultural Labor." *University of Pennsylvania Journal of Labor & Employment Law* 1 (1998): 487–510.

Magdoff, Fred, John Bellamy Foster, and Frederick H. Buttel. "An Overview." In *Hungry for Profit: The Agribusiness Threat to Farmers, Food, and the Environment,* edited by Fred Magdoff, John Bellamy Foster, and Frederick H. Buttel. New York: Monthly Review Press, 2000.

Majka, Linda C., and Theo J. Majka. *Farm Workers, Agribusiness, and the State.* Philadelphia: Temple University Press, 1982.

Mapes, Kathleen. *Sweet Tyranny: Migrant Labor, Industrial Agriculture, and Imperial Politics.* Urbana: University of Illinois Press, 2009.

Markowitz, Gerald E., and David Rosner. *Deceit and Denial: The Deadly Politics of Industrial Pollution.* Berkeley: University of California Press, 2002.

Marquardt, Steve. "'Green Havoc': Panama Disease, Environmental Change, and Labor Process in the Central American Banana Industry." *American Historical Review* 106 no. 1 (2001).

———. "Pesticides, Parakeets, and Unions in the Costa Rican Banana Industry, 1938–1962." *Latin American Research Review* 37, no. 2 (2001): 3–36.

Marshall, Woodville K. "Provision Ground and Plantation Labor in Four Windward Islands: Competition for Resources During Slavery." In *Cultivation and Culture: Labor and the Shaping of Black Life in the Americas,* edited by Ira Berlin and Philip Morgan, 203–20. Charlottesville: University of Virginia, 1993.

May, Stacy, and Galo Plaza. *The United Fruit Company in Latin America.* Washington, DC: National Planning Association, 1958.

Mayer, Don, and Kyle Sable. "Yes! We Have No Bananas: Forum Non Conveniens and Corporate Evasion." *International Business Law Review* 4 (2004): 130–64.

McConnell, Grant. *The Decline of Agrarian Democracy.* Berkeley: University of California Press, 1953.

McWilliams, Carey. *Factories in the Field: The Story of Migratory Farm Labor in California.* Boston: Little, Brown, 1939.

Meléndez Aguirre, Denis H. "El expediente de La Marcha Sin Retorno." Managua: Centro Alexander Von Humboldt and Centro de Información y Servicios de Asesoría en Salud, 2006, http://www.cieets.org.ni/media/contenido/attachments/DMA_Expediente_Marcha_sin_Retorno_090311.pdf.

Melosi, Martin. *Effluent America: Cities, Industry, Energy, and the Environment.* Pittsburgh: University of Pittsburgh Press, 2001.

Melucci, Alberto. "The New Social Movements: A Theoretical Approach." *Social Science Information* 19 (1980): 199–226.

Miller, Laurel E. "Forum Non Conveniens and State Control of Foreign Plaintiff Access to U.S. Courts in International Tort Actions." *University of Chicago Law Review* 58 (1991): 1369–92.

Moberg, Mark, and Steve Striffler, eds. *Banana Wars: Power, Production and History in the Americas.* Durham, NC, and London: Duke University Press, 2003.

Moody, Kim. *Workers in a Lean World: Unions in the International Economy.* Haymarket Series. London and New York: Verso, 1997.

Mora Castellanos, Eduardo. "Obreros, pesticidas, salud y relaciones de fuerza en los bananales del Caribe costarricense." *Ambien-Tico* 33–34 (1995).

Mora Solano, Sindy. "Costa Rica en la década de 1980: Estrategias de negociación política en tiempos de crisis." *Intercambio* 4, no. 5 (2007): 165–83.

Murphy, Michelle. "Chemical Regimes of Living." *Environmental History* 13, no. 4 (2008): 695–703.

———. *Sick Building Syndrome and the Problem of Uncertainty: Environmental Politics, Technoscience, and Women Workers.* Durham, NC: Duke University Press, 2006.

Murray, Douglas L. *Cultivating Crisis: The Human Cost of Pesticides in Latin America.* Austin: University of Texas Press, 1994.

Nash, Linda. *Inescapable Ecologies: A History of Environment, Disease, and Knowledge.* Berkeley: University of California Press, 2003.

National Agricultural Chemicals Associations. *Open Door to Plenty.* Washington, DC, 1958.

Nayar, N. M. "The Bananas: Botany, Origin, Dispersal." In *Horticultural Reviews,* 117–64. New York : John Wiley & Sons, 2010.

Ngai, Mae M. *Impossible Subjects: Illegal Aliens and the Making of Modern America.* Politics and Society in Twentieth-Century America. Princeton, NJ: Princeton University Press, 2004.

Nixon, Rob. *Slow Violence and the Environmentalism of the Poor.* Cambridge, MA: Harvard University Press, 2011.

O'Bannon, J H. "Worldwide Dissemination of Radopholus Similis and Its Importance in Crop Production." *Journal of Nematology* 9, no. 1 (1977): 19–25.

Olson, William A., Robert T. Habermann, Elizabeth K. Weisburger, Jerrold M. Ward, and John H. Weisburger. "Brief Communication: Induction of Stomach Cancer in Rats and Mice by Halogenated Aliphatic Fumigants." *Journal of the National Cancer Institute*, no. 51 (1973): 1993–95.

Orantes, C. M., R. Herrera, M. Almaguer, E. G. Brizuela, C. E. Hernández, H. Bayarre, J. C. Amaya, D. J. Calero, P. Orellana, R. M. Colindres, M. E. Velázquez, S. G. Núñez, V. M. Contreras, and B. E. Castro. "Chronic Kidney Disease and Associated Risk Factors in the Bajo Lempa Region of El Salvador: Nefrolempa Study, 2009." *MEDICC Review* 13, no. 4 (2011): 14–22.

Pellow, David N. *Resisting Global Toxics: Transnational Movements for Environmental Justice*. Urban and Industrial Environments. Cambridge, MA: MIT Press, 2007.

Penagos, H. G. "Contact Dermatitis Caused by Pesticides among Banana Plantation Workers in Panama." *International Journal of Occupational and Environmental Health* 8, no. 1 (2002): 14–18.

Peraza, S., C. Wesseling, A. Aragón, R. Leiva, R. A. García-Trabanino, C. Torres, K. Jakobsson, C. G. Elinder, and C. Hogstedt. "Decreased Kidney Function among Agricultural Workers in El Salvador." *American Journal of Kidney Disease* 59, no. 4 (2012): 531–40.

Pérez Sáinz, Juan Pablo. *From the Finca to the Maquila: Labor and Capitalist Development in Central America*. Boulder, CO: Westview Press, 1999.

Pérez Vargas, Víctor. "Los inconvenientes del 'Forum Non Conveniens.'" *Revista de Ciencias Jurídicas* 100 (Feb–Apr 2003): 61–84.

Petersen, Roger A. *The Legal Guide to Costa Rica*. Miami: Centro Legal R & M, S.A, 1994.

Petras, James F., and Henry Veltmeyer. *Social Movements in Latin America: Neoliberalism and Popular Resistance*. Social Movements and Transformation. New York: Palgrave Macmillan, 2011.

Petryna, Adriana. *Life Exposed: Biological Citizens after Chernobyl*. In-Formation Series. Princeton, NJ: Princeton University Press, 2002.

Polakoff, Erica, and Pierre LaRamée. "Grass-Roots Organizations." In *Nicaragua without Illusions : Regime Transition and Structural Adjustment in the 1990s*, edited by Thomas W. Walker. Wilmington, DE: SR Books, 1997.

Polanyi, Karl. *The Great Transformation: The Political and Economic Origins of Our Time*. 2nd Beacon Paperback ed. Boston, MA: Beacon Press, 2001.

Potashnik, G., and I. Yanai-Inbar. "Dibromochloropropane (DBCP): An 8-Year Reevaluation of Testicular Function and Reproductive Performance." *Fertility and Sterility* 47, no. 2 (1987): 317–23.

Pulido, Laura. *Environmentalism and Economic Justice*. Tucson: University of Arizona Press, 1996.

Putnam, Lara. *The Company They Kept: Migrants and the Politics of Gender in Caribbean Costa Rica, 1870–1960*. Chapel Hill: University of North Carolina Press, 2002.

Raffensperger, Carolyn, and Joel A. Tickner. *Protecting Public Health and the Environment: Implementing the Precautionary Principle.* Washington, DC: Island Press, 1999.

Ramírez, A L, and C M Ramírez. "Esterilidad masculina causada por la exposición laboral al nematicida 1,2-Dibromo-3-Cloropropano. [Male Sterility Caused by Occupational Exposure to the Nematicide 1,2-Dibromo-3-Chloropropane]." *Acta médica costarricense* 23, no. 2 (1980): 219–22.

Raynolds, Laura T., Douglas L. Murray, and John Wilkinson. *Fair Trade: The Challenges of Transforming Globalization.* London and New York: Routledge, 2007.

———. "The Global Banana Trade." In *Banana Wars: Power, Production, and History in the Americas,* edited by Steve Striffler and Mark Moberg, 23–47. Durham, NC, and London: Duke University Press, 2003.

Rebello, Justin. "U.S. District Court in Calif. Rules against Dow Chemical in Pesticide Exposure Case." *Lawyers Weekly USA,* December 17 2007.

Revista centroamericana de ciencias de la salud. "Salud ocupacional en el sector bananero centroamericano: Salud ocupacional en trabajadores del Valle de la Estrella—Standard Fruit C. Costa Rica." 4, no. 9 (1978): 9–67.

Revista Envío. "El Nemagón en el banquillo: Acusan los bananeros." March 1998.

Robertson, Cassandra Burke. "Forum Non Conveniens on Appeal: The Case for Interlocutory Review." *Southwestern Journal of International Law* 18 (2011): 445–73.

Robinson, Ian. "Does Neoliberal Restructing Promote Social Movement Unionism? U.S. Developments in Comparative Perspective." In *Unions in a Globalized Environment: Changing Borders, Organizational Boundaries, and Social Roles,* edited by Bruce Nissen. Armonk, NY: M.E. Sharpe, 2002.

Roelofs, Cora. *Preventing Hazards at the Source.* Fairfax, VA: American Industrial Hygiene Association, 2007.

Rogge, Malcolm J. "Towards Transnational Corporate Accountability in the Global Economy: Challenging the Doctrine of Forum Non Conveniens in in Re: Union Carbide, Alfaro, Sequihua, and Aguinda." *Texas International Law Journal* 36 (2001): 299–317.

Rosenberg, Beth, and Charles Levenstein. "Unintended Consequences: Impacts of Pesticide Bans on Industry, Workers, the Public, and the Environment." Lowell, MA: Toxics Use Reduction Institute, University of Massachusetts, 1995.

Rosenthal, Erika. "The DBCP Pesticide Cases: Seeking Access to Justice to Make Agribusiness Accountable in the Global Economy." In *Agribusiness and Society: Corporate Responses to Environmentalism, Market Opportunities, and Public Regulation,* edited by Kees Jansen and Sietze Vellema. London and New York: Zed Books, 2004.

Rosner, David, and Gerald Markowitz, eds. *Dying for Work: Workers' Safety and Health in Twentieth-Century America.* Bloomington and Indianapolis: Indiana University Press, 1989.

———. "The Trials and Tribulations of Two Historians: Adjudicating Responsibility for Pollution and Personal Harm." *Medical History* 53 (2009): 271–92.

Royal Dutch Shell PLC. *Annual Report and Form 20-F for the Year Ended December 31, 2006*. London: Royal Dutch Shell, 2007.

Royo, Antoni. "La ocupación del Pacífico Sur costarricense por parte de la compañía bananera (1938–1984)." *Diálogos* no. 2 (2003). http://historia.fcs.ucr.ac.cr/articulos/2003/zonasur.htm.

Russell III, Edmund P. "Speaking of Annihilation: Mobilizing for War against Human and Insect Enemies: 1914–1945." *Journal of American History* 82 no. 4 (1996): 1505–29.

———. "The Strange Career of DDT: Experts, Federal Capacity, and Environmentalism in World War II." *Technology and Culture* 40, no. 4 (1999): 770–96.

———. *War and Nature: Fighting Humans and Insects with Chemicals from World War I to Silent Spring*. Studies in Environment and History. Cambridge and New York: Cambridge University Press, 2001.

Saint Dahl, Henry. "Forum Non Conveniens, Latin America and Blocking Statutes." *University of Miami Inter-American Law Review* 35, no. 1 (2003/2004): 25–63.

———. *McGraw-Hill's Spanish and English Legal Dictionary*. New York: McGraw-Hill, 2004.

Saldaña-Portillo, María Josefina. *The Revolutionary Imagination in the Americas and the Age of Development*. Durham, NC, and London: Duke University Press, 2003.

Sandifer, S. H., R. T. Wilkins, C. B. Loadholt, L. G. Lane, and J. C. Eldridge. "Spermatogenesis in Agricultural Workers Exposed to Dibromochloropropane (DBCP)." *Bulletin of Environmental Contamination and Toxicology* 23, no. 4–5 (1979): 703–10.

Sarat, Austin. "At the Boundaries of Law: Executive Clemency, Sovereign Prerogative, and the Dilemma of American Legality." *American Quarterly* 57, no. 3 (2005): 611–31.

Sass, R. "Agricultural 'Killing Fields': The Poisoning of Costa Rican Banana Workers." *International Journal of Health Services* 30, no. 3 (2000): 491–514.

Sassen, Saskia. *Losing Control? Sovereignty in an Age of Globalization*. New York: Columbia University Press, 1996.

Schoonover, Thomas David. *The United States in Central America, 1860–1911: Episodes of Social Imperialism and Imperial Rivalry in the World System*. Durham, NC: Duke University Press, 1991.

Schoultz, Lars. *Beneath the United States: A History of U.S. Policy toward Latin America*. Cambridge, MA: Harvard University Press, 1998.

Scott, Alan. *Ideology and the New Social Movements*. London: Unwin Hyman, 1990.

Seidman, Gay. *Beyond the Boycott: Labor Rights, Human Rights, and Transnational Activism*. New York: Russell Sage Foundation, 2007.

Sellers, Christopher. "Body, Place, and the State: The Makings of an 'Environmentalist' Imaginary in the Post-World War II U.S." *Radical History Review*, no. 74 (1999): 31.

———. *Hazards of the Job: From Industrial Disease to Environmental Health*. Chapel Hill: University of North Carolina Press, 1999.

Sepúlveda Malbrán, Juan Manuel, ed. *Las organizaciones sindicales centroamericanas como actores del sistema de relaciones laborales.* San José, Costa Rica: Oficina Internacional de Trabajo (International Labor Organization), 2003.

Shell Chemical Corporation. "El banano: Sus plagas, enfermedades y malezas." 1959.

———. "Proceedings of the Shell Nematology Workshop: St. Louis, MO." 1957.

———. "Proceedings of the Shell Nematology Workshop: Portland, OR." 1959.

Siegel, Charles S., and David S. Siegel. "The History of DBCP from a Judicial Perspective." *International Journal of Occupational and Environmental Health* 5, no. 2 (1999): 127–35.

Siqueira, Carlos Eduardo. *Dependent Convergence: The Struggle to Control Petrochemical Hazards in Brazil and the United States.* Amityville, NY: Baywood Publishing Company, 2003.

Sklair, Leslie. "Social Movements and Global Capitalism." *Sociology* 29 no. 3 (1995): 495–512.

Slocum, Karla. *Free Trade and Freedom: Neoliberalism, Place, and Nation in the Caribbean.* Ann Arbor: University of Michigan Press, 2006.

Slutsky, M., J.L. Levin, and B.S. Levy. "Azoospermia and Oligospermia among a Large Cohort of DBCP Applicators in Twelve Countries." *International Journal of Occupational and Environmental Health* 5 (1999): 116–22.

Smith, C., K. Kerr, and A. Sadripour. "Pesticide Exports from U.S. Ports, 2001–2003." *International Journal of Occupational and Environmental Health* 14, no. 3 (2008): 176–86.

Smith, Erin Foley. "Right to Remedies and the Inconvenience of *Forum Non Conveniens*: Opening U.S. Courts to Victims of Corporate Human Rights Abuses." *Columbia Journal of Law & Social Problems* 44 (2010): 145–92.

Soil Fumigation Handbook. Midland, MI: Dow Chemical Company, 1958.

Soluri, John. "Accounting for Taste: Export Bananas, Mass Markets, and Panama Disease." *Environmental History,* no. 3 (2002): 386–410.

———. *Banana Cultures: Agriculture, Consumption, and Environmental Change in Honduras and the United States.* Austin: University of Texas Press, 2005.

Stahler-Sholk, Richard. "Review: Sandinista Economic and Social Policy: The Mixed Blessings of Hindsight." *Latin American Research Review* 30, no. 2 (1995): 235–50.

Stephens, Beth. *International Human Rights Litigation in U.S. Courts.* Boston and Leiden: Martinus Nijhoff Publishers, 2008.

Stoll, Steven. *The Fruits of Natural Advantage: Making the Industrial Countryside in California.* Berkeley: University of California Press, 1998.

Strasser, Susan. *Satisfaction Guaranteed: The Making of the American Mass Market.* New York: Pantheon Books, 1989.

Strauss, Marcy. "Toward a Revised Model of Attorney-Client Relationship: The Argument for Autonomy." *North Carolina Law Review* 65 (1987): 315–49.

Szasz, Andrew. *Shopping Our Way to Safety: How We Changed from Protecting the Environment to Protecting Ourselves.* Minneapolis: University of Minnesota Press, 2007.

Sze, Julie. *Noxious New York: The Racial Politics of Urban Health and Environmental Justice.* Urban and Industrial Environments. Cambridge, MA: MIT Press, 2007.

Teitelbaum, Daniel. "The Toxicology of 1,2-Dibromo-3-Chloropropane (DBCP)." *International Journal of Occupational and Environmental Health* 5, no. 2 (1999): 122–26.

Thrupp, Lori Ann. "Pesticides and Policies: Approaches to Pest-Control Dilemmas in Nicaragua and Costa Rica." *Latin American Perspectives* 15, no. 4 (2004): 37–70.

———. "Sterilization of Workers from Pesticide Exposure: The Causes and Consequences of DBCP-Induced Damage in Costa Rica and Beyond." *International Journal of Health Services* 21 no. 4 (1991): 731–57.

Tomich, Dale. "'Une Petite Guinée': Provision Ground and Plantation in Martinique, 1830–1843." In *Cultivation and Culture: Labor and the Shaping of Black Life in the Americas,* edited by Ira Berlin and Philip Morgan, 203–20. Charlottesville: University of Virginia, 1993.

Torkelson, T.R., S.E. Sadek, V.K. Rowe, J.K. Kodama, H.H. Anderson, G.S. Loquvam, and C.H. Hine. "Toxicological Investigations of 1,2-Dibromo-3-Chloropropane." *Toxicology and Applied Pharmacology* 3 (1961): 345–39.

Torres, C., A. Aragón, M. González, I. López, K. Jakobsson, C.G. Elinder, I. Lundberg, and C. Wesseling. "Decreased Kidney Function of Unknown Cause in Nicaragua: A Community-Based Survey." *American Journal of Kidney Disease* 55, no. 3 (2010): 485–96.

Torres-Rivas, Edelberto. *History and Society in Central America.* Translations from Latin America Series. Austin: University of Texas Press, 1993.

Trost, Cathy. *Elements of Risk: The Chemical Industry and Its Threat to America.* New York: Times Books, 1984.

Tucker, Richard P. *Insatiable Appetite: The United States and the Ecological Degradation of the Tropical World.* Berkeley: University of California Press, 2000.

United Fruit Company Department of Research. "Problems and Progress in Banana Disease Research." Boston: United Fruit Company, 1958.

United States General Accounting Office. "Report to the Chairman, Environment, Energy, and Natural Resources Subcommittee, Committee on Government Operations, House of Representatives: Export of Unregistered Pesticides Is Not Adequately Monitored by EPA." 1989.

Van Wendel de Joode, B.N., I.A.M. De Graaf, C. Wesseling, and H. Kromhout. "Paraquat Exposure of Knapsack Spray Operators on Banana Plantations in Costa Rica." *International Journal of Occupational and Environmental Health* 2, no. 4 (1996): 294–304.

Wald, Sarah. "Visible Farmers/Invisible Workers Locating Immigrant Labor in Food Studies." *Food, Culture, and Society* 14, no. 4 (2011): 567–86.

Walker, Thomas W. *Revolution & Counterrevolution in Nicaragua.* Boulder, CO: Westview Press, 1991.

Walker, Thomas W., and Christine J. Wade. *Nicaragua : Living in the Shadow of the Eagle.* 5th ed. Boulder, CO: Westview Press, 2011.

Ward, Jerrold M., and Robert T. Habermann. "Pathology of Stomach Cancer in Rats and Mice Induced with the Agricultural Chemicals Ethylene Dibromide and Dibromochloropropane." *Bulletin of the Society of Pharmacological and Environmental Pathologists* 2, no. 2 (1974): 10–11.

Wargo, John. *Our Children's Toxic Legacy: How Science and Law Fail to Protect Us from Pesticides.* New Haven, CT: Yale University Press, 1998.

Warning Labels: A Guide for the Preparation of Warning Labels for Hazardous Chemicals, Manual L-1. Washington, DC: Manufacturing Chemists Association of the United States, 1949.

Wehunt, E J, D J Hutchison, and D I Edwards. "Reaction of Musa Acuminate to Radopholus Similis." Abstract. *Phytopathology* 55, no. 10 (1965): 1082.

Weideman, M. "Toxicity Tests in Animals: Historical Perspectives and New Opportunities." *Environmental Health Perspectives* 101, no. 2 (1993): 222–25.

Weir, David, and Mark Schapiro. *Circle of Poison: Pesticides and People in a Hungry World.* San Francisco: Institute for Food and Development Policy, 1981.

Weir, David, Mark Schapiro, and Terry Jacobs. "The Boomerang Crime." *Mother Jones,* November 1979, 40–48.

Wesseling, Catharina, Anders Ahlbom, Canila Antich, Ana Cecilia Rodríguez, and Roberto Castro. "Cancer in Banana Plantation Workers in Costa Rica." *International Journal of Epidemiology* 25, no. 6 (1996): 11.

Wesseling, Catharina, Daniel Antich, Christer Hogstedt, Ana Cecilia Rodríguez, and Anders Ahlbohm. "Geographical Differences of Cancer Incidence in Costa Rica In Relation to Environmental and Occupational Pesticide Exposure." *International Journal of Epidemiology* 28, no. 3 (1999): 365–74.

Wesseling, Catharina, Aurora Aragón, Luisa Castillo, et al., "Hazardous Pesticides in Central America." *International Journal of Occupational and Environmental Health* 7, no. 4 (2001): 287–94.

Wesseling, Catharina, Luisa Castillo, and Carl-Gustav Elinder. "Pesticide Poisonings in Costa Rica." *Scandinavian Journal of Work, Environment, and Health* 19, no. 4 (1993): 227–35.

Wesseling, Catharina, Jennifer Crowe, Christer Hogstedt, Kristina Jakobsson, Rebekah Lucas, and David H. Wegman. "The Epidemic of Chronic Kidney Disease of Unknown Etiology in Mesoamerica: A Call for Interdisciplinary Research and Action." *American Journal of Public Health* 103, no. 11 (2013): 1927–30.

Wesseling, Catharina, Christer Hogstedt, Anabelle Picado, and Leif Johansson. "Unintentional Fatal Paraquat Poisonings among Agricultural Workers in Costa Rica: Report of 15 Cases." *American Journal of Industrial Medicine* 32, no. 5 (1997): 433–41.

Wesseling, Catharina, Matthew Keifer, Anders Ahlbohm, Robert McConnell, Jai-Dong Moon, Linda Rosenstock, and Christer Hogstedt. "Long-Term Neurobehavioral Effects of Mild Poisonings with Organophosphate and N-Methyl Carbamate Pesticides among Banana Workers." *International Journal of Occupational and Environmental Health* 8 no. 1 (2002): 27–34.

Wesseling, Catharina, Robert McConnell, Timo Partanen, and Christer Hogstedt. "Agricultural Pesticide Use in Developing Countries: Health Effects and Research Needs." *International Journal of Health Services* 27 no. 2 (1997): 273–308.

Wesseling, Catharina, Berna van Wendel de Joode, Matthew Keifer, Leslie London, Donna Mergler, and Lorann Stallones. "Symptoms of Psychological Distress and Suicidal Ideation among Banana Workers with a History of Poisoning by Organophosphate or N-Methyl Carbamate Pesticides." *Occupational and Environmental Medicine* 67, no. 11 (2010): 778–84.

Wesseling, Catharina, Berna van Wendel de Joode, and Patricia Monge. "Pesticide-Related Illness and Injuries among Banana Workers in Costa Rica: A Comparison between 1993 and 1996." *International Journal of Occupational and Environmental Health* 7 (2001): 90–97.

West, Irma. "Occupational Disease in California Attributed to Pesticides and Agricultural Chemicals." *Archives of Environmental Health* 1 (1960): 40–46.

———. "Occupational Diseases of Farmworkers." *Archives of Environmental Health* 9 (1964): 92–98.

Whorton, D., R. M. Krauss, S. Marshall, and T. H. Milby. "Infertility in Male Pesticide Workers." *Lancet* 2, no. 8051 (1977): 1259–61.

Williams, Phillip J. "Dual Transitions from Authoritarian Rule: Popular and Electoral Democracy in Nicaragua." *Comparitive Politics* 26, no. 2 (1994): 169–85.

Williams, Robert G. *Export Agriculture and the Crisis in Central America.* Chapel Hill: University of North Carolina Press, 1986.

Wilson, Bruce M. *Costa Rica: Politics, Economics, and Democracy.* Boulder, CO: Lynne Rienner Publishers, 1998.

Wong, Mitchell M. "Forum Non Conveniens: Circumstances after 'Sinochem.'" *New York Law Journal* 237, no. 58 (2007): 4–5.

Wright, Angus Lindsay. "Rethinking the Circle of Poison: The Politics of Pesticide Poisoning among Mexican Farm Workers." *Latin American Perspectives* 13 (1990): 26–59.

Yonay, Ehud. "The Nematode Chronicles." *New West*, May 1981, 66–74+.

Yozell, Emily. "The Castro Alfaro Case: Convenience and Justice—Lessons for Lawyers in Transcultural Litigation." In *Human Rights, Labor Rights, and International Trade*, edited by Lance A. Compa and Stephen F. Diamond. Philadelphia: University of Pennsylvania Press, 1996.

Zamora, Augusto. "Relaciones con USA: Camino de doble vía." *Revista Envío*, October 1996.

Zavon, M. R., and C. A. Wilzbach. "Radiation Control Activities in a Local Health Department." *Public Health Reports* 74, no. 5 (1959): 439–40.

Ziegler, J. "Health Risk Assessment Research: The OTA Report." *Environmental Health Perspectives*. 101, no. 5 (1993): 402–06.

Zierler, David. *The Invention of Ecocide: Agent Orange, Vietnam, and the Scientists Who Changed the Way We Think About the Environment.* Athens: University of Georgia Press, 2011.

UNPUBLISHED SOURCES

Barrantes Cascante, Ramón. "Posición de Foro Emaús ante el Tribunal Internacional del Pueblo sobre violaciones a los derechos humanos y ambientales en las plantaciones bananeras de Costa Rica." Paper presented at the International People's Tribunal, New York, April 26 1998.

Boix Bornay, Vicent. "DBCP: Un artefacto químico que sigue estallando." 2005.

Bucheli, Marcelo. "Multinational Corporations and the Politics of Vertical Integration: The Case of the Central American Banana Industry in the Twentieth Century." n.p., n.d., http://www.economia.unam.mx/amhe/pdfs/banano_bucheli.pdf. Reprint.

Cordero Ulate, Allen. "Nuevas desigualdades; nuevas resistencias: El caso de los ex trabajadores bananeros costarricences afectados por los agroquímicos." Paper presented at the XXVIII International Congress of the Latin American Studies Association, Rio de Janeiro, Brazil, June 11–14 2009.

Davis, Frederick Rowe. "Pesticides and Toxicology: Episodes in the Evolution of Environmental Risk Assessment (1937–1997)." PhD diss., Yale University, 2001.

Delp, Linda, Marisol Arriaga, Guadalupe Palma, Haydée Urita, and Abel Valenzuela. "NAFTA's Labor Side Agreement: Fading into Oblivion? An Assessment of Workplace Health & Safety Cases." Los Angeles: UCLA Labor Center, 2004, www.labor.ucla.edu/publications/reports/nafta.html.

Hendler, Scott. "Bend It Like Beckham: Forum Manipulation and Abuse of the Foreign Sovereign Immunities Act by Multinational Corporations." Inter-American Bar Association/Inter American Federation of Lawyers Annual Conference (2004).

Murray, Douglas L. "The Politics of Pesticides: Corporate Power and Popular Struggle over the Regulatory Process." PhD diss., University of California Santa Cruz, 1983.

Taylor, A. L. "Nematocides and Nematicides—History." http://flnem.ifas.ufl.edu/HISTORY/nematicide_his.htm.

INDEX

Page numbers in italics indicate figures.

CACM (Central American Common Market), 52

CAFTA (Central American Free Trade Agreement), 8, 170, 184, 189–90, 192, *193*, 220, 277n26

Calderón, Rafael Ángel, 146, 150

California, and transnational litigation, 110, 119, 202–8, 212

California Department of Food and Agriculture (CDFA), 97–98

California Occupational Safety and Health Administration (CalOSHA), 78

California Rural Legal Assistance, 95–96, 98

Calvosa, Carmelo, 95–96

Canales Martínez, et al. v. Dow Chemical Company, et al. (2002), 165–66, 282n116

capitalism, 3–5, 231

Carazo, Rodrigo, 146, 277n26

carbofuran (Furadan), 70–71, 103

Carmona López, Lino, 169

Carson, Rachel, 27, 45

Carter, Jimmy, 101–2, 266n155

Carter, Michael, 198–201, 203, 210

Cassity, Henry, 78–80

Castle & Cooke Foods. *See* Dole Food companies (Dole Food Company; formerly Standard Fruit Company)

Castro Alfaro case: *amicus curiae* or friend of the court briefs and, 121; compensation and, 126–27; defendant lawyers and, 118; defendants and, 126; dissenting opinion and, 121–24, 125–26, 128, 272n62; FNC and, 117–19, 121–26, 128, 177, 272n58; majority opinion and, 121, 124–26, 272n58; mobility of corporations and, 117, 121, 124–25; mobility of plaintiffs and, 117, 121, 123–24, 125, 272n62; plaintiff lawyers and, 117, 119–21, 127, 128; plaintiffs, 126, 271n52; representation and, 119–21, 126–27; settlements of litigation and, 126–28; translation and, 119–21, 126–27

CDFA (California Department of Food and Agriculture), 97–98

Central America, and stereotypes of Latin America, 187–88, 189, 199, 204–9, 289n73. See also *Tellez/Laguna* case

Central American banana production and

trade: overview of, 51–53; agribusiness and, 55; agricultural workers' power in context of, 52; banana unions and, 69; "boomerang crime" and, 101, 268n176; European Union and, 9; health effects history of, 6, 9, 217; historiography of, 8–9; independent growers model and, 52, 68–69, 251n28; independent producers and, 52, 68–69, 177, 251n28; internal immigrant workers and, 51; land reform and, 51–52; nematode and, 53–58, 55; pesticide history and, 9–10, 51, 52; U.S. imperialism and, 8–9, 51; women agricultural workers and, 52, 251n30. See also *afectados, los* (the affected); Central America and Central American courts; *specific countries*

Central American banana production and trade, and DBCP: overview and history of, 1, 4, 47–48, 58–61, 72, 216; agricultural workers' exposure and, 61–66; agricultural workers' resistance and, 69–70, 71; agricultural workers versus corporations' control/power and, 52, 66–69, 71–72; application of DBCP and, 68–69; benefit/risk equation and, 89; children and, 63–64, 153, 279n61; DBCP and, 58–60; DBCP registration and, 47–48, 56–58; enforcement of national state pesticides policy and, 61, 80, 260n42; families' exposure and, 63–64; inequalities in exposures/protections and, 47, 78–81, 83–86, 90; "innocently unaware" of hazards and, 66–67; marketing of DBCP and, 55, 59; pesticide alternatives and, 70–72; pesticides policy of national states and, 10, 61, 80, 93–95, 104, 254n95, 260n42, 268n180; protective equipment/measures and, 61, 64–65, 84–85, 92–93, 254n96, 256n126; research on DBCP use and, 56, 58–59; warning labels in Spanish and, 65–66; women agricultural workers and, 62, 63–64, 153. *See also* Costa Rican banana production and trade, and DBCP; Guatemalan banana production and trade, and DBCP; health effects, of DBCP;

COSIBA-CR (*Coordinadora de Sindicatos Bananeros de Costa Rica*), 147, 149, 168–69, 278n41

Costa Rica: accountability of national state for DBCP use in, 151, 155–56; globalization and, 150–51; neoliberalism and, 146–47, 150; social welfare policies of, 146; Solidarity Associations in, 147, 151; sterility in culture of, 115–16, 144; unions and, 95, 143, 146–49. *See also* Costa Rican banana production and trade; Costa Rican national movement, of DBCP-affected workers

Costa Rican banana production and trade: overview of, 51; application techniques for DBCP and, 59–60; banana unions and, 95, 147; DBCP use in, 59–60, 104, 217; enforcement of pesticides policy and, 61, 260n42; exposure to DBCP and, 61–64; health effects of DBCP in context of, 3, 91–92, 101, 108, 115–17, 145, 152–54, 157–58, 176, 235n2, 279n59, 269n6; independent producers and, 69; nematode and, 53–54, 58; nematode research and, 55; non-sterility health effects of DBCP and, 145, 152–54, 157–58, 279n59; notification procedures for banned pesticides and, 91–92; pesticides policy of national state and, 10, 61, 93–95, 104, 254n95, 260n42; power of agricultural workers versus corporations and, 69; protective equipment/measures and, 64–65, 92–93; resistance against DBCP use and, 69; salary for workers and, 133, 274n101; settlements of litigation and, 142–44, 152, 279n57; sterility health effect of DBCP and, 3, 91–92, 101, 108, 235n2, 269n6; strikes and, 52, 147; transnational litigation and, 3, 106, 107–109, 110–114, 116–117, 119–121, 134, 142–43. *See also* Costa Rica; Costa Rican national movement, of DBCP-affected workers

Costa Rican courts: appeals process in *Abarca*, 139, 275n121; Bustamante Code and, 138; Civil Code and, 138, 165; FNC or location in context of, 137–39, 165,

275n120; judicial traditions and, 137–39, 275n120; litigation by DBCP-affected workers and, 3, 114, 137; mobility of corporations in DBCP litigation in, 139; sovereignty and, 137–39, 275n120; transnational corporations and DBCP cases in, 139

Costa Rican national movement, of DBCP-affected workers: overview of, 14, 144, 145–46, 172–73, 221–22, 223; azoospermia or absence of sperm in semen and, 144, 178–79; banana unions in context of, 146, 148–49; births/conception before/after exposure to DBCP in context of litigation and, 144, 171; children's health effects and, 153–54, 279n61; compensation and, 3, 108, 145, 151, 153–57, 159–72, 166–67, 169, 171–72, 269n6, 279n61, 282n109; CONATRAB and, 148–51, 156–60, 164–65, 167–70, 172, 278nn40–41, 279n44, 281n78, 282n109; COSIBA-CR and, 147, 149, 168–69, 278n41; death of *los afectados* in context of, 171; *Defensoría de los Habitantes* and, 150–56, 160–61, 164, 168, 280n78; democracy and, 15, 110, 146, 149, 221, 226, 231; human rights and, 147, 150, 155, 164; Interinstitutional Commission and, 156–57, 280n78; Law 8130 and, 145, 159–60, 163–65, 167–71; medical definition of health effects of DBCP and, 3, 145, 151, 153, 155–56, 160–62, 167, 169, 279n61; Memorandum of Understanding and, 159–62, 164, 167–68, 172, 282n109; men's health effects and compensation in context of, 151, 153–54, 159, 162–63, 169, 171–72, 279n61; national state and, 150–56; national state compensation and, 108, 157, 160, 163–72, 269n6, 282n109; non-sterility health effect compensation payment and, 160, 162, 169, 171–72; oligospermia or lower-than-normal sperm count and, 144; penalties against corporations in context of, 151, 155, 157, 173; public protests and, 149, 156–58, 166–68, 278n41, 281n86, 281n90; settlements and, 142–144, 148, 151–52, 165,

NAFTA (North American Free Trade Agreement), 190, 229
Najlis, Michele, 194, 211
Nanne, Henry, 92–93
Nash, Linda, 65, 256n126
National Academy of Science, 31, *31*, 33
National Agricultural Chemicals Association (NACA), 28–33, *31*, 41
National Labor Relations Act of 1935, 49
national states: compensation for DBCP health effects and, 3, 108, 145, 151, 153–72, 269n6, 279n61, 282n109; corporations and, 4, 8, 15, 26, 28–29, 30, 39–40, 45, 51, 74, 87–88, 92–94, 97–98, 99–102, 151, 155, 157, 173, 189–90, 222–23; court access for international plaintiffs and, 127–128, 139–41, 165–166, 174–75, 202, 212–13, 228; globalization and, 5–6, 109, 150–51, 174, 231; health and welfare of population and, 4–5, 155; inequalities in exposures/protections and, 4, 6, 27–28, 47–48, 74, 81–84, 86–87, 90–91, 94, 96–99, 103, 215, 218; neoliberalism and, 4–5, 146–47, 150, 176, 234n12; pesticide regulation and, 4, 5, 10, 26–29, 37, 40, 61, 93–95, 104, 254n95; pesticide-injured workers and, 6, 78, 142–165, 167–173, 174–175, 178, 182–186, 189–92, 195, 215, 221–222; trade agreements and, 170, 190, 197–98, 223–24, 300n3
Nemacur (fenamiphos), 70–71
Nemagon: overview and history of, 19, 86; agribusiness and, 16, 22–24, 55, 240nn48–49; agricultural abundance in marketing of, 55; animal testing and, 3, 7, 12, 34–37, 42–43, 45–46, 65, 246n166; application techniques and, 44, 58–59; exposure of agriculture workers to, 38–39; Florida citrus groves and, 38, 44; health effects of, 41, 42–46, 176, 178, 180, *185*, 198–99; registration of, 38, 39, 45–46, 56; "safe use" of, 41, 43–44, 46, 75; Spanish-language warning labels and, 65; Standard Fruit and, 58–60; testicular damage in animals and, 35–36, 40, 44–45, 73, 74; toxicological research as concurrent with

marketing and, 33–35, 46; U.S. agriculture and, 48; warning labels and, 35, 40–41, 65. *See also* Shell Chemical company (Shell)
nematode: agribusiness and, 16, 22–25, *26*, 54–55, 240nn48–49; chloropicrin and, 18, 20; description of, 1, 18, 22; EDB and, 18, 55; Hawaiian Islands pineapple industry and, 17–19; pesticide alternatives to DBCP and, 70–72, 78; transnational history of banana production and, 53–54; transnational research and, 55–56
neoliberalism, 4–5, 109, 146–47, 150, 176, 234n12
Nicaraguan banana production and trade: overview of, 47, 51; agricultural workers' exposure to DBCP and, 61–62; DBCP use and, 1, 7, 60, 217; enforcement of pesticides policy and, 61, 260n42; independent producers and, 177; "informal" imperialism and, 8; pesticides policy and, 10, 61, 260n42, 268n180; health effects of DBCP and, 3, 176–77, 178–80, 186, 190–91, *191*, 235n2; transnational litigation and, 3, 14, 177–79, 181–82, 186–88, 195–97, 289n62, 289n64, 293n128. *See also* Nicaraguan national movement, of DBCP-affected workers
Nicaraguan courts: FNC and, 137, 139–40, 179, 199, 202–4, 207, 213; ICSID and, 197–98, 293n128; litigation against transnational corporations and, 178, 182, 186, 188, 196–97, 289n62, 289n64, 293n128; litigation under Law 364 and, 186–88, 191–92, 195–97, 293n128; litigation verdicts against Dole in, 186, 195–96, 207; litigation verdicts against Dow in, 186, 196, 212; litigation verdicts against Oxy in, 196, 212; litigation verdicts against Shell and, 186, 196, 212; representation and, 188; trademarks embargo and, 197–98, 293n128. *See also* Nicaraguan banana production and trade; Nicaraguan national movement, of DBCP-affected workers; Special Law to Process Lawsuits Filed by People

Serrano, Germán, 164, 167, 283n125
settlements, of litigation, 126–28, 140,
    142–44, 152, 224, 279n57
Shell Chemical Company (Shell): *Abarca*
    case and, 137–40, 282n115; *Erazo* case
    and, 129–30; health effects for produc-
    tion workers and, 78; ICSID and,
    197–98, 293n128; litigation verdicts
    against, 182, 186, 196, 212; Nicaraguan
    litigation, 186–87, 196–97, 212; *Sibaja*
    case and, 106, 111, 114; trademarks
    embargo and, 196–97. *See also*
    Nemagon
Sibaja case, 106, 109–14, 268n1, 269nn9–
    10, 270n22
Siegel, Charles S., 140, 268n1, 269n10,
    271n52
*Sindicato de Trabajadores de Plantaciones
    Agrícolas* (SITRAP), 163, 170
SITRAP (*Sindicato de Trabajadores de
    Plantaciones Agrícolas*), 163, 170
Sloan, M. J., 56–57
slow violence (structural violence), 9, 179,
    237n1
Smith, Shearn, 118
Soluri, John, 10, 54, 254n99
Sparks, Mark, 210–11, 300n225
Special Law to Process Lawsuits Filed by
    People Affected by the Use of Pesticides
    Manufactured with DBCP (Law 364):
    Nicaraguan court verdicts and, 186–88,
    191–92, 195–97, 293n128; Nicaraguan
    national movement and, 178–96, 198–
    200, 202, 204, 207–8, 212–14, 223, 226;
    trade agreements and, 197–98, 223–24,
    300n3; transnational litigation in con-
    text of, 179, 186, 188, 196–97, 289n62,
    289n64, 293n128; U.S. opposition to,
    182–84, *185*, 186, 188–90, *193*; litigation
    verdicts against Dole and, 186, 195–96,
    207; litigation verdicts against Dow
    and, 186, 196, 212; litigation verdicts
    against Oxy and, 196, 212; litigation
    verdicts againt Shell and, 186, 196, 212
Stahler-Sholk, Richard, 176
Standard Fruit Company (now Dole Food
    companies; Dole Food Company). *See*
    Dole Food companies (Dole Food

Company; formerly Standard Fruit
    Company)
State Department, United States, 15,
    101, 187, 189–90, 206–7, 222–23,
    266n155
sterility, and DBCP health effects: over-
    view and statistics, 1, 13, 233n2;
    azoospermia or absence of sperm in
    semen and, 36, 143–44, 178–79, 207,
    259n28, 297n191; banana production
    and, 3, 91–92, 101, 108, 235n2, 269n6;
    cultural understanding of, 115–16, 144;
    oligospermia or lower-than-normal
    sperm count and, 142–44, 149, 178–79,
    259n28, 276n5, 297n191; production
    workers and, 1, 73, 76–78, 87, 125, 144,
    162, 172, 199, 201, 259n8
Stockholm Convention, 229
structural violence (slow violence), 9, 179,
    237n1

Taylor, A. L., 22–24
Taylor, Paige, 68
Téllez, José Adolfo, 1
Téllez/Laguna case (*Tellez v. Dole* [2007]),
    1, 202–3, 208–11, 224, 297n190,
    298n196
Texas courts, and transnational litigation
    for DBCP-affected workers: *amicus
    curiae* or friend of the court briefs and,
    121; defendant lawyers and, 118; defend-
    ants and, 126; dissenting opinion and,
    121–24, 125–26, 128, 272n62; FNC and,
    117–19, 121–26, 128–29, 177, 272n58,
    273n83; majority opinion and, 121,
    124–26, 272n58; mobility of corpora-
    tions and, 117, 121, 124–25; mobility of
    plaintiffs and, 117, 121, 123–24, 125,
    272n62; plaintiff lawyers and, 117,
    119–21, 127, 128; plaintiffs and, 117, 121,
    123–24, 125, 126, 271n52, 272n62; repre-
    sentation and, 119–21, 126–27; settle-
    ment of litigation and, 126–28, 199;
    translation and, 119–21, 126–27
Thrupp, Lori Ann, 4, 254n98, 260n42,
    268n180
Torkelson, Ted R., 34, 42
Toruño, Socorro, 204–5, 207, 208

American relations and, 8–9, 51–52, 179, 237n1

Weir, David, 4, 101, 103, 108–10, 125
Wesseling, Catharina, 161
Wintemute, Glenn A., 100
Wong, Mitchell M., 302n25
World Bank, 176, 197, 266n155

World Health Organization, 101, 104, 115, 276n5
Wright, Agnus, 4, 6, 74

Yozell, Emily, 119–21, 127

Zavon, Mitchell, 36–39, 43, 77–78